Compendium of Conifer Diseases

Edited by

Everett M. Hansen
Oregon State University
Corvallis

Katherine J. Lewis
University of Northern British Columbia
Prince George

APS PRESS
The American Phytopathological Society

Front cover photograph by Everett Hansen
Back cover photograph by Katherine Lewis

Reference in this publication to a trademark, proprietary product,
or company name by personnel of the U.S. Department of Agriculture
or anyone else is intended for explicit description only and does not
imply approval or recommendation to the exclusion of others
that may be suitable.

Library of Congress Catalog Card Number: 97-74232
International Standard Book Number: 0-89054-183-3

© 1997 by The American Phytopathological Society

All rights reserved.
No portion of this book may be reproduced in any form, including
photocopy, microfilm, information storage and retrieval system,
computer database or software, or by any means, including
electronic or mechanical, without written permission from the publisher.

Copyright is not claimed in any portion of this work written by
U.S. government employees as a part of their official duties.

Printed in the United States of America on acid-free paper

The American Phytopathological Society
3340 Pilot Knob Road
St. Paul, MN 55121-2097, USA

Preface

This compendium is the cumulative effort of many forest pathologists from most parts of the world where conifers are important. Each author was asked to address the topic globally but was allowed to emphasize the local situation that was most familiar. The authors of individual sections are recognized at the end of each contribution, and their affiliations are given below.

We have attempted to make this compendium useful to a wide range of people, from students in forestry and pathology to practitioners and researchers who require a broad view of the field. Each article concludes with a short list of references to provide access to the relevant literature. General references and texts on forest pathology are listed in the introduction to this volume.

The editors gratefully acknowledge the following individuals, who prepared sections of the compendium or contributed slides or figures:

Don Barrett, Wood End, Oxford, England

George M. Blakeslee, School of Forest Resources and Conservation, University of Florida, Gainesville

James W. Byler, U.S. Department of Agriculture, Forest Service, Coeur d'Alene, Idaho

Brenda Callan, Pacific Forestry Centre, Canadian Forest Service, Victoria, British Columbia

Gary A. Chastagner, Research and Extension Center, Washington State University, Puyallup

Teresa Coutinho, Department of Microbiology and Biochemistry, University of the Orange Free State, Bloemfontein, South Africa

Claude Delatour, Laboratorie de Pathologie Forestiere, INRA, Seichamps, France

L. David Dwinell, U.S. Department of Agriculture, Forest Service, Athens, Georgia

Gregory Filip, Department of Forest Science, Oregon State University, Corvallis

W. Forstreuter, Philipps-Universitat, Fachbereich Biologie, Morphologie und Systematik der Pflanzen, Marburg, Germany

Peter D. Gadgil, Forest Research Institute, Rotorua, New Zealand

Brian Geils, U.S. Department of Agriculture, Forest Service, Ft. Collins, Colorado

John Gibbs, Forest Authority, Research Division, Alice Holt Lodge, Surrey, England

Donald J. Goheen, U.S. Department of Agriculture, Forest Service, Central Point, Oregon

Thomas C. Harrington, Department of Plant Pathology, Iowa State University, Ames

Magnus Hellgren, Department of Forest Mycology and Pathology, Swedish University of Agricultural Sciences, Uppsala

Paul E. Hennon, U.S. Department of Agriculture, Forest Service, Juneau, Alaska

Yasuyuki Hiratsuka, Tattori Mycological Institute, Japan, and Northern Forestry Centre, Canadian Forest Service, Edmonton, Alberta

Ian A. Hood, New Zealand Forest Research Institute, Rotorua, New Zealand

Richard Hunt, Pacific Forestry Centre, Canadian Forest Service, Victoria, British Columbia

John L. Innes, Swiss Federal Institute for Forest Snow and Landscape Research, Birmensdorf

Risto Jalkanen, Finnish Forest Research Institute, Rovaniemi

Glen A. Kile, CSIRO, Division of Forestry, Parkes, New South Wales, Australia

Kari Korhonen, Finnish Forest Research Institute, Vantaa

E. George Kuhlman, U.S. Department of Agriculture, Forest Service, Athens, Georgia

Paul D. Manion, College of Environmental Science and Forestry, State University of New York, Syracuse

M. D. Mehrotra, Division of Forest Pathology, Forest Research Institute, Dehra Dun, India

Tatanya I. Morozova, Siberian Institute of Physiology and Biochemistry of Plants, Irkutsk, Russia

Kenneth M. Old, CSIRO, Division of Forestry, Parkes, New South Wales, Australia

D. Omdal, Washington Department of Natural Resources, Olympia

C. Parker

Robert F. Patton, Department of Plant Pathology, University of Wisconsin, Madison

Roger Peterson, U.S. Department of Agriculture, Forest Service, Ft. Collins, Colorado

Harry Powers, U.S. Department of Agriculture, Forest Service, Athens, Georgia

Zdenka Prochazkova, Forestry and Game Management Research Station, Uherske Hradiste, Czech Republic

Pablo Rosso, Department of Botany and Plant Pathology, Oregon State University, Corvallis

L. F. Roth, Department of Botany and Plant Pathology, Oregon State University, Corvallis

Robert F. Scharpf, Placerville, California

C. G. (Terry) Shaw III, U.S. Department of Agriculture, Forest Service, Juneau, Alaska

Wayne A. Sinclair, Department of Plant Pathology, Cornell University, Ithaca, New York

Halvor Solheim, Norwegian Forest Research Institute, As, Norway

Glen R. Stanosz, Department of Plant Pathology, University of Wisconsin, Madison

Jan Stenlid, Department of Forest Mycology and Pathology, Swedish University of Agricultural Sciences, Uppsala

Jeffrey Stone, Department of Botany and Plant Pathology, Oregon State University, Corvallis

Jack R. Sutherland, Applied Forest Science and Pacific Forestry Centre, Victoria, British Columbia

Frank H. Tainter, Department of Forest Resources, Clemson University, Clemson, South Carolina

Walter G. Thies, U.S. Department of Agriculture, Forest Service, Corvallis, Oegon

Borys Tkacz, U.S. Department of Agriculture, Forest Service, Flagstaff, Arizona

Kim Von Weissenburg, University of Helsinki, Helsinki, Finland

D. Wiens, Rancho Santa Ana Botanic Garden, Claremont, California

A. Dan Wilson, U.S. Department of Agriculture, Forest Service, Stoneville, Mississippi

Michael J. Wingfield, Department of Microbiology, University of the Orange Free State, Bloemfontein, South Africa

Stephen Woodward, Department of Forestry, University of Aberdeen, Aberdeen, Scotland

Contents

Introduction
- 1 Purpose and Scope
- 1 Diseases of Conifers

Part I. Diseases of Forest Trees

- **4 Root Diseases**
- **4 *Phytophthora* Diseases**
- 4 Littleleaf Disease
- 6 Port Orford Cedar Root Disease
- **7 Rhizina Root Disease**
- **8 *Leptographium* Diseases**
- 8 Black-Stain Root Disease
- 9 Other *Leptographium* Species Associated with Conifer Roots
- **11 Root Decays**
- 12 Annosum Root Rot
- 13 Armillaria Root Disease
- 14 Laminated Root Rot
- 16 Red-Brown Butt Rot
- 16 Tomentosus Root Rot
- 17 Brown Root Rot
- 18 Other Root-Rot Fungi
- **18 Stem and Branch Diseases**
- 18 Blue-Stain Fungi and Bark Beetles
- 19 Pine Wilt
- **20 Stem Decays**
- **21 Wound Decays**
- 21 Red Rot
- 21 Amylostereum Rot
- 22 Yellow Brown Top Rot
- 23 Brown Crumbly Rot
- 23 Other Wound Decays
- **24 True Heartrots**
- 24 Red Ring Rot
- 25 Rust-Red Stringy Rot
- **26 Stem Rusts**
- **26 Rusts of Soft Pines**
- 26 White Pine Blister Rust
- 27 Other Soft Pine Stem Rusts
- **27 Rusts of Hard Pines**
- 27 Fusiform Rust
- 29 Eastern Gall Rust
- 29 Western Gall Rust
- 30 Resin Top Disease
- 31 Comandra Blister Rust
- 31 Stalactiform Blister Rust
- 32 Sweetfern Blister Rust
- 32 Other Hard Pine Stem Rusts
- **33 Limb Rusts**
- **33 Pine Twist Rust**
- **34 Gymnosporangium Stem Rusts**
- **36 Parasitic Plants**
- 36 Dwarf Mistletoes
- 38 Leafy Mistletoes and Other Parasitic Plants
- **40 Cankers and Twig Blights**
- 41 Larch Canker
- 42 Sphaeropsis Shoot Blight and Canker
- 43 Scleroderris Canker
- 45 Pitch Canker
- 46 Atropellis Cankers
- 47 Other Canker Diseases
- **49 Cone and Seed Diseases**
- 49 Spruce Cone Rust
- 50 Southwestern Pine Cone Rust
- 50 Southern Pine Cone Rust
- 50 Cone and Seed Fungi
- **51 Foliage Diseases**
- 51 Needle and Broom Rusts
- **53 Needle Blights and Needle Casts**
- 54 Rhabdocline Needle Casts
- 55 Swiss Needle Cast
- 57 Brown Spot Needle Blight
- 57 Dothistroma Needle Blight
- 59 Cyclaneusma Needle Cast
- 59 Other Foliage Diseases of Pines
- 61 Foliage Diseases of Other Conifers
- **64 Abiotic Diseases**
- **66 Decline Diseases**
- 67 Recent Forest Declines in Europe
- 69 Northeastern Subalpine Red Spruce Decline
- 70 Yellow-Cedar Decline
- 71 Pole Blight
- 72 Austrocedrus Decline

Part II. Diseases in the Forest

- 73 Europe
- 74 Fennoscandia
- 76 Boreal Forest of North America
- 77 Eastern Siberia and the Russian Far East
- 79 Coastal Western North America
- 80 Inland West of North America
- 82 Southern United States
- 83 India
- 85 Australia and New Zealand
- 86 Africa

Part III. Diseases in Special Settings
88 Forest Tree Nurseries
88 Christmas Tree Plantations
89 Horticultural Landscapes

93 **Host Index**

95 **Index**

Color Plates (following page **48**)

Introduction

Purpose and Scope

This compendium differs from others in this series in several ways. It covers a larger and more diverse group of hosts than other compendia and deals with diseases of conifers growing both in forests managed for the production of forest products and in wild, unmanaged stands. We have thus tried to present both the economic and the ecological impacts of conifer diseases. In order to stay within the strict format of the compendium series, we have focused on the most significant diseases of the more important genera. We have concentrated on diseases of conifers as forest trees, only briefly addressing diseases in ornamental and landscape, nursery, and Christmas tree settings. Finally, we have tried to place the importance of conifer pathology in the larger context of forestry in the world. A series of essays discusses the significance of conifers and their diseases in different geographical regions. The intent is to identify the subset of diseases that have regional significance and that make a real difference in the way forests grow and the values gained from them.

Diseases of Conifers

The importance of conifers in the world extends far beyond their use as raw material for forest products. They are key components of forested ecosystems that cover a large portion of the northern hemisphere and key contributors to the global carbon cycle and ecosystem-level processes. They provide habitat for many other organisms. Diseases cause important economic losses in conifer forests, but they are also important regulators of forest structure and composition as well as of decomposition and nutrient cycling processes.

The causal agents of disease in coniferous trees and angiosperms span the same diverse array of organisms and abiotic stresses, but the relative importance of the various groups differs markedly. Viruses, bacteria, and nematodes are all present in and on conifers but are seldom associated with pathogenesis. Viruses may be more widespread than presently recognized (witness the recent demonstration of tomato mosaic virus in red spruce trees in the northeastern United States), but because of problems with extracting them from conifer tissue, they may remain undetected. Few bacterial diseases are known, perhaps because of the generally acidic pH of conifer environments. Several species of nematodes can reach damaging populations in conifer nurseries, but nematodes generally do not cause problems in the forest. The dramatic exception is the pinewood nematode, which is very destructive in Japan and China.

Most diseases of conifers are caused by fungi and the group of parasitic plants called mistletoes. Especially in the drier parts of western North America, dwarf mistletoes are often the most conspicuous and damaging forest pathogens. The center of diversity for both dwarf mistletoes and their Pinaceae hosts lies in this region, and mistletoes and conifers have coevolved. These obligate parasites distort and divert tree growth to their own benefit but kill trees only slowly.

The ubiquitous "fungus roots," or mycorrhizae, of trees make fungi parts of the trees themselves. This symbiosis, however, is just the middle ground of a spectrum of nutritional relationships between fungi and conifers, each as essential to the forest as mycorrhizae. Saprophytic fungi, at one end of the scale, decompose the dead needles, cones, twigs, and branches shed continuously in the forest. Without them, many of the nutrients in the system would be tied up in an ever-deepening litter layer.

Pathogenic fungi depend on living trees for at least a portion of their nutrition. Many have saprophytic ability as well or live on the dead parts of live trees, but others are obligate parasites that thrive on vigorous trees. These fungi have coevolved with their hosts and exist in an often delicate balance with them. Just as the saprophytic and symbiotic fungi play essential roles in the forest, so the pathogens manage many ecosystem interactions. In a forest without disease, succession would be slowed and plant and animal diversity lowered, with special losses among cavity-dwelling birds and mammals. On the other hand, without disease, the conifer forest of western Oregon would still have western white pine as an important component and larch could be productively grown throughout northern Europe. Disease, caused mostly by fungi, results in more economic destruction in forests than insects and fire combined. In Oregon and Washington, an estimated 13% of the annual tree growth is lost to disease.

Biology and Pathology

Fungi pathogenic to conifers are many and diverse. They include members of Oomycetes and especially Ascomycetes and Basidiomycetes and attack trees in all their parts and at all life stages from seed to senescence. Their parasitism ranges from nonspecific and crude attacks by wound pathogens and some toxin producers to the intimate and prolonged cell-to-cell contact of the obligate parasites within their hosts.

Oomycetes are primitive organisms more closely related to algae than to other fungi. Species of *Phytophthora* and *Pythium* are pathogenic on conifers. Most have wide host ranges and elaborate life cycles, producing spores that swim and hormones that diffuse through the watery milieu in which they live. The pathogenic ascomycetes are a more diverse group. Most spread by rain-splashed or windblown spores. Many are foliar pathogens, and another group induces stem cankers.

From the forest pathologist's point of view, there are two important groups of basidiomycetes, the rusts and the wood-rotters. The tree rusts have highly involved life cycles and highly evolved host-parasite relationships. Rusts, like most other obligate parasites, have evolved intimate cell-to-cell relationships with their hosts and may persist many years without seriously damaging the tree. The higher basidiomycetes, the hymenomycetes (or pore and gill fungi), have life cycles that are much less elaborate than those of the rusts, but in some cases, they maintain an even longer-lasting relationship with their hosts. These include the wood-decomposing fungi that have the enzymes to break down and metabolize cellulose and lignin.

Ecology: Disease in the Forest

Pathogenic fungi and the diseases they cause are important factors in the interactions between natural forest ecosystems. Individual trees in a natural forest are often not healthy, but the forest as a whole is seldom threatened. Death and disability from disease come to individuals in forest populations as they do to humans, most frequently as an accompaniment to old age. Wood decay is one of the dominant influences on the structure of old-growth forests. Both the true heartrot fungi, which enter through natural infection courts, and wound-decay fungi, which infect fire scars or wounds on fallen trees, increase in individual trees and in a stand as a whole with time. From 5 to 60% of the gross wood volume of old-growth Douglas-fir stands in Oregon is decayed, 82% of that by *Phellinus pini*. Decay is probably as important as any other single influence in pushing western forest stands toward the usual climax composition of western hemlock. Although hemlock may regenerate under the Douglas-fir canopy, the trees stay in the understory until the overstory fir stand begins to break up from the cumulative effects of age, principally decay, and the associated wind breaks and bark beetle attack.

Root-rot fungi also change the nature of the natural forest. They are the oldest and largest organisms in the forest, outliving the trees themselves and surviving from one tree generation to the next in the roots and stumps of infected trees. If we can adjust our thinking to the spatial and temporal scale of the forest, root rot might appear as a wave of change washing across and back again. *Phellinus weirii*, for example, spreads vegetatively from infected to healthy roots through the forest at a rate of about 34 cm per year.

Healthy, young-growth conifer forests are often very monotonous places, with a species diversity approaching that of a corn field. But in slowly expanding root rot infection centers, the stand is opened by death of susceptible conifers, and various herbaceous and woody angiosperms become established in the light. Animal diversity around infection centers also increases with the new sources of food and shelter. Disease- and shade-tolerant trees may seed in and grow through the brush, but several species, which require more light and bare soil for seedling establishment, do not. In the absence of susceptible tree roots in the infection center, the pathogen dies out of the originally killed root systems over a period of 40–50 years while spreading within the surrounding stand. A rough balance is established between the fungus and the forest as the disease slowly moves, leaving a different plant and animal community behind. When catastrophic fire renews the forest, the pathogen may reappear in a mosaic from some old infected snags and other formerly infected areas may be healthy.

Bark beetles are involved in the disease cycle. Blue-stain fungi (*Ceratocystis* and *Ophiostoma* spp.) help bark beetles kill trees. When conditions are favorable, often after drought or windstorms, bark beetle populations may become enormous in stressed or windthrown trees and then healthy trees are attacked and killed. However, most often it is the weakened or recently killed trees that sustain the initial attacks in new outbreaks and maintain the year-to-year endemic populations, and root-disease fungi such as *Armillaria ostoyae*, *Leptographium wageneri*, and *Heterobasidion annosum* provide a steady source of these trees. New beetle attacks are strongly correlated with preexisting root disease centers in western North America and no doubt elsewhere as well.

Pathogenic fungi are important in the response of the forest to external changes as well as to its internal dynamics. Plant disease is the consequence of interactions between the host plant, the pathogenic fungus, and the environment. A change in any one factor, such as the virulence of the pathogen or the susceptibility of the tree, can influence disease severity. These interactions are among the mechanisms by which a forest community adapts to the changing environment in which it grows. Infrequent weather extremes, such as early frosts, may damage individual trees with low tolerance to cold. In many cases, trees weakened but not killed by the effects of weather are attacked by canker fungi. Canker- and needle-disease fungi seem to specialize in removing poorly adapted individuals from a stand or at least in reducing their reproductive potential. Especially in times of climatic change, disease may be an important factor in maintaining the fitness of the tree population.

Disease in Disrupted Environments

The forest is a resilient community capable of changing through time in response to changing conditions. Fungi are the instruments of much of this dynamism. When environmental change is too extreme or too rapid, however, the fungal response can damage or even destroy the forest. Disruptions to the forest environment by human activity have created most of the damaging disease problems that engage forest pathologists today. The simple practice of leaving stumps rather than snags has created a new problem with root-rot fungi. *H. annosum* is a root and butt rot fungus whose spores normally infect trees through wounds, but fresh stump surfaces provide ideal infection courts. Since stumps seem to be an inevitable by-product of forest harvest, we have allowed infection by this fungus to reach a level much greater than that in the undisturbed forest.

Rust fungi respond to different environmental changes. Fusiform rust (*Cronartium quercuum* f. sp. *fusiforme*), for example, was a mycological curiosity on pines and oaks throughout the nineteenth century in the southeastern United States. This natural, balanced, host-pathogen relationship was upset by intensive forest management. Silvicultural and genetic innovations that increased pine growth also led to increased infection by this obligate parasite.

The consequence of extreme environmental change is often a damaging disease epidemic. This has happened repeatedly where tree species or pathogenic fungi are introduced into a new environment. Native host-parasite combinations have a genetic buffer that tends to limit damage, even when the environment is disturbed. Movement of a tree or fungus beyond its area of adaptation removes the buffer, sometimes with disastrous consequences. The move need not be far. Several attempts to grow ponderosa pine in plantations west of the Cascade Mountains in Oregon have failed because of epidemics of *Lophodermella morbida*, a needle cast fungus, even though ponderosa pine grows naturally in pure stands less than 50 miles to the east. The fungus is rare under the dry conditions of the main pine range; but in this new, wetter environment, it has been destructive.

Trees moved far beyond their native range often meet the same end. In 1957, F. R. Moulds wrote an article published in the *Journal of Forestry* (volume 55, pages 563–566) entitled "Exotics Can Succeed in Forestry as in Agriculture" and cited the greatly increased growth of Monterey pine in East Africa, Chile, and New Zealand compared with that of its economically unsatisfactory form in its native California. One reason given for success was the escape from native diseases with the move to a new environment. Within a few years, however, *Dothistroma pini*, an innocuous needle fungus on native pines in North America, had devastated the exotic plantations in the southern hemisphere. The title remains accurate, however. Monterey pine is still grown commercially in New Zealand with the aid of repeated fungicide sprays and the selection of resistant trees—just like in agriculture.

(Adapted in part from Hansen, E. M., 1977, Fungi, forests, and man, pages 105-124 in: Mushrooms and Man, T. Walters, ed., Linn-Benton Community College, Albany, OR)

Selected References

Allen, E. A., Morrison, D. J., and Wallis, G. W. 1996. Common Tree Diseases of British Columbia. Natural Resources Canada, Canadian Forest Service, Victoria, British Columbia.

Bakshi, B. K. 1976. Forest Pathology: Principles and Practice in Forestry. Controller of Publications, Delhi, India.

Browne, F. G. 1968. Pests and Diseases of Forest Plantation Trees. Clarendon Press, Oxford.

Butin, H. 1995. Tree Diseases and Disorders: Causes, Biology, and Control in Forest and Amenity Trees. D. Lonsdale and R. Strouts, eds. and trans. Oxford University Press, Oxford.

Commonwealth Mycological Institute. 1980. Distribution Maps of Plant Diseases. 3rd ed. Commonwealth Mycological Institute, Kew, England.

Farr, D. F., Bills, G. F., Chamuris, G. P., and Rossman, A. Y. 1989. Fungi on Plants and Plant Products in the United States. American Phytopathological Society, St. Paul, MN.

Funk, A. 1981. Parasitic microfungi of western trees. Can. For. Serv. Pac. For. Res. Cent. Inf. Rep. BC-X-222.

Funk, A. 1985. Foliar fungi of western trees. Can. For. Serv. Pac. For. Res. Cent. Inf. Rep. BC-X-265.

Gibson, I. A. S. 1979. Diseases of Forest Trees Widely Planted as Exotics in the Tropics and Southern Hemisphere. Vol. 2, The Genus *Pinus*. Commonwealth Mycological Institute, Kew, England, and Commonwealth Forestry Institute, University of Oxford, Oxford.

Hepting, G. H. 1971. Diseases of forest and shade trees of the United States. U.S. Dep. Agric. For. Serv. Agric. Handb. 386.

Hiratsuka, Y. 1987. Forest tree diseases of the prairie provinces. Can. For. Serv. North. For. Cent. Inf. Rep. NOR-X-286.

International Union of Forest Research Organizations (IUFRO). 1988. Proc. Working Group Meet. S2.06.01, Root and Butt Rots; S2.06-10, Rusts of Pine; S2.06.02, Canker Diseases; S2.07.09, Diseases and Insects in Nurseries. L. Jukka, ed. En Bok om Skogens Hälsa, Samerka Ab, Helsinki, Finland.

Ivory, M. H. 1987. Diseases and disorders of pines in the tropics: A field and laboratory manual. Overseas Dev. Adm. Oxford For. Inst. Overseas Res. Publ. 31.

Manion, P. D. 1991. Tree Disease Concepts. Prentice Hall, Englewood Cliffs, NJ.

Myren, D. T., ed. 1994. Tree Diseases of Eastern Canada. Natural Resources Canada, Canadian Forestry Service, Science and Sustainable Development Directorate, Ottawa.

Peace, T. R. 1962. Pathology of Trees and Shrubs. Clarendon Press, Oxford.

Phillips, D. H., and Burdekin, D. A. 1982. Diseases of Forest and Ornamental Trees. MacMillan, London.

Riffle, J. W., and Peterson, G. W., tech. coords. 1986. Diseases of trees in the Great Plains. U.S. Dep. Agric. For. Serv. Gen. Tech. Rep. RM-129.

Scharpf, R. F. 1993. Diseases of Pacific Coast conifers (revised). U.S. Dep. Agric. For. Serv. Agric. Handb. 521.

Sinclair, W. A., Lyon, H. H., and Johnson, W. T. 1987. Diseases of Trees and Shrubs. Cornell University Press, Ithaca, NY.

Tainter, F. H., and Baker, F. A. 1996. Principles of Forest Pathology. John Wiley & Sons, New York.

Part I. Diseases of Forest Trees

Root Diseases

Phytophthora Diseases

Phytophthora is a genus in the class Oomycetes. These water molds produce swimming zoospores to seek out and infect the fine roots of their hosts in water-saturated soils. They are not true fungi but are more closely related to some algae. *Phytophthora* species are important agricultural pathogens, and they attack conifers when they are grown in agricultural situations (in nurseries or as Christmas trees, in seed orchards, or in old field sites) more frequently than in forests. *Pythium*, another genus of water mold, is important in nurseries as a cause of damping-off.

The symptoms on the tree reflect destruction of the absorptive root system. Aboveground, the entire crown is affected, showing symptoms of chronic water stress (reduced growth, chlorosis, and needle loss) as soils dry. In some host-*Phytophthora* combinations, the pathogen grows up the roots through the phloem tissue and girdles the tree, resulting in rapid death. In other cases, infection is confined to the fine roots and results in a chronic disease.

It is often difficult to isolate *Phytophthora* from diseased trees. Isolation from freshly infected tissues, selective media, or baiting is usually required. In practice, diagnosis is usually based on root symptoms and site conditions. Disease is often associated with poorly drained soils or seasonal flooding.

Zoospores initiate infections. They are chemotactic and follow gradients of root exudates in saturated soils to find susceptible rootlets. The swimming spore encysts behind the root tip, germinates, and penetrates the root cortex. Enzymes kill cells, and the pathogen absorbs released nutrients for its own growth. As roots die, *Phytophthora* often forms resting spores (chlamydospores or oospores) with thick cell walls. These spores allow it to survive dry periods and may be released into the soil as roots decompose. The resting spores may survive 5 years or longer, germinating to form zoospores when soil conditions are again favorable.

Little is known about *Phytophthora* species in forest situations. *P. cinnamomi* has been introduced into many ecosystems around the world; littleleaf disease in the southeastern United States is perhaps the best known example on conifers. *P. lateralis* was introduced into the mixed coniferous forests of the western United States and causes a lethal root disease of Port Orford cedar. These examples illustrate the danger that *Phytophthora* spp. pose to coniferous forests. *P. gonapodyides*, on the other hand, is readily recovered from some forest streams in Europe and western North America but has not been associated with a disease of trees in the forest. Much more work is needed to understand the ecology and pathology of *Phytophthora* spp. in forest situations.

Littleleaf Disease

Other name: Phytophthora root rot
Causal organism: *Phytophthora cinnamomi* Rands in poorly drained, eroded soils
Hosts: shortleaf pine (*Pinus echinata*) and to a lesser degree, other southern pines
Distribution: the southern Piedmont Region of the southeastern United States

Littleleaf is a root disease complex involving a soil-inhabiting fungus, *Phytophthora cinnamomi*, and a host predisposed by unfavorable soil conditions. It represents a classic example of how land abuse and changes in forest ecosystems have led directly to the development of a new disease.

Land clearing for agriculture in the southeastern United States accelerated after 1790. Repeated cropping of cotton, with attendant soil loss, eventually produced impoverished farmlands that were subsequently abandoned. Much of this idle land, reduced to subsoil from sheet and gully erosion, was naturally colonized by shortleaf and other southern pines after it was abandoned for the last time during the early 1900s, resulting in overstocked, unmanaged stands. Approximately 20–50 years later, disease was found to be most severe in stands on these infertile, old field sites.

From about 1850 to 1875, before the introduction of the chestnut blight fungus, chestnut and chinquapin were killed across much of their southern ranges in the United States. *P. cinnamomi* has been implicated as the cause. It appears that the pathogen was well established prior to the major episodes of land abuse that lead to soil loss, land abandonment, and host revegetation, which finally led to the development of littleleaf disease.

Importance and Geographic Distribution

In the comprehensive disease survey of 1952, the impact of littleleaf disease on annual growth accounted for a sawtimber loss of 146 million board feet. The disease ultimately affected approximately 6 million hectares, or about 35% of the commercial shortleaf pine area. On 2.4 million hectares, it was regarded as the most serious limitation to the sustained management of shortleaf pine. The disease was prevalent in the commercial range of shortleaf pine, an eight-state area that includes the Piedmont Plateau (Alabama through Virginia), the Appalachian Plateau (eastern Kentucky and Tennessee), and north-central Mississippi in the Atlantic Coastal Plain. Distribution of the disease was correlated with stand age, poorly drained soils, and the degree of erosion. Today, littleleaf is less of a problem because of the decreased importance of shortleaf pine and its replacement by more resistant species.

Causal Organism

Early failures in isolating *P. cinnamomi* directly from roots are readily explained by the ephemeral nature of root pathogenesis. The fungus attacks and quickly kills only the succulent root tips of the pine host. Saprophytic dominance and survival of the pathogen within the root tissues is short lived because of the suppressing effect of competitive saprophytic colonization by soil microorganisms. *P. cinnamomi* was first isolated from soil by using apples as bait and later from diseased rootlets with selective media. Today, serological assay kits in which enzyme-linked immunosorbent assay (ELISA) is used are commercially available to detect *P. cinnamomi* in soil and plant samples.

Other soil organisms influence root pathogenesis. The abundance of plant-parasitic nematodes, the sheath (*Hemicycliophora vidua*) and spiral (*Helicotylenchus dihystera* and *H. erythrinae*) nematodes in particular, in diseased stands was twice that found in healthy stands of shortleaf pine. Species such as these may be important factors that contribute to overall root injury in the littleleaf complex. Naturally occurring mycorrhizae also deter *P. cinnamomi* on shortleaf pine, physically and chemically. Complete resistance of shortleaf pine mycorrhizae to infection by zoospores of *P. cinnamomi* was shown, whereas nonmycorrhizal roots were completely susceptible.

P. cinnamomi is identified in culture by its relatively rapid growth in petaloid or camellioid patterns on agar media. Its vegetative hyphae give rise to microscopic, globose to pyriform chlamydospores that form singly or in clusters of three to 10. Long-term survival of *P. cinnamomi* in soil and host tissue is likely by means of chlamydospores. Asexual reproduction occurs in the presence of free water with the development of stalked, terminally borne, lemon-shaped sporangia.

Epidemiology

Symptoms of littleleaf arise as a result of nitrogen deficiency in the tree. This deficiency is associated with a reduction in the absorptive capacity of the rootlets caused by the primary attack by *P. cinnamomi*. Other soil factors, specifically poor aeration, low fertility, and periodic moisture stress, are also injurious to the rootlets and may prevent root rejuvenation and tree recovery.

The influences of the soil environment on the host, the pathogen, and the host-pathogen interaction are the primary determinants of disease expression. Excessive soil moisture and the consequent poor aeration are particularly important when temperature is favorable. High moisture favors mobility of the pathogen, while poor aeration is debilitating to the host and even influences zoospore attraction. In the long term, poor soils favor the disease primarily by reducing the host's vigor and its ability to regenerate normal roots after local infection of short duration.

Diagnosis

The early symptom of littleleaf is rootlet mortality. As root loss progresses without rootlet regeneration, the secondary symptoms of crown decline become visible. Initially, only a slight yellowing and dwarfing of new foliage (symptoms typical of nutrient or water deficiencies) may be evident. Within a few years, however, the symptoms advance to the diagnostic littleleaf condition (Fig. 1). The crown is thin; shoot growth is stunted; and the branches, with only terminal tufts of needles to support, often assume an ascending habit. The sparse foliage that remains becomes chlorotic and may be reduced in length from the normal 8–13 cm to approximately 1 cm. As the foliage declines, crown dieback advances and radial growth is drastically reduced.

Final symptoms may include the abnormally heavy production of small cones, which persist and contain mostly abortive seeds. As the crown dies, sprouts of normal vigor and foliage color develop profusely along the lower part of the stem. Ordinarily, afflicted trees succumb within 6 years after the onset of visible symptoms; however, some may die as quickly as 1 year, and others may languish in gradual decline for 12 years or longer. During the past several decades, however, these stressed trees have often attracted the southern pine beetle and may be killed before the littleleaf disease complex can fully develop.

Control Strategy

In the southeastern United States, the fungus is nearly ubiquitous in the soil, and its presence must be assumed in any disease-management strategy in the Piedmont. The chances of a shortleaf pine stand developing littleleaf disease can be judged on-site by using a simple soil-rating scale (Fig. 2) based on four readily measurable soil characteristics. When high-risk sites have been identified, the forester is often advised to convert the stands to less susceptible species. Risk prediction is probably the single most important control strategy today.

Experience has shown that loblolly pine may survive the littleleaf complex somewhat better than shortleaf pine and grow into the small- to medium-sized sawlog class, thus producing a much more valuable product. Lateral roots of loblolly pine are less susceptible to infection than those of shortleaf pine.

Nutrient amendment by soil fertilization is the only successful example of littleleaf therapy to date. The application of 5-10-5 commercial fertilizer at the rate of 2,273 kg/ha together with 454 kg of ammonium sulfate prevents symptom development in healthy trees and improves the condition of trees with littleleaf during the early stages of disease development. Soil rehabilitation is an alternative that can ultimately lead to avoidance of the pathogen. Silviculturally converting the site to

Fig. 1. Healthy shortleaf pine (left) and a tree with typical symptoms of littleleaf disease (right). (Reprinted from Zak, 1957)

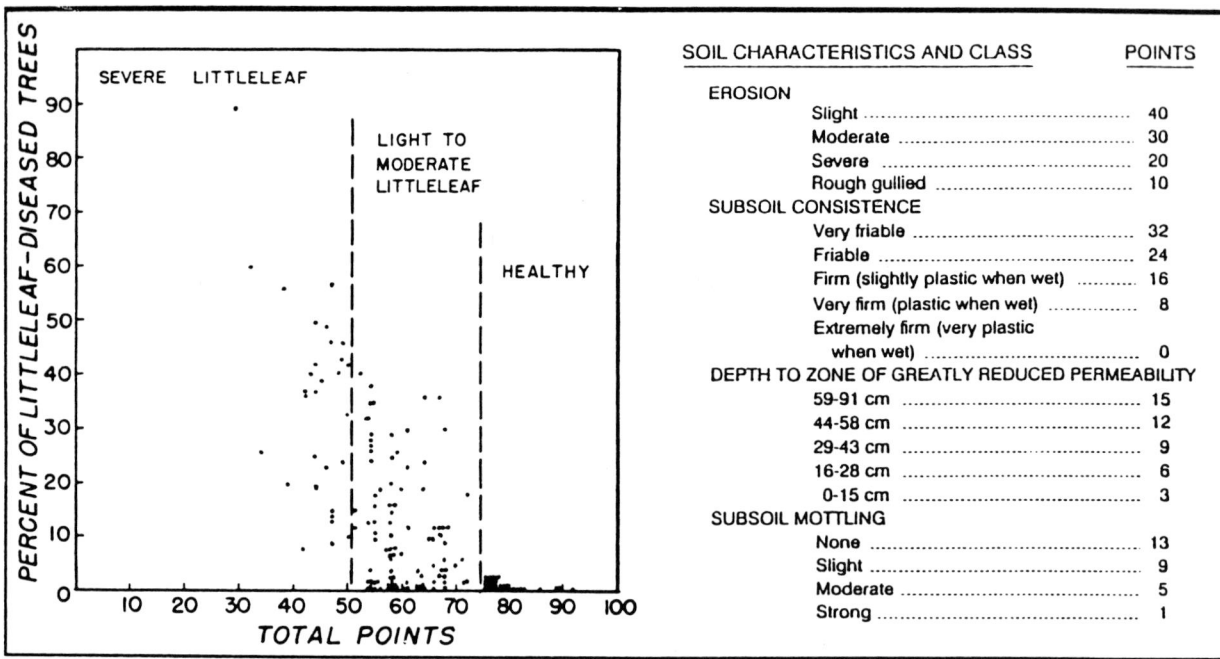

Fig. 2. Soil-rating scale for calculating the chance that a site will develop littleleaf disease. The higher the numerical index, the lower the risk of littleleaf. (Reprinted from Campbell et al, 1953)

soil-building species of hardwoods or even grass can renew the productivity of littleleaf sites while providing cover and protection from erosion.

Field observations and greenhouse inoculation trials indicate that individual shortleaf pines exhibit natural resistance to littleleaf disease. Presently, shortleaf pine from open-pollinated parents in seven states are being tested for long-term growth response on sites with severe littleleaf symptoms in Georgia, South Carolina, and Virginia. Evaluation of these plantings after 16 years has shown that shortleaf pine can be successfully selected for adaptability to littleleaf sites. Screening of tissue cultures derived from commercial seed collections suggests a widely diverse population with a high proportion of very resistant (tolerant) genomes.

Selected References

Campbell, W. A., Copeland, O. L., and Hepting, G. H. 1953. Managing shortleaf pine in littleleaf disease areas. U.S. Dep. Agric. For. Serv. Southeast. For. Exp. Stn. Pap. 25.

Copeland, O. L., Jr., and McAlpine, R. G. 1955. The interrelations of littleleaf, site index, soil, and ground cover in Piedmont shortleaf pine stands. Ecol. 36:635-641.

Oak, S. W., and Tainter, F. H. 1988. How to identify and control littleleaf disease. U.S. Dep. Agric. For. Serv. South. Reg. Prot. Rep. R8-12.

Zak, B. 1957. Littleleaf of pine. U.S. Dep. Agric. For. Serv. For. Pest Leafl. 20.

Zarnoch, S. J., Ruehle, J. L., Belanger, R. P., Marx, D. H., and Bryan, W. C. 1996. Growth and crown vigor of 25-year-old progenies on a littleleaf disease site. U.S. Dep. Agric. For. Serv. Res. Paper SE-289.

Zentmyer, G. A. 1980. *Phytophthora cinnamomi* and the Diseases It Causes. Monogr. 10. American Phytopathological Society, St. Paul, MN.

(Prepared by F. Tainter)

Port Orford Cedar Root Disease

Other names: Phytophthora root rot, cedar root rot
Causal organism: *Phytophthora lateralis* Tucker & Milbrath
Hosts: *Chamaecyparis lawsoniana* and *Taxus brevifolia*
Distribution: southwestern Oregon and northern California and ornamental plantings of Port Orford cedar in western Canada and the western United States

Port Orford cedar (POC), also known as Lawson's cypress, is very susceptible to this introduced pathogen. Pacific yew is killed more slowly and usually only in areas of high disease hazard. Other hosts are not known, and the origin of the pathogen has not been determined. Port Orford cedar root disease was first reported in 1923 in nurseries near Seattle, Washington, where ornamental cultivars of POC were growing. The disease eliminated POC as a commercial ornamental tree. In 1952, it was first reported on POC in the tree's native range and has now spread through most of the forest areas where the tree grows. The annual loss during 1980–1989 was estimated at 2,100 m^3 and valued at \$2,300,000. Ecological consequences of the loss of POC are evident in sensitive ecosystems.

Symptoms and Diagnosis

POC trees of all ages are killed rapidly, usually within 1 year of the first visible crown symptoms. *Phytophthora lateralis* colonizes the root phloem rapidly, growing to the root crown and then up the stem. A sharp demarcation between reddish, infected phloem below and white, healthy tissue above (Plate 1) is visible until the stem starts to dry out or until secondary bark beetles (*Phloeosinus* sp.) create extensive larval galleries in the dying inner bark.

Crown symptoms on yew develop more slowly, and several years can elapse between infection and death. The demarcation between living and dead phloem is visible in trees with advanced symptoms, but infected tissue is seldom found far above ground.

Field diagnosis in both cedar and yew is based on the distinctive demarcation between infected and healthy tissue in dying trees. The pathogen can be isolated from recently killed tissues, and commercial ELISA kits can be used to detect *P. lateralis* up to several years after tree death.

Disease Cycle

P. lateralis is a cool-climate species. It is active through the mild, wet winters of the area and inactive, or even dies, during the warm, dry summer months. It is carried upslope in mud and debris and washes downslope in water. In the native forests where POC grows, uphill transport is primarily along roads on vehicles and road-maintenance and logging equipment, while downslope movement occurs in streams and through overland flow during periods of heavy winter rains. Zoospores infect POC roots within 24 hr of contact.

Chlamydospores and oospores (*P. lateralis* is homothallic) form within 2 weeks in infected roots and remain infective for at least 6 years. Resting spores are transported in mud on vehicles during wet weather. If soil is later dropped near healthy cedar roots, the cycle may begin again as spores germinate to produce sporangia that release zoospores during periods of soil saturation. Zoospores swim only a few millimeters but can be carried much greater distances in flowing water. They can remain motile for several hours. Little infection occurs without saturated soil conditions.

Effects on the Forest

The consequences of infection are dramatic. Infected POC invariably die, and the extensive system of logging roads has provided access for the pathogen to most of the POC range. Many cedars survive, however, within the general area of introduction. POC grows in most locales as a minor component of the conifer forest, although it may be the predominant species in local areas, especially in wet places and on ultramafic soils (derived from old, metamorphosed rock high in several heavy metals). Mortality approaches 100% along infested streams and averages perhaps 25% adjacent to infested roads. Trees growing even a short distance away from these inoculum sources often remain healthy, however. Mortality levels away from roads and streams depend on the frequency and intensity of other soil-moving mechanisms, principally timber harvest and mining activity but also animals.

Ecological effects are most evident in habitats where POC is the predominant tree species. The sensitive riparian habitat is often dramatically altered with death of the POC. With death of the overstory, several species of evergreen shrubs often assume dominance, resulting in changes in the structure and microenvironment of both the terrestrial and the aquatic communities.

Distinctive plant community associations including many rare, endemic species are found on the ultramafic soils characteristic of much of the region. POC is especially important as a dominant conifer on these soils if subsurface water is available. Disease incidence has not been high on these sites, primarily because timber volumes are low and therefore opportunities for introduction of the pathogen are reduced. Where it is present, however, the pathogen has a dramatic effect by killing what is often the only overstory species growing on the site.

Disease Management

The U.S. Forest Service has an active and aggressive program to stop the further spread of the pathogen, reduce inoculum levels in infested areas, and assure the continued viability of POC as a forest species in the region. All forest management activities in areas where POC grows are evaluated for their potential effects on the POC resource, and activities are modified or appropriate mitigating measures are undertaken as necessary. Principal actions include road-use restrictions such as wet-season or permanent closure, sanitation through equipment washing, and inoculum reduction by removing POC from the roadside areas where they would be most vulnerable (Plates 2 and 3).

POC trees vary in susceptibility to the pathogen, and trees that die more slowly than most have been identified. Resistance is genetically controlled, and trees are being propagated for use in a breeding program. It is hoped that trees with useful resistance will be produced from controlled crosses between selected parents.

Selected References

Hansen, E. M., and Hamm, P. B. 1996. Survival of *Phytophthora lateralis* in infected roots of Port Orford cedar. Plant Dis. 80:1075-1078.

Hansen, E. M., Hamm, P. B., and Roth L. F. 1989. Testing Port-Orford-cedar for resistance to *Phytophthora*. Plant Dis. 73:791-794.

Kliejunas, J. 1994. Port-Orford-cedar root disease. Fremontia 22:3-11.

Zobel, D. B., Roth, L. F., and Hawk, G. M. 1985. Ecology, pathology, and management of Port-Orford-cedar (*Chamaecyparis lawsoniana*). U.S. Dep. Agric. For. Serv. Gen. Tech. Rep. PNW-184.

(Prepared by E. Hansen)

Rhizina Root Disease

Other names: tea break fungus, group dying, rotmurkla (Swedish), la rhizina (French), Wurzellorchel (German), kuplamörsky (Finnish)

Causal organism: *Rhizina undulata* Fr.:Fr. (syn. *Rhizina inflata* (Schaeff.) P. Karst.)

Hosts: many conifer species

Distribution: widely distributed around the world

Rhizina undulata is dependent on fires (e.g., forest fires, slash burning, and camp fires), which heat and thereby activate ascospores resting in the ground. In acidic soil, the hyphae from germinating ascospores may colonize roots of fresh or living coniferous stumps and live seedlings and trees. Yellowish white mycelial strands soon cover the roots, and the fungus penetrates the bark through lenticels, which it covers with tufts of white mycelium, and attacks the inner bark and cambial zone where necrotic areas appear.

Fruit bodies (apothecia) occur scattered on the burned ground or in "fairy rings" around fire sites early in the growing season. They are first visible as small, white balls and when about 6 mm in diameter, as flat, brown "buttons" with cream-colored rims (Plate 4). The young fruit bodies are tightly attached to the mineral soil or burned litter. They grow to a diameter of 3–8 cm, and their centers rise from the substrate, forming fleshy, convex, undulating, cushionlike structures with the typical cream-colored rims. Later in the growing season, several apothecia may fuse, forming irregular clumps up to 25 cm wide, which then turn black (Plate 5), lose the rim coloration, and start to deteriorate to the point that they can hardly be found the next growing season. The lower side is loosely connected to the substrate with fragile but numerous rhizomorphs or hyphal strands (Plate 6). The fruit bodies are often very numerous and on favorable sites can cover nearly half the soil surface. Yellow rhizomorphs can be found in the soil. In the boreal zone, the fungus persists in a parasitic and expansive stage for 2–3 years after a fire, but in other areas (e.g., the British Isles), it may persist up to 7 years. Viable ascospores remain in a dormant stage in the soil for tens of years. A saprophytic stage (on roots of recently cut stumps) may precede the parasitic stage by a few years.

Symptoms and Diagnosis

Needles of seedlings turn brown, and leader growth is stunted within a few weeks of infection during the early part of the growing season. No other symptoms occur on aboveground parts, but the roots are heavily infected and covered with yellowish white mycelium. Diseased mature trees have prolific cone production, reduced shoot growth, wilting leaders and lateral branches, copious resin flow on the lower part of the stems up to several meters, and yellow and later brown needles.

Even green needles can be shed suddenly. Trees die within 2–5 years after infection. In the boreal zone, mature trees are seldom or never attacked; but in other climate zones, infection of mature trees appears to be common. Aboveground symptoms are only indicative; diagnosis is based on the vicinity of fruit bodies (within a meter or so but not always present) and fire sites (within tens of meters) as well as the presence of fungal strands and lesions on the roots.

Effects on Tree and Forest

The effect on seedlings is dramatic and fast, and often nearly all seedlings planted immediately after a burn will die within a month or two (in Scandinavia, Siberia, and North America). In Scandinavia, England, Scotland, and Northern Ireland, seedlings on 60–70% of the burned areas can be heavily attacked in any given year (Plate 6). Mortality may exceed 70%. In western North America, reports on seedling mortality from the late 1960s and early 1970s are conflicting, but where planting is done immediately after a burn, which in turn has come immediately after a clear-cut, the fungus may cause severe local losses.

Seedlings of Douglas-fir (*Pseudotsuga menziesii*) are susceptible, but pole-stage trees are resistant or immune. *R. undulata* has caused death of old-growth western hemlock (*Tsuga heterophylla*) in Washington and on nearly mature red spruce (*Picea rubens*) in Vermont. Death in nurseries in the United States has been reported from Maine and Maryland. In central and southern Europe, mature conifers and especially trees 20–30 years old (pole stage) die at the edges of successively enlarging gaps (another name for the disease is "group dying"). The gaps can be up to 40 m across in stands of more susceptible hosts such as Sitka spruce (*Picea sitchensis*), but they are restricted to fire sites and their vicinity. Mature trees on the edges of burned areas up to several tens of meters (maximum 100 m) into the stand can be killed by the fungus, while seedlings usually die within only a few meters from the edge of the burn. Secondary damage caused by wind throw and sun scald is common around the edges of such gaps, extending them for several years after the fungus has disappeared.

Predisposing Factors

In addition to the absolute requirement of fire, conditions associated with Rhizina root disease include acidic soil and light sandy or peat soils rather than heavier clay and silty soils. Also, the previous crop must have been predominantly conifers. Fresh conifer roots have to be available within a year or two after the fire has activated the spores in the soil. The susceptibility of planting stock appears to be greater (or a larger portion is killed) than that of naturally regenerated trees.

Disease Management

Disease management is related to fire control and camping habits. In areas where the fungus is known to occur, camp fires should be avoided or separated from surrounding soil and conifer roots by ditches. Correspondingly, an expanding gap can be stabilized or arrested by maintaining a ditch around it for up to 7 years. Prescribed and slash burning should be avoided. However, because burning has beneficial silvicultural and ecological effects, other controls are often advised. Regeneration by sowing seed rather than planting may prevent damage. Also, timing of burning and planting is important. By delaying burning up to 3 years after clear-cutting or planting up to 3 years after burning, damage to seedlings is reduced. In Swaziland, a delay after fire of only 3 months was sufficient; but in South Africa, this was not adequate. The seedling mortality caused by broadcast burning is greater than that caused by burning slash gathered into a few heaps. Because of the erratic occurrence of the fungus on burned sites, specific precautionary measures are often neglected, but the damage to planting stock and pole-size stands as well as to mature trees can be considerable.

Selected References

Jalaluddin, M. 1967. Studies on *Rhizina undulata*. II. Observations and experiments in East Anglian plantations. Trans. Br. Mycol. Soc. 50:461-472.

Lundquist, J. E. 1993. Assessing Rhizina root disease hazard: A case study. S. Afr. For. J. 164:51-53.

Phillips, D. H., and Young, C. W. T. 1976. Group dying of conifers. Br. For. Comm. Leafl. 65.

Thies, W. G., Russell, K. W., and Weir, L. C. 1977. Distribution and damage appraisal of *Rhizina undulata* in western Oregon and Washington. Plant Dis. Rep. 61:859-862.

Thompson, J. H., and Tattar, T. A. 1973. *Rhizina undulata* associated with disease of 80-year-old red spruce in Vermont. Plant Dis. Rep. 57:394-396.

(Prepared by K. Von Weissenberg)

Leptographium Diseases

Leptographium spp. are conidial fungi with morphological features and sporulation behavior that facilitate dispersal by insects. Conidia, formed in a sticky drop at the end of a rigid conidiophore, and sometimes long-necked perithecia are formed in insect galleries beneath the bark of dead or dying trees. The spores are commonly carried by bark beetles. Sexual stages, where known, are in the genus *Ophiostoma* and are relatives of *O. ulmi*, the fungus that causes Dutch elm disease.

Some species, including *L. wageneri* on conifers, cause wilt diseases by colonizing the vascular system and disrupting water transport. These fungi progress vertically within the tree, well beyond the point of introduction. Most *Leptographium* spp., however, colonize ray parenchyma cells preferentially; colonization is primarily radial in the stem. These are the blue-stain fungi, commonly associated with bark beetles, and are discussed in the section on stem diseases. Another small group of *Leptographium* spp. is associated with root-feeding beetles and is discussed below.

Black-Stain Root Disease

Causal organism: *Leptographium wageneri* (Kendrick) M. J. Wingfield var. *wageneri* T. C. Harrington & F. W. Cobb; var. *pseudotsugae* T. C. Harrington & F. W. Cobb; and var. *ponderosum* (T. C. Harrington & F. W. Cobb) T. C. Harrington & F. W. Cobb (teliomorph *Ophiostoma wageneri* (D. J. Goheen & F. W. Cobb) T. C. Harrington; syns. *Verticicladiella wageneri* Kendrick and *Ceratocystis wageneri* D. J. Goheen & F. W. Cobb)

Hosts: *Pseudotsuga menziesii, Pinus ponderosa, Pinus monophylla, Pinus contorta, Tsuga mertensiana*, and others

Distribution: western United States and Canada (damage is most frequent in interior areas of southwest Oregon and northern California)

Black-stain root disease (BSRD) is caused by three varieties of *Leptographium wageneri*: var. *wageneri* on pinyon pines, var. *ponderosum* on ponderosa and lodgepole pines and mountain hemlock, and var. *pseudotsugae* on Douglas-fir (Plate 7). The varieties are host specific in the forest, although cross infections are observed in artificial inoculations. The disease was first described from pines in 1938 and from Douglas-fir in 1967. Awareness as well as disease incidence has increased rapidly, and today BSRD is one of the diseases of greatest concern in western forests. On Douglas-fir especially, disease severity is closely related to forest silvicultural activities, and the observed increase in damage in recent decades is apparently the result of intensified forestry operations. The conversion of large areas of old, natural forest to young, relatively uniform, dense stands of predominantly Douglas-fir and the practice of precommercial thinning have increased since the 1950s. The insect vectors create a potential for rapid disease increase and explain the interactions with forest operations. BSRD is found only in western North America, but both exotic and native conifers in Europe are susceptible, and insects that might function as vectors are present.

Symptoms and Diagnosis

Aboveground symptoms of BSRD are similar to those caused by other root-disease fungi, except that they develop more rapidly. Sapling Douglas-fir trees may die within 3 years of the appearance of the first visible symptoms. Loss of old needles occurs first, followed by reduced terminal growth, chlorosis, sometimes a crop of "stress cones," and death.

By the time crown symptoms are obvious, the characteristic black stain can usually be observed in the xylem at the soil line. In tangential view, the stain appears as broad streaks extending from the roots up the stem (Plate 8). In cross section, the stain appears in crescents and follows the annual rings. This pattern of staining contrasts with the radial sectors or more diffuse staining associated with other *Leptographium* spp. The stain fades after the tree is dead and may be masked by other stains from secondary fungi.

Diagnosis can be based on the distinctive stain pattern in fresh material. Confirmation is possible by microscopic examination of the hyphae in the xylem. Hyphae of *L. wageneri* are confined to xylem tracheids; they are not found in parenchyma cells (Plate 9). The sectored pattern associated with blue-stain fungi, in contrast, results from their colonization of ray parenchyma cells. The black-stain fungus can be isolated from stained wood in recently infected material. Isolation is difficult from areas of older infection and is usually not possible from trees that have been dead for a year or longer. The fungus grows on most media; the addition of cycloheximide to the medium reduces the growth of unrelated fungi.

Disease Cycle

L. wageneri can grow a few centimeters from root to root through the soil and through root grafts. Insects, however, provide its primary means of dispersal. In Douglas-fir, a root-feeding bark beetle, *Hylastes nigrinus* Mann., and two root and crown weevils, *Steremnius carinatus* Mann. and *Pissodes fasciatus* LeConte, are known vectors of the pathogen (Plate 10). Vectoring by similar insects is strongly suspected in the pines as well.

These insects breed in the dying roots of trees severely stressed by a variety of agents, including BSRD. They do not have an obligate relationship with the fungus, but they are very efficient at finding black-stain-infected roots on trees, even before crown symptoms are visible to the human eye. They overwinter as pupae in galleries excavated at the interface between xylem and phloem and emerge as adults in the spring. Adults fly (or in the case of *S. carinatus*, crawl), apparently following olfactory cues released from stressed trees. They land, burrow down to the roots, and feed on root phloem. If feeding wounds etch the xylem and if the insect is carrying the fungus, a new infection may result.

Effects on the Forest

Trees are killed by BSRD in irregularly expanding infection centers that can reach many hectares in size in pinyon woodlands. In Douglas-fir plantations, most infection centers become inactive while they are still small, but occasionally spread continues at a meter or more a year until trees are about 30 years old. Expansion usually slows as Douglas-fir trees mature, but mortality can continue for many years. Pines of all ages may be killed, although infected older trees are often attacked by other bark beetles that mask the underlying black-stain infection.

Disease incidence in Douglas-fir is highest next to roads, in plantations that have been precommercially thinned and that have suffered severe soil disturbance during the harvest of the previous stand, and in areas of disturbed soil drainage. Trees growing under these conditions are often stressed and attractive to the insect vectors of BSRD. After precommercial thinning, the insects respond to volatile chemicals released from the cut trees. Predisposing stand factors are not as well characterized for BSRD on pines but include high soil moisture and pure stands of pine on sites formerly occupied by other species.

BSRD is locally very damaging; mortality levels of up to 50% can occur in individual plantations of Douglas-fir. In most plantations, however, mortality does not exceed 1%. High mortality is associated with severely disturbed sites. If other conifer species are not present, then infection centers fill with hardwood brush. If hemlock or other conifers are present, they may occupy the site. In pinyon-juniper woodlands of the arid southwestern United States, pinyon pine is largely eliminated from the large infection centers.

Disease Management

At present, active management of BSRD is practiced only in Douglas-fir, and actions are preventative. Hazard rating is basic. If there is no BSRD in the vicinity, special precautions are not necessary. If BSRD is in surrounding stands, practices to reduce soil disturbance and tree stress and to avoid attracting the insect vectors are important. Harvest of the mature stand should minimize soil compaction, skid trails, and landings. This is best accomplished with various cable suspension systems. If ground-based skidding of logs is necessary, careful operation, appropriate equipment, and planned skid trails can minimize disturbance. Nonhost species can be planted in areas of greatest hazard, such as on landings and along roads.

Douglas-fir can be planted at a spacing wide enough that precommercial thinning is not necessary, if survival and establishment of the seedlings are satisfactory. If thinning must be done, nonhost species should be planted in areas of greatest hazard and near already established infections. The time of the year when thinning is done affects the attraction of vector insects to the slash. The *Hylastes* beetle has a single flight period during late spring or early summer. If thinning is finished after the flight but early enough in the summer that the slash has time to dry before winter, it will not be attractive to the insects the next spring. Thinning during the fall, winter, or early spring, however, leaves slash green and attractive for the next flight period.

Selected References

Goheen, D. G., and Hansen, E. M. 1993. Effects of pathogens and bark beetles on forests. Pages 175-196 in: Beetle-Pathogen Interactions in Conifer Forests. T. D. Schowalter and G. M. Filip, eds. Academic Press, San Diego, CA.

Harrington, T. C., and Cobb, F. W., Jr., eds. 1988. Leptographium Root Diseases on Conifers. American Phytopathological Society, St. Paul, MN.

(Prepared by E. Hansen)

Other *Leptographium* Species Associated with Conifer Roots

A few *Leptographium* species have been associated with conifer mortality, primarily as associates of root-feeding bark beetles (Scolytidae) and weevils (Curculionidae) that attack living trees. Except for *L. wageneri*, discussed above, *Leptographium* spp. are saprophytes or weak pathogens and are not believed capable of causing disease independently. They may, however, exacerbate the damage caused by the phloem-feeding insects.

Disease Cycle

Unlike stem-feeding bark beetles, which produce aggregating pheromones and attack trees en masse, root-feeding bark beetles and weevils do not tend to aggregate before they attack. Because the individual attacks cause damage over only a limited area, they usually do not kill large trees, but some of these insects have been noted as significant pests of newly planted seedlings and young trees. Other beetle species are capable of attacking larger trees during successive years, resulting in death of portions of the root stem or root collar area and, eventually, death of the entire tree. Some of the damage is caused by egg laying and larval mining, but much of the damage caused by root-feeding bark beetles results from the adults feeding on the inner bark of the root and root collar.

Pines and other members of Pinaceae generally respond to wounds caused by insects by producing resin, and a number of fungi are found in such resin-soaked tissues. Prominent among this mycoflora are species of *Leptographium* and other fungi with *Ophiostoma* teleomorphs. Some of these fungi are weak pathogens, as indicated by wound inoculations, and they appear to further damage roots already damaged by the insect grazing and egg laying. Among the insect-associated *Leptographium* spp. tested for pathogenicity, *L. terebrantis* S. J. Barras & T. J. Perry is the most virulent. Other *Leptographium* spp. produce more limited lesions.

Symptoms and Diagnosis

Common root symptoms of trees attacked by bark beetles or weevils and the associated *Leptographium* spp. include copious resin production and discoloration of the phloem and xylem. Substantial quantities of resin may exude from the site of insect damage. When it crystallizes, the resin may encrust the soil and other debris around the root and thus obscure the evidence of insect feeding. Phloem tissue surrounding the insect damage may darken as it dies, and the xylem tissue is frequently soaked with resin. Many *Leptographium* spp., especially *L. procerum* and *L. terebrantis* (Fig. 3), appear to grow well in resin-soaked tissues and may exacerbate the damage by inducing further resin production (Plate 11). Melanized hyphae are typically abundant in the ray parenchyma tissue, and the tissue colonized by the *Leptographium* species may be stained blue gray to black in a wedge-shaped pattern. Smaller trees may die as the resin-soaked xylem is occluded and no longer conducts water or as the trees are girdled by dying phloem tissue. In larger trees, individual roots can be killed or small patches of the xylem and phloem may die, leaving partially healed scars or "cat faces" at the bases of the trees. The patches tend to coalesce after insect attacks in successive years, and the entire tree may die.

Although *Leptographium* spp. are generally easy to isolate and the cultures are usually quickly identifiable to the genus level, identification to species is difficult. There are no keys to aid identification of the conidial states of these fungi. Many *Leptographium* spp. associated with conifer roots are morphologically similar and usually occur in association with other *Leptographium* spp. Many species await description.

Effects on Tree and Forest

Considerable losses associated with the insect damage described above have been noted in plantations of young seedlings or saplings, but such losses appear to be minimal in natural forests. Because the insects typically breed in stumps or other logging debris, most losses have been noted in cleared forests replanted with seedlings. In southern Africa, problems with root-feeding bark beetles developed in pine plantings after slash burning was discontinued because of problems with *Rhizina* infections of the seedlings. Christmas tree plantations typically generate stumps through selective harvesting, and these plantations frequently have problems with root-feeding weevils. Damage to larger trees by root-feeding bark beetles

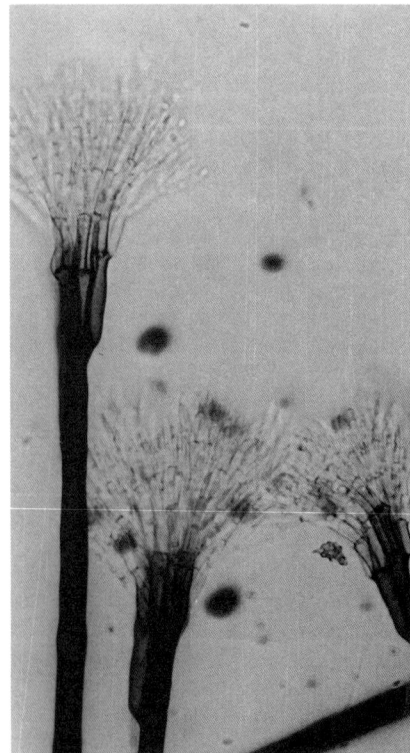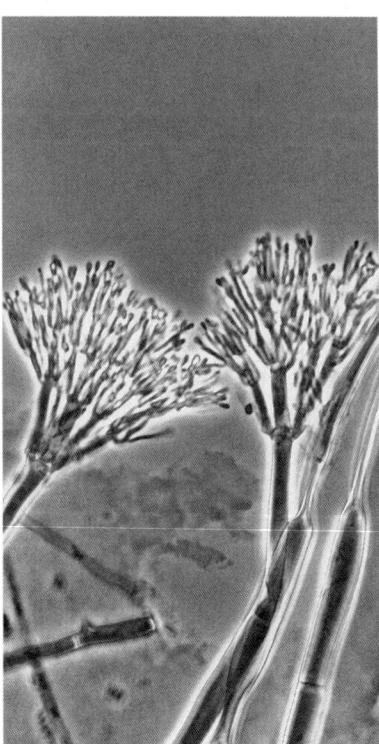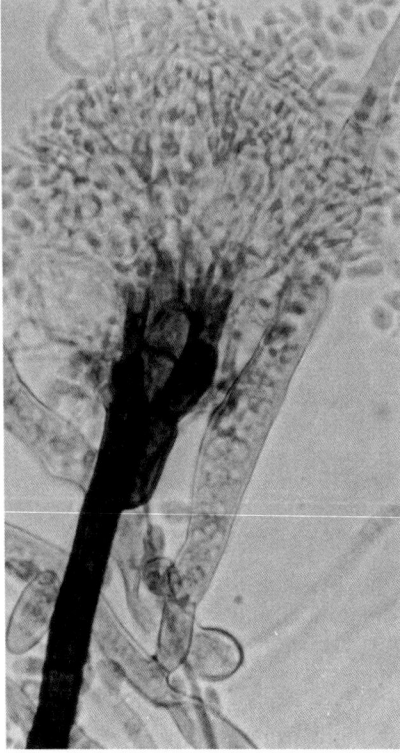

Fig. 3. Conidiophores of *Leptographium procerum* (left), *L. wageneri* (center), and *L. terebrantis* (right) from pine roots.

and weevils is often restricted to trees under stress or those planted outside their native habitat.

Pines are the most commonly affected conifers, but spruce and Douglas-fir can also be hosts of beetle-*Leptographium* complexes. *L. procerum* has been associated with a syndrome called white pine root decline (or procerum root disease) on eastern white pine (*Pinus strobus*) in Christmas tree plantations in eastern North America (Plate 11). The fungal infection apparently follows damage to the roots and root collar caused by root-feeding weevils (*Hylobius pales* and *Pissodes nemorensis*), which serve as vectors of the fungus. *L. serpens* has been associated with dying pines in Italy and South Africa and has been reported from the southeastern United States, but it is not believed to be a primary pathogen. In South Africa, the fungus is associated with root damage caused by bark beetles in the genera *Hylastes* and *Hylurgus*. Red pine decline in the Lake States has been associated with root-feeding weevils, bark beetles, and at least one *Leptographium* sp. (*L. terebrantis*), but some as yet undefined predisposing factor seems to be involved in this complex syndrome of tree mortality. *L. terebrantis* is most commonly associated with *Dendroctonus valens* and *D. terebrans*, which are North American root- and stump-feeding bark beetles capable of killing large pines over a period of years through repeated attack of the large roots and root collar area. Trees under attack by *D. valens* and *D. terebrans* are often attacked and killed by stem-feeding bark beetles.

Disease Management

Control of the damage caused by beetle-*Leptographium* complexes must center on control of the insects. Local population explosions of the beetles after clear-felling of forests may be largely unavoidable. Uprooting or burning breeding material (stumps or slash) by may be feasible in some situations. Direct chemical control of the insects in stumps or treatment of newly planted seedlings would rarely be cost effective. Trap logs may be used to lure newly emerged adults away from planted seedlings. If seedling losses are high, replanting may be necessary after the beetle population has subsided. Planting can be delayed to avoid the peak of the beetle population, but subsequent weed-management problems would probably negate the value of such a management tool. High-value saplings, such as Christmas trees, may be protected with insecticides. If larger pines are killed, it is likely that some predisposing factors are involved. Selection of species that are better adapted or more vigorous may avoid future problems on the site.

Selected References

Furniss, R. L., and Carolin, V. M. 1977. Western forest insects. U.S. Dep. Agric. For. Serv. Misc. Pub. 1339.

Harrington, T. C. 1993. Diseases of conifers caused by species of *Ophiostoma* and *Leptographium*. Pages 161-172 in: *Ceratocystis* and *Ophiostoma*: Taxonomy, Ecology, and Pathogenicity. M. J. Wingfield, K. A. Seifert, and J. F. Webber, eds. American Phytopathological Society, St. Paul, MN.

Harrington, T. C., and Cobb, F. W., Jr., eds. 1988. Leptographium Root Diseases on Conifers. American Phytopathological Society, St. Paul, MN.

Klepzig, K. D., Raffa, K. F., and Smalley, E. B. 1991. Association of an insect-fungal complex with red pine decline in Wisconsin. For. Sci. 37:1119-1139.

Nevill, R. J., and Alexander, S. A. 1992. Root- and stem-colonizing insects recovered from eastern white pines with procerum root disease. Can. J. For. Res. 22:1712-1716.

Wingfield, M. J. 1986. Pathogenicity of *Leptographium procerum* and *Leptographium terebrantis* on *Pinus strobus* seedlings and established trees. Eur. J. For. Pathol. 16:299-308.

Wingfield, M. J., Harrington, T. C., and Crous, P. W. 1994. Three new *Leptographium* species associated with conifer roots in the United States. Can. J. Bot. 72:227-238.

(Prepared by T. C. Harrington and M. J. Wingfield)

Root Decays

Root-decay fungi belong to a community of organisms that live belowground most their lives. They are root, not soil, inhabitors and are primarily basidiomycetes. The specialized ability of root-decay fungi to penetrate and colonize healthy root systems enables them to avoid competition for substrate with saprophytic soil organisms. They exist at endemic levels in most forested ecosystems, although forest-management activities have caused increases in the incidence and severity of the diseases they cause.

Root-decay fungi cause dysfunction in the root system by killing cambial tissue, disrupting the transport system, and decomposing cellulose, causing loss of structural integrity. Symptoms expressed by an infected tree are chlorotic (yellowish) and thin foliage, distress cone crops, and reduced leader length. Infected trees eventually die standing or are blown down because of root decay. The length of time from symptom expression to death varies from 1 to more than 40 years, depending on host and environmental factors and the species of root-rot fungus.

Penetration of the root bark is the first step in the disease cycle. Most fungi rely primarily on direct penetration, although methods and strategies differ between species. Ectotrophic mycelium (mycelium that grows on the root surface), which is produced by many of the root-disease fungi, directly penetrates root bark and enables the fungus to spread from tree to tree across root contacts. After penetration of the bark and infection of the cambium, the fungus begins the colonization phase by killing cambial tissue and spreading along the cambium or by growing into the heartwood of the root and then spreading longitudinally. Colonization of the cambium, sapwood, or heartwood is determined by the fungal species and by the host's response to infection. For example, resinous species, such as pine, can respond to infection by producing resins in the sapwood that inhibit growth of some fungi in this tissue.

Most of the root-decay fungi cause white rots, although a few cause brown rots. White rot fungi produce an exocellulase enzyme that breaks down cellulose and a collection of enzymes and compounds that break down lignin. Cellulose and lignin are degraded simultaneously, and the resultant sugar units are absorbed by the fungus for nutrition. The decayed wood appears pitted because these lytic enzymes function in the vicinity of the hyphae, and it is white because of the removal of the pigmented lignin. Brown rot fungi produce both exocellulases and endocellulases but have no ability to break down lignin. The brown color is the result of the oxidized lignin, and the typical crumbly texture of brown rot is caused by loss of cellulose.

Once a sufficient food base is garnered to support the energy-demanding reproductive phase, most root-decay fungi produce an aboveground basidiocarp, either on the ground directly above a colonized root or on the lower stem of the host tree. The epidemiological role of basidiospores varies among species. Root-decay fungi attack and invade only living tissue, but once the parasitic phase is complete and the tree is killed, the fungus becomes saprophytic and expands within the dead root system. The rate of saprophytic colonization is usually faster than that of parasitic colonization because the tree cannot actively respond to infection. Colonization of the dead root system and lower stem becomes a race between the root-decay fungus and saprophytic organisms. Many of the root-decay fungi can remain within dead host roots for decades, depending on the size of the root system, the species of fungus, and the rate of root system deterioration caused by environmental conditions and the action of saprophytes. The colonized root system provides a substantial food base from which the fungus can infect new hosts through root contacts or the specialized hyphal structures produced by some root-decay fungi—and the disease cycle begins anew. Root-to-root spread of these fungi results in outwardly expanding infection centers. The centers can be very discrete and reach several hectares in size, or they can have less discrete edges and include only a few trees.

Root systems of stumps and dead trees provide a large food base for the fungus and therefore create high inoculum potential. Stand regeneration policies generally require establishment of seedlings (often of the same host species) within 2–3 years after harvest. Seedlings that develop root contacts with colonized stump roots may become infected, resulting in an accelerated rate of disease expansion compared with that in unmanaged forests. Other forest management practices, such as thinning, also may cause more rapid expansion of root disease pockets through the increase in inoculum potential and subsequent infection of residual trees.

Root-decay fungi are natural and important components of many forest ecosystems. They contribute to plant and animal diversity by creating a wide variety of tree ages and stand structures and under natural conditions are in overall balance with other ecosystem components. However, forest management activities have upset the natural balance, resulting in increased incidence and severity of root disease. This increase and the greater demands on forests for fiber production have caused root decays to become a major concern in forest management.

Management of root decays has emphasized avoiding or minimizing further spread through the establishment of nonhost species, avoidance of partial cutting in root rot sites, and elimination of inoculum from the ground by stump removal and similar techniques. The ability to identify symptoms and signs of root-decay fungi and knowledge of their biology and ecology are necessary to properly manage root diseases within economic and ecological boundaries.

(Prepared by K. Lewis)

Annosum Root Rot

Other names: rotticka (Swedish); le fomes, pourriture rouge, maladie du rond (French); Wurzelschwamm, Wurzelfäule (German); podredumbre roja del pino (Spanish)

Causal organism: *Heterobasidion annosum* (Fr.:Fr.) Bref. (syns. *Fomes annosus* (Fr.:Fr.) Cooke., *Fomitopsis annosa* (Fr.:Fr.) P. Karst., *Polyporus annosus* Fr., and *Trametes radiciperda* Hart; anamorph *Spiniger meineckellus* (A. Olson) Stalpers)

Hosts: many conifers, especially species of *Abies, Juniperus, Larix, Picea, Pinus, Pseudotsuga,* and *Tsuga* and less often, broad-leaved trees

Distribution: widely distributed in the northern hemisphere; Australasia (as *H. araucariae* P. K. Buchanan); locally in South Africa and South America

Symptoms and Diagnosis

Heterobasidion annosum is a polypore that causes root and butt rot of many coniferous trees. It is a white rot fungus that degrades primarily wood lignin. The advanced decay is soft and spongy and often contains characteristic white cellulose pockets and black specks. In some trees, particularly pines, the fungus colonizes only the roots, causing strong resin exudation and rapid death of the tree. However, in most conifers, the decay rises into the stem as a heartrot (Plate 12), and the tree may stay alive for decades without any external symptoms. The disease tends to occur in patches that enlarge progressively. Because of the high mortality, such disease centers are especially conspicuous in pine stands (Plate 13).

Other root-rot fungi may cause similar symptoms, and a positive diagnosis of annosum root rot in the field is often possible only where fruit bodies are found. They are mostly perennial and develop usually in concealed places under the roots of dead trees or in hollow stump cavities (Plate 14). Small, developing, or abortive fruit bodies often appear as characteristic pustules on the bark, especially on pine species. Conidiophores, which develop in pure cultures or on the surface of infected wood under moist conditions, are an identifying characteristic of *H. annosum*.

Effects on Tree and Forest, Ecological Role

H. annosum is probably the most destructive fungal pathogen of managed coniferous forests in the northern hemisphere. From the ecological point of view, this root rot increases species diversity in coniferous forests, particularly by favoring the regeneration of broad-leaved trees in the canopy gaps that it causes.

Taxonomic Status

Although so far classified under a single species name, *H. annosum* is in fact a complex of several intersterile groups that probably represent separate species. Because these groups have only recently been discovered, their distribution areas are not well known. Three intersterile groups have been identified from Europe (P, S, and F) and two from North America (P and S). They also show small morphological differences. Practically nothing is known about the intersterile groups of *H. annosum* in Asia.

The *H. annosum* that occurs in Australasia is quite different from the *H. annosum* found in the northern hemisphere and has recently been described as a new species: *H. araucariae* P. K. Buchanan. This fungus is weakly pathogenic and grows on species of *Agathis, Araucaria,* and *Pinus*. Local occurrences of *H. annosum* have also been reported from South Africa and South America, but their taxonomic status is unclear. The European P (pine) group is generally a fungus of pine forests, but it does not show much host specialization. In addition to pine species, spruce, juniper, larch, and other conifers are susceptible. Even broad-leaved trees can be attacked when they grow with pines (Plate 15). Excluding the areas north of latitude 64°, the distribution of the P type seems to cover almost all of Europe.

The main hosts of the European S (spruce) group are Norway spruce and Siberian fir. With spruce stumps as a base, it can also attack Scot's pine saplings. Older pines, however, are very resistant to this fungus, and the annosum root rot in mature pine stands is always caused by the P type. The S type is the main cause of butt rot of Norway spruce within the natural distribution area of this tree, except in the northernmost areas where the fungus is rare. When cultivated in western Europe outside its natural distribution, Norway spruce is attacked mostly by the P type. The S type so far has not been found in the British Isles.

The F (fir) group occurs in *Abies* forests (Plate 16). It causes damage to *A. alba* and other species of *Abies* in the Italian and Balkan peninsulas but is less common in the *A. alba* forests of central Europe. Seedlings of *Pseudotsuga*, growing close to infected true fir stumps, are also susceptible, but older trees seem to be resistant.

Like its European counterpart, the P group in North America attacks mostly pine species. It is distributed across the continent and is especially harmful in southeastern pine plantations. In laboratory experiments, the North American and European P groups show a high degree of interfertility. However, they differ in their biochemical properties, and it is unclear whether they can produce vigorous hybrids.

The North American S group causes root and butt rot on *Abies* and other nonpine genera in western areas of the continent from California to Alaska. In the laboratory, some isolates of the North American S group are interfertile with isolates from two European groups, F and S, and others are not. It is not known whether the hybrids would be competitive in nature.

Because the intersterile groups of *H. annosum* on different continents differ from each other, the risk of accidentally introducing new pathogens should be considered in intercontinental timber transport.

Disease Cycle

H. annosum produces both asexual and sexual spores. The asexual spores (conidia) are not airborne, and their role in the infection process is not well known. The fruiting bodies of *H. annosum* produce large numbers of basidiospores when the weather is warm and the soil is not too dry. Despite the hidden position of the fruiting bodies, the basidiospores are distributed effectively into the air. They settle on the bark of trees and on stump surfaces and fall on forest soil where they are washed down by rain water and carried by burrowing animals. The spores infect freshly exposed wood. They usually infect a living tree through wounded roots or less often through stem wounds.

Once established in a stand, whether through stumps from thinning or wounds on trees, the fungus grows along the root system and attacks neighboring trees through root contacts and grafts. Spreading at a rate of 0.2–2 m per year, a fairly large disease center up to 30 m or more in diameter can form from one fungal individual (genet). However, most genets of *H. annosum* in managed forests are small, indicating the importance of spore infection in the spread of this fungus.

The occurrence and spread of *H. annosum* is exacerbated by human activities in the forest, particularly harvesting operations. Thinning operations cause injuries to the roots of residual trees and fresh stump surfaces, both of which provide infection courts. Occasional stress, for example, that caused by drought or air pollutants, may facilitate root infection. The rate of colonization of the root system is very dependent on the soil type and is particularly rapid in former agricultural soil and in other soils with a high pH. Peat soils are almost free of the disease.

Disease Management

In healthy forests, it is important to prevent infection of cut stumps or tree wounds by spores. The number of thinnings should be kept to a minimum, and the operations should be carried out carefully in order to minimize root damage. Thinnings should be done during the season when the number of spores in the air is small: winter in the north and dry, hot summers in the south. During high-risk seasons, infection by spores can be controlled by treating the fresh conifer stump surfaces with chemical or biological agents (urea, borax, or spores of *Phlebiopsis gigantea*).

Spread of the fungus in a stand can be slowed by planting with wide spacing, growing mixed stands, and avoiding the cultivation of susceptible trees on hazardous sites. In a stand where the disease is already present, it is frequently worthwhile to reduce the rotation time.

Mechanical removal of the stumps or rotation with a resistant tree species can significantly reduce inoculum levels on infested sites. If such operations are not possible, natural regeneration or sowing is generally better than planting. Slash burning or stump treatment with a competing fungus at clear-felling may reduce infection in the next tree generation. Total eradication of *H. annosum* in managed forests is difficult, but with proper management, the economic losses can be kept at an acceptable level.

Selected References

Hodges, C. S. 1969. Modes of infection and spread of *Fomes annosus*. Annu. Rev. Phytopathol. 7:247-266.

Johansson, M., and Stenlid, J., eds. 1994. Proc. Int. Conf. Root Butt Rots, 8th. Parts 1 and 2. IUFRO Working Party S2.06.01. Swedish University of Agricultural Science, Uppsala, Sweden.

Negrutskii, S. F. 1986. Kornevaya gubka (*Heterobasidion annosum*). 2nd ed. Agropromizdat, Moscow.

Otrosina, W. J., and Scharpf, R. F., tech. coords. 1989. Proceedings of the symposium on research and management of annosus root disease (*Heterobasidion annosum*) in western North America. U.S. Dep. Agric. For. Serv. Gen. Tech. Rep. PSW-116.

Worrall, J. J., and Harrington, T. C. 1992. *Heterobasidion*. Pages 86-90 in: Methods for Research on Soilborne Phytopathogenic Fungi. L. L. Singleton, J. D. Mihail, and C. M. Rush, eds. American Phytopathological Society, St. Paul, MN.

(Prepared by K. Korhonen)

Armillaria Root Disease

Other names: shoestring root disease, honey mushroom, mushroom root rot; Hallimasch (German); pourridié a armillaire (French); honungsskivling (Swedish)

Causal organism: *Armillaria ostoyae* (Romagnesi) Herink (syns. *A. obscura* Schaeff.:Fr., *A. mellea* var. *obscura* Gillet, and *Armillariella ostoyae* (Romagnesi) Herink)

Hosts: many conifer and hardwood species in forests, plantations, and amenity plantings worldwide

Distribution: worldwide in boreal, temperate, and tropical forests; present in various hardwood and coniferous forests across the northern hemisphere

Causal Organism

We now recognize that a number of distinct species were earlier lumped under the name *Armillaria mellea*. K. Korhonen began the process of distinguishing individual species from the complex on the basis of mating compatibility. A numerical nomenclature of biological species was developed in North America (NABS I-XI). *A. ostoyae* is the most common pathogenic species on conifers in the northern hemisphere, but several other species cause butt rot or kill young or weakened trees. *A. mellea* is primarily a pathogen of hardwoods but may also cause disease on conifers in other management situations, such as that of Christmas trees.

Symptoms and Diagnosis

Woody plants infected with *A. ostoyae* express an array of symptoms, including reduction in shoot growth, changes in foliar characteristics, crown dieback, stress-induced reproduction, basal stem indicators such as resinosis (Plate 17), and death. Generally, symptom expression and rate of development relate to the position of attack and the rate of destruction of the host root system. If the disease progresses rapidly or the host is small, then not all symptoms may become evident before the tree dies. Similarly, symptom development in conifers is often more pronounced on vigorous hosts.

Many of these symptoms are nonspecific and may be induced in trees affected by prolonged drought or attacked by bark beetles or other root pathogens. Thus, to confirm Armillaria root disease, the root collar and lower bole of the tree must be examined for signs specific to the fungus, including mycelial fans (Plate 18), rhizomorphs, basidiomes (Plate 19), and decay. *A. ostoyae* can also be confirmed by culturing from the host.

Disease Cycle

The critical points in the disease cycle are 1) infecting, killing, and colonizing a food base; 2) surviving in the food base; and 3) spreading from the food base to a susceptible host. The food base generally consists of infected stumps and roots, which also can serve as avenues of spread. To persist in a given area, as they can for centuries, root-disease fungi require a continual supply of new food bases. Root contacts between infected and adjacent roots or the attachment of rhizomorphs to host root systems enable *Armillaria* spp. to spread and persist at a given site. *Armillaria* spp. also disperse by airborne basidiospores produced in the basidiomes; however, the epidemiological role of basidiospores is not clear. Historically, their role has been considered minimal, but recent studies indicate that basidiospores may tolerate harsh environments and may be a factor in disease spread. Their role in forests with an established population of *Armillaria* spp. may be less important than that in first-rotation forests or shelter belts established in areas with previously little or limited tree cover. The time from infection to mortality depends on several factors, including the inoculum potential of the fungus and the species, vigor, and age of the host. Trees less than 15 years old may be killed within a relatively short time, whereas others may be able to check the spread of the fungus by producing callus and adventitious roots. Trees that succumb to the fungus are used as new food bases.

Effects on Tree and Forest, Ecological Role

Armillaria spp. are endemic to many forests, living on roots and lower stems of conifers, broad-leaved trees, and associated understory vegetation. As natural components of the mycoflora of native forests, *Armillaria* spp. cause endemic disease. Depending on the environmental or biological conditions, disease may be present to a greater or lesser degree in a particular place. As such, disease caused by *Armillaria* spp. varies considerably in time and space.

As saprophytes, the species are major decomposers in forests and decayers of wood in service. In many forests, the role of *Armillaria* as a decomposer is its most conspicuous activity. *Armillaria* causes a typical white rot, and in many tree species, both sapwood and heartwood may be decomposed.

In addition to parasitic and saprophytic activities, some species of *Armillaria* behave as mycotrophic associates in several achlorophyllous taxa in Orchidaceae. In eastern North America, *A. mellea* is commonly observed as a mycoparasite on the agaric *Entoloma arbortivum*.

Predisposing Factors

There are about 40 species of *Armillaria* worldwide, some of which act as virulent pathogens and others as opportunists that act selectively on trees stressed by lack of light or damaged by machinery, defoliation (e.g., by insects, fungi, or frost), freezing, drought, or pollution. *A. ostoyae* is generally considered an aggressive pathogen and is capable of infecting and killing apparently vigorous trees.

Predisposing stresses may be more important for disease development in relatively resistant species than in the more susceptible species. In general, hardwoods are considered more resistant to death caused by Armillaria root disease than coniferous species in northern temperate forests.

Disease Management

There currently are no easily applied, cost-effective methods for controlling Armillaria root disease in forests, although numerous strategies are employed to limit the threat. These strategies include silvicultural options, avoidance of hazardous sites, direct inoculum reduction, and a host of biological control methods.

Significant mortality often occurs in forests where the composition and structure of tree species has been altered by management activities. Thus, the most significant silvicultural consideration for natural forests is to maintain the composition of the local tree species and their relative "natural" densities. Suitable species must be matched with suitable sites to minimize disease hazard. "Off-site" planting may predispose the host to disease.

Biological control of *Armillaria* spp. has for the most part been unsuccessful. Two features of *Armillaria* spp. make control by introduced organisms difficult: 1) the inoculum already exists in portions of the woody substrate and thus maintains a positional advantage over introduced fungi; and 2) the fungus spreads rapidly in the cambial zone of freshly killed trees. The inability to maintain effective populations of organisms antagonistic to *Armillaria* under field conditions is a major factor limiting successful biological control in forest environments.

The longest-standing recommendation for the control of Armillaria root disease is the removal of inoculum from the soil. This action entails removing diseased trees and their root systems and other colonized woody material. Although the ecology of many *Armillaria* spp. is not completely understood, there are three situations in which removal of stumps and roots appears to be effective: 1) when the pathogen causes a primary disease from existing inoculum but does not continue to spread from secondary inoculum; 2) when secondary inoculum is important in the disease process but occurs in distinct patches, thus allowing for careful removal in specific areas; and 3) when secondary inoculum is important but the crop can be managed on a relatively short rotation, such as in intensively managed fruit orchards.

Selected Reference

Shaw, C. G., III, and Kile, G. A. 1991. Armillaria root disease. U.S. Dep. Agric. For. Serv. Agric. Handb. 691.

(Prepared by D. Omdal and C. G. Shaw III)

Laminated Root Rot

Other name: yellow laminated root rot
Causal organism: *Phellinus weirii* (Murrill) R. L. Gilbertson (syns. *Poria weirii* (Murrill) Murrill, *Fomitiporia weirii* Murrill, and *Inonotus weirii* (Murrill) Kotlaba & Pouzar). The Asian fungus variously known as *Phellinus* or *Inonotus sulphurascens* or *P. heinrichii* may be conspecific.
Hosts: all species of Pinaceae native to northwestern North America, particularly Douglas-fir (*Pseudotsuga menziesii*), mountain hemlock (*Tsuga mertensiana*), and grand fir (*Abies grandis*). The cedar form causes butt rot of western red cedar (*Thuja plicata*).
Distribution: northwestern United States and Canada; possibly Russia and Japan

Causal Organism

There are at least two distinct species lumped under the name *Phellinus weirii*. The original collection was associated with a butt rot of western red cedar. The fungus that causes root rot of Douglas-fir and other conifers is different, but the nomenclature is still confused. The relationship of North American isolates to similar fungi in Asia is also unclear. The following discussion is based on the fungus that causes laminated root rot of Douglas-fir and associated conifers in North America.

Symptoms and Diagnosis

Laminated root rot is often first detected when openings caused by fallen trees appear in a forest stand. As roots decay, a tree dies standing or is robbed of structural support and is windthrown. Trees tend to fall in a random pattern, and the upturned roots are short and stubby because of the extensive decay (Plate 20). Disease centers may vary from a few trees to a hectare or more. Standing dead or symptomatic trees are typically present around the periphery and scattered within an infection center (Plate 21).

Crown symptoms, which may be quite variable, include reduced leader growth; short, chlorotic foliage; and distress cone crops. These symptoms, usually not seen until at least half the root system is affected, may not appear until 15 years after initial infection and may be present for 10 years or longer before a tree dies.

The disease can be identified in living suspect trees by examining the root collar and major roots for the characteristic gray white superficial (ectotrophic) mycelium (Plate 22). The reddish brown, wiry, setal hyphae (0.3 mm long) can be seen with a hand lens scattered in the superficial mycelium or in pieces of wood with advanced decay.

Laminated root rot is distinguished by the characteristic appearance of the decayed wood. Early stages of decay appear as reddish brown to chocolate brown irregular patches or crescent-shaped stains, usually in the outer heartwood, on fresh stump tops or cross sections of major roots. As the disease progresses, short, oval pits about 1×0.5 mm appear and become more numerous. Later, the wood tends to separate into layers along annual rings, creating sheets of yellowish, pitted wood; hence, the name laminated root rot (Fig. 4).

The fruiting bodies, which are rare during most years, are inconspicuous, resupinate, pore-covered crusts that usually form on the undersides of fallen trees and uprooted stumps near the forest floor. When young, they are a light gray brown with white, sterile margins. Later, they turn a uniform chocolate brown.

Disease Cycle

Laminated root rot can persist from generation to generation on a site as long as susceptible tree species are present. When infected trees die, are harvested, or burn, the pathogen can continue to live saprophytically in stumps and large roots for 50 years or longer. Infection in a young stand begins when roots of young trees contact infested stumps and roots from the preceding stand. The pathogen spreads little, if at all, by windblown spores and does not appear to grow through the soil. Virtually all spread is by mycelia across root contacts or within roots.

As the fungus advances along a tree's roots, the roots distal to the fungus are killed. Some seedlings may be infected and killed during the first year. The infection spreads between living trees through root contact. Initial mortality involves scattered individuals close to old, infested stumps. When the stand is 15–20 years old, roots of adjacent trees make contact, the pathogen advances to new hosts, and disease centers appear. These centers expand at an average radial rate of about 30 cm per year and eventually become understocked openings of various sizes that contain dead standing trees, windthrows, and occasionally unaffected trees. When the stand is replaced, the process starts over.

Effects on Forests, Ecological Role

Laminated root rot is the most important single natural disturbance causing long-term change in the forest ecosystems of the northwestern United States and Canada, and it is one of the most difficult to manage. While Douglas-fir is the most economically important host, nearly all species of commercially important conifers can be affected. It is estimated that the disease annually reduces forest fiber production by about 4.4 million cubic meters. Centers commonly fill in with hardwoods, shrubs, and resistant conifers. Susceptible conifers may regenerate in the centers, but they commonly contact inoculum at an early age and die. Through the formation of centers or "gaps" in homogeneous Douglas-fir stands, this disease greatly increases the structural diversity and number of species, thus providing big-game forage, small-animal habitat, and a source for minor forest products. The impact of this disturbance is greater in stands west of the crest of the Cascade Mountains (west side) than it is farther east, where many other agents are active in causing change.

Predisposing Factors

Within its natural range, laminated root rot does not appear to be restricted by topography, climate, or soil conditions. Disease incidence is greatest on middle to upper slopes, which may reflect distribution of the most susceptible hosts. Disease incidence is also greater in stands with a high density of the most susceptible hosts. Recent work has shown that plant associations in disease centers differ from those in uninfected parts of the stand. These associations may be useful indicators of disease.

Disease Management

Mitigating strategies involving root removal during harvest (e.g., push-over logging, in which roots come up as the tree is pushed over), postharvest stump removal, inactivation of inoculum (e.g., by fumigation), and regeneration with less susceptible species are available and are often best implemented at the time of final harvest. A survey of the intensity and distribution of the disease is an essential element in the selection of management alternatives.

Selected References

Hadfield, J. S., Goheen, D. J., Filip, G. M., Schmitt, C. L., and Harvey, R. D. 1986. Root Diseases in Oregon and Washington Conifers. U.S. Department of Agriculture Forest Service, Pacific Northwest Region, Portland, OR.

Thies, W. G. 1984. Laminated root rot: The quest for control. J. For. 82:345-356.

Thies, W. G., and Sturrock, R. N. 1995. Laminated root rot in western North America. U.S. Dep. Agric. For. Serv. Gen. Tech. Rep. PNW-349.

Wallis, G. W. 1976. *Phellinus* (*Poria*) *weirii* root rot: Detection and management proposals in Douglas-fir stands. Can. For. Serv. Tech. Rep. 12.

Fig. 4. Advanced decay caused by laminated root rot. (Courtesy U.S. Department of Agriculture, Forest Service)

(Prepared by W. G. Thies)

Red-Brown Butt Rot

Other names: velvet top fungus; Kiefern Braunporling (German)

Causal organism: *Phaeolus schweinitzii* (Fr.:Fr.) Pat. (syn. *Polyporus schweinitzii* Fr.:Fr.)

Hosts: principally species of *Abies, Cedrus, Larix, Picea, Pinus,* and *Pseudotsuga.* Isolated incidences of its occurrence have also been reported on hardwoods, including *Eucalyptus, Fraxinus, Prunus,* and *Ulmus* spp.

Distribution: throughout the world, particularly in the coniferous ecosystems of North America and Europe; also in the South Pacific on hardwoods

Symptoms and Diagnosis

Phaeolus schweinitzii causes an extensive, brown, cubical rot in scattered trees of old-growth forests, plantations, or parklands. It is a major cause of butt rot in Douglas-fir and other commercial conifer timber species in the northern hemisphere. The decay is normally confined to the nonreactive inner wood of roots or stems. Unless basidiocarps are present, evidence of infection is often not apparent until felling or the trees collapse as a result of the extensive decay caused by the fungus (Plate 23). Basidiocarps are velvety in texture and a dark rusty brown with yellow rims when active (Plate 24). Bracket forms grow from damaged wood at the bases and on aboveground parts of trees. More circular structures with short stalks may arise from decaying roots around the tree base. The brown, cubical rot can be confused with that caused by *Sparassis crispa,* which may be present in similar habitats. In the absence of sporocarps, a useful distinguishing feature is that active mycelium of *P. schweinitzii* is yellow, whereas that of *S. crispa* is white. Sometimes growth and identification of this fungus in culture is the only certain method of diagnosis.

Disease Cycle and Predisposing Factors

It is now known that in areas where basidiocarps exist, there is normally a persistent fungal infestation of the soil. It seems likely that the annual deposition of basidiospores helps to maintain these soil infestations and should be regarded as a potential source of infection in tree roots (Plate 25). In Great Britain, where the origin of the distinct genotypes (strains) of the fungus found in different trees in the same plantation has been investigated, it has been concluded that root infection from spore-generated inoculum in the soil could explain the genetic diversity of the fungus present in different trees. For example, infection caused by spores in the soil can be compared with that caused by *Phellinus weirii,* in which groups of trees all contain a single genotype (strain) of the fungus as a result of direct mycelial transfer between roots of adjacent trees.

Despite the nearly ubiquitous presence of *P. schweinitzii* in many of the forest soils examined in Great Britain, successful invasion of trees is not common, which suggests some predisposing agent. Field and laboratory work has strongly suggested that *Armillaria ostoyae* (*Armillariella obscura*) is a predisposing factor in the infected Sitka spruce plantations in Great Britain. An association with *A. ostoyae* was also observed in Idaho when root systems of Douglas-fir were examined, but there *A. ostoyae* was considered secondary to *P. schweinitzii.*

Physical damage to aboveground parts of trees has also been implicated as a route of infection. For example, the mechanical damage caused by hauling has been identified in Bulgaria as the main factor in the infection of *Pinus peuce* and has resulted in butt rot that extends as far as 17 m up the stem. In North America, wounds from storm damage are regarded as major avenues of infection. Some authors consider *P. schweinitzii* to be responsible for more damage than any other single wood-decaying fungus in North America.

Disease Management

In Great Britain, where damage is almost exclusively confined to susceptible species growing on sites with a previous history of old, mixed woodland where *A. ostoyae* is common, up to 25% of the crop has been found to be infected when it was clear-felled. If crops such as Sitka spruce continue to be raised on such sites, it seems unlikely that any reduction in disease incidence can be expected. In upland sites in northern Great Britain with no recent history of natural woodland, disease occurrence is rare, despite the presence of *P. schweinitzii* in the soil. On such sites, it is possible that if the buildup of *A. ostoyae* can be prevented as successive crops are harvested, disease from *P. schweinitzii* might be avoided.

Selected References

Barrett, D. K. 1985. Basidiospores of *Phaeolus schweinitzii*: A source of soil infestation. Eur. J. For. Pathol. 15:417-425.

Barrett, D. K., and Greig, B. J. W. 1985. The occurrence of *Phaeolus schweinitzii* in the soils of Sitka spruce plantations with broad-leaved or non-woodland histories. Eur. J. For. Pathol. 15:412-417.

Barrett, D. K., and Uscuplic, M. 1971. The field distribution of interacting strains of *Polyporus schweinitzii* and their origin. New Phytol. 70:581-598.

Blakeslee, G. M., and Oak, S. W. 1980. Residual naval stores stumps as reservoirs of inoculum for infection of slash pines by *Phaeolus schweinitzii*. Plant Dis. 64:167.

(Prepared by D. Barrett)

Tomentosus Root Rot

Causal organism: *Inonotus tomentosus* (Fr.:Fr.) S. Teng (syns. *Polyporus tomentosus* Fr.:Fr., *Coltricia tomentosa* (Fr.:Fr.) Murrill, and *Onnia tomentosa* (Fr.:Fr.) P. Karst.)

Hosts: a broad range of conifers but most frequently spruce (*Picea*) and pine (*Pinus*) species

Distribution: boreal forests throughout the northern hemisphere. In North America, *I. tomentosus* root rot is most frequent in central interior forests of British Columbia and in Ontario and Quebec. It is also found in spruce or mixed conifer forests in the United States, particularly in Wisconsin and the Pacific Northwest, and has been reported from Scandinavia, Siberia, Europe, and northern Asia.

Symptoms and Diagnosis

Inonotus tomentosus causes a slowly developing dysfunction of the root system. Symptoms exhibited by infected trees, including thin foliage, reduced leader growth, and chlorosis, are similar to those caused by other root diseases. Aboveground symptoms are generally not obvious until the fungus is well established in the root system and are therefore not reliable for diagnostic purposes. Eventually, the tree dies standing or is windthrown. Windthrown trees have short, stubby roots (Plate 26) in which the pits caused by *I. tomentosus* are obvious, particularly when the roots are viewed from the ends. The disease occurs in small clumps of trees (often only two or three trees per clump), which gradually coalesce into larger areas.

During the early stages of decay, a dark pink to pinkish brown stain is evident in the roots, normally in the heartwood (Plate 27). However, positive diagnosis should not be made without finding the pits that are characteristic of advanced decay. During the early stages of decay, the pits are very small and often slightly yellowish. As decay progresses, they expand into a honeycomb pattern of white pits. In extremely advanced decay, the pits no longer contain mycelium, but the honeycomb pattern remains. Fruiting bodies normally appear from mid-July to September, but their production is weather dependent.

Therefore, trees without fruiting bodies may still be infected (Plate 28).

Disease Cycle

Little is known about the infection process of tomentosus root rot. The fungus can spread via root contact, and it is thought that the fungus is able to directly penetrate the bark of small roots (less than 5 cm in diameter). Ectotrophic mycelium has been observed on roots of various sizes and appears on or in the outer bark. This mycelium may play a role in spread of the fungus. There is also evidence that the pathogen is spread by basidiospores. When isolates of the fungus from neighboring clumps of infected trees are paired on culture media, they form a barrier zone and are vegetatively incompatible. This indicates that the isolates have different genotypes and therefore that these infections probably originated from basidiospores rather than vegetatively by root contact.

Once infection does occur, the fungus spreads distally and proximally in the roots and eventually infects the root collar, where it colonizes the heartwood in the butt. The fungus usually grows longitudinally in the heartwood; radial spread into the sapwood is limited until the tree is extremely stressed or dead. At this point, the spread is more radial than longitudinal. Once a sufficient food base is colonized, the fungus may produce epigeous fruiting bodies, which are connected to colonized roots.

Effects on Tree and Forest, Ecological Role

Trees infected with *I. tomentosus* may be killed outright or be predisposed to windthrow. They also grow more slowly than healthy trees and often contain significant amounts of decayed wood. The magnitude of loss varies with the level of infection within any one stand. Results of one study indicated that severely infected trees (more than 75% of the root system infected) suffered a 20% reduction in growth. Losses to butt cull were determined to be up to 25% of the total log volume. Spruce regeneration established on sites with previous histories of tomentosus root rot becomes infected by root contact with infected stumps. On severely infected sites, the losses to the disease can be substantial and result in insufficient stocking. As a result, the presence of tomentosus root disease on a site can limit the selection of species for regeneration.

In natural ecosystems, this fungus is very important in causing small openings, thereby creating more structural and compositional diversity in the forest. As a killer of trees, it provides dead standing and downed woody material, which provides habitat for many animals.

Predisposing Factors

It is thought that *I. tomentosus* is able to infect vigorous trees as easily as stressed trees. However, once the fungus is in the tree, death may occur earlier in trees that are stressed by other factors as well. Tomentosus root disease itself is thought to predispose trees to attack by some bark beetles, root collar weevils, and other agents that typically inhabit stressed trees.

Disease Management

Management of tomentosus root rot should begin during the planning and preharvest stages of forest management. Once a stand is regenerated, the options for management are more limited. In general, management strategies fall into three categories: 1) establishing an alternative species that is less likely to become infected, either because the species is less susceptible (e.g., *Abies* spp.) or its rooting habit results in slower spread (e.g., pine); 2) establishing a mix of host and nonhost species that decreases the potential for root contacts because of differences in rooting habit; and 3) removing inoculum from the soil. Because it is quite expensive and also site dependent, the third strategy is not currently practical. It is also recommended that thinning of trees in young stands or partial cutting of mature stands be avoided in areas where *I. tomentosus* is present. It is possible that thinning or partial cutting will increase the amount of inoculum in cut stumps (because of the increased radial spread of the fungus in the roots) and therefore increase the rate of spread to surrounding residual trees.

Selected References

Lewis, K. J., and Hansen, E. M. 1991. Survival of *Inonotus tomentosus* in stumps and subsequent infection of young stands in north central British Columbia. Can. J. For. Res. 21:1049-1057.

Lewis, K. J., Morrison, D. J., and Hansen, E. M. 1992. Spread of *Inonotus tomentosus* from infection centers in spruce forests in British Columbia. Can. J. For. Res. 22:68-72.

Whitney, R. D. 1977. *Polyporus tomentosus* root rot of conifers. Can. For. Serv. Great Lakes For. Res. Cent. For. Tech. Rep. 18.

(Prepared by K. Lewis)

Brown Root Rot

Causal organism: *Phellinus noxius* (Corner) G. H. Cunningham (syn. *Fomes noxius* Corner)

Hosts: species of *Agathis, Araucaria, Cupressus,* and *Pinus* (and many angiosperm tree species)

Distribution: pantropical in Africa, Asia, Australia, and the South Pacific

Symptoms and Diagnosis

Foliage of infected trees becomes yellow and then brown and in older trees may be cast (Plate 29). Resinosis on the stem occurs near ground level. A brown or black fungal crust envelops root surfaces and may extend up the stem as a "stocking" approximately 60 cm above the soil level (Plate 30). Decayed wood is discolored pale yellow or red brown with thin, brown zone lines. Advanced decay is a light, dry, friable, honeycomb pocket rot with associated diffuse, brown hyphae. The fruiting body is a woody poroid bracket or resupinate crust.

Disease Cycle

Primary infection occurs when roots come in contact with colonized stump roots of the previous cover. The fungus grows along the root surface and also penetrates the bark and cambium, eventually reaching and girdling the root collar. Soon thereafter, crown symptoms appear. Death is rapid in young plants. Older trees die more slowly or are uprooted because of extensive root decay (Plate 31). Infection between neighboring trees results in mortality centers that expand as the plantation ages. Basidiospores may initiate a small number of new infection centers through stump colonization.

Mortality losses in conifer plantations average up to 10%, although some local occurrences may be severe (up to 50% within 1 ha). *Phellinus noxius* occurs naturally as a decay fungus in native forests. The disease is favored by warm, moist conditions and is therefore less prevalent at high altitudes where conditions are cool and on dry ridge sites. Some resistance develops with age, but mortality occurs in trees of all ages. Dominant and suppressed trees appear equally susceptible.

Disease Management

Removal of colonized tree stumps prior to planting reduces disease levels but is costly. Stump inoculation with harmless decay fungi has shown potential for disease control. Specimen trees may be treated by pruning off diseased roots and isolated by using buried root barriers.

Selected References

Bolland, L. 1984. *Phellinus noxius*: Cause of a significant root-rot in Queensland hoop pine plantations. Aust. For. 47:2-10.

Bolland, L., Tierney, J. W., Winningham-Martin, S. M., and Ramsden, M. 1989. Investigations into the feasibility of biological control of *Phellinus noxius* and *Poria vincta* in Queensland hoop pine plantations. Pages 72-82 in: Proc. Int. Conf. Root Butt Rots, 7th. IUFRO Working Party S2.06.01. D. J. Morrison, ed. Forestry Canada, Pacific Forestry Centre, Victoria, B.C.

(Prepared by I. Hood)

Other Root-Rot Fungi

Ceriporiopsis rivulosa (Berk. & M. A. Curtis) R. L. Gilbertson & Ryvarden (syn. *Poria albipellucida* D. Baxter) causes a laminated and pitted wood decay that is very similar to that caused by laminated root rot (*Phellinus weirii*). *C. rivulosa* is most common in western North America, although it also occurs in the eastern United States and south into Central America. Dead conifers and living western red cedar and redwood are the main hosts. The basidiocarp is annual and resupinate; it is initially white and turns pale brown with age.

Inonotus dryadeus (Pers.:Fr.) Murrill causes a white rot in the roots and butts of true firs. It is found in firs along the Pacific coast from British Columbia to Mexico, although it is most commonly found from northern California south. It produces large, sessile, brown, and slightly tomentose basidiocarps at the bases of infected trees and sometimes on the roots.

Oligoporus sericeomollis (Rom.) Pouz. (syn. *Poria asiatica*) (Plate 32) is a brown rot fungus found throughout conifer forests of North America. Its primary host species is cedar, although it attacks a wide variety of both coniferous and deciduous hosts. It is also important in the decay of dead conifer wood. In living trees, it is most frequently found in the heartwood, where it causes a brown, cubical rot. The basidiocarp is annual, resupinate, white, and very thin.

Perenniporia subacida (Peck) Donk causes feather rot. This widely distributed fungus is most common in forests at low elevations and causes a white, stringy rot in the roots and butts of conifers. It is usually found on dead trees but also causes root rot of living trees, particularly hemlock and true firs. Basidiocarps are perennial, effused, and ivory colored.

Oligoporus balsameus (Peck) R. L. Gilbertson & Ryvarden (syns. *Polyporus balsameus* Peck and *Postia balsameus* (Peck) Jülich) causes balsam butt rot, a common, brown, cubical decay of the roots and butt of balsam fir (*Abies balsamea*) and other species of *Abies* and *Picea* in the northern coniferous forests of North America. It is also found in Europe. Decay is associated with basal wounds and frost cracks on trees.

Rigidoporus lineatus (Pers.) Ryvarden and *Ganoderma australe* (Fr.:Fr.) Pat. are pathogens of *Pinus kesiya*, which was introduced into New Guinea for commercial purposes. Its susceptibility to the root rots caused by these fungi and others is the reason that it is no longer planted in New Guinea. Both fungi cause a white root rot with numerous zone lines. Wilt symptoms develop quickly and are followed by tree death. *R. lineatus* has an annual, pileate basidiocarp that is sessile or has a slight stipe. It is zonate and pinkish brown with a bright orange red pore surface when fresh.

Rigidoporus vinctus (Berk.) Ryvarden (syn. *Poria vincta* (Berk.) Cooke) (Plate 33) is found in Australia, frequently in conjunction with *Phellinus noxius*. It can cause significant losses in plantations of hoop pine (*Araucaria cunninghamii*). In North America, it is primarily a white rot of dead hardwoods. The basidiocarps are annual to perennial, resupinate, and highly variable in morphology. The color ranges from pale brown to pink, and the pores are quite small.

Sparassis crispa (Wulfen:Fr.) Fr. (syn. *S. radicata* Weir) is known as the cauliflower fungus (Plate 34). In Europe, North America, and Japan, it causes a root rot of old trees of many conifer species in the family Cantharellaceae. The basidiomes are large, white, and coralloid and grow on perennial stalks from decayed roots. The decay is a brown, cubical rot, and yellowish mycelial fans form beneath the bark.

In tropical forests, there are several fungi that cause root diseases of pines, particularly trees that are stressed because of overstocking, drought, or poor site conditions. In South Africa, *Sphaeropsis sapinea* (Fr.:Fr.) Dyko & Sutton in Sutton causes a root rot on *Pinus elliottii* and *P. taeda*. Two related fungi, *Botryosphaeria dothidea* (Moug.:Fr.) Ces. & De Not. and *Lasiodiplodia theobromae* (Pat.) Griffon & Maubl., cause root rots of *P. taeda* in Hawaii and Venezuela, respectively. *Pseudophaeolus baudoni* causes a root rot of *P. caribaea* in South Africa, but the infection centers are very small and in only one region.

Selected References

Gilbertson, R. L., and Ryvarden, L. 1987. North American Polypores, vols. 1 and 2. Fungiflora, Oslo.

Morrison, D. J., ed. 1989. Proc. Int. Conf. Root Butt Rots, 7th. IUFRO Working Party S2.06.01. Forestry Canada, Pacific Forestry Centre, Victoria, B.C.

(Prepared by K. Lewis)

Stem and Branch Diseases

Blue-Stain Fungi and Bark Beetles

Causal organisms: Blue-stain fungi of several genera, mainly in Ophiostomatales, are involved. The most important genera are *Ceratocystis* Ellis & Halst. and *Ophiostoma* Syd. & P. Syd. and the anamorph genus *Leptographium* Lagerberg & Melin.
Hosts: most conifers
Distribution: throughout the range of the hosts

Ophiostomatales are fungi with long-necked perithecia, evanescent asci, and hyaline ascospores that lack pores or slits. There currently is little agreement on the taxonomic status of the genera in Ophiostomatales, so the use of the term "ophiostomatoid fungi" to cover these species is common.

Vectors and Hosts

Blue-stain fungi are dispersed by bark beetles of the genera *Dendroctonus, Ips, Scolytus,* and others. Only a few species of bark beetles are able to successfully attack living, healthy trees. The most-studied and well-documented are *D. frontalis* Zimmerman (the southern pine beetle) associated with *Ophiostoma minus* (Hedgc.) Syd. & P. Syd. in attacking southern pines; *D. ponderosae* Hopkins (the mountain pine beetle) associated with *O. clavigerum* (Robinson-Jeffrey & Davidson) T. C. Harrington in attacking mainly lodgepole pine; and *I. typographus* L. (the Eurasian spruce bark beetle) associated with *Ceratocystis polonica* (Siemaszko) Moreau (syn. *Ophiostoma polonicum* Siemaszko) in attacking Norway spruce.

Symptoms and Diagnosis

An early symptom of bark beetle attack is sawdust around entrance holes and at the base of a tree. Wilting symptoms on the foliage may be visible within 2 weeks after attack (e.g., by the southern pine beetle). In colder climates, the foliage may stay green up to 1 year after successful attack.

In susceptible trees, little or no resinosis is observed in the bark around beetle galleries. Resistant trees may contain the beetles and the fungi introduced by them in resin-soaked lesions. Occluded sapwood can be observed in wedges just beneath the bark beetle galleries. Later, these wedges may be colonized by blue-stain fungi (Plate 35). During successful attacks, all sapwood may be occluded and blue stained 2–3 weeks after infestation, depending on factors such as tree species and temperature. For some bark beetles, a particular associated blue-stain fungus (e.g., *C. polonica* for *I. typographus* and *O. clavigerum* for *D. ponderosae*) is always at the leading edge of fungal penetration of the sapwood. The fungal infection following southern pine beetle infestations is more complex. A variety of fungal species seems to be involved, not only blue-stain species.

Disease Cycle

The fruiting structures of blue-stain fungi associated with bark beetles are obviously adapted for insect dispersal, and the only way for spores to reach new host material is via beetle vectors. Most bark beetles have numerous pits or pockets on their exoskeletons. These pits may be filled with a variety of fungal spores before the beetles leave the old galleries. The pits act like mycangia and are referred to as such by some authors.

When blue-stain fungi are introduced into the bark of conifers, they start to grow in both the phloem and sapwood. In the phloem, blue stains develop mostly longitudinally. Transverse fungal growth is rather slow and does not usually match the rate of larval tunneling. Fungal growth in sapwood is mainly along the medullary rays and in other living cells of sapwood, such as the resin-duct epithelial cells. Blue-stain fungi live primarily on readily available nutrients, such as carbohydrates, in the cells. Typically, wedge-shaped occluded areas arise beneath the beetle galleries.

Fruiting structures of blue-stain fungi develop at the barkwood interface. As the beetles exit this tissue, their movement in the galleries causes the sticky fungal spores to adhere to their bodies. The beetles may also feed on blue-stain fungi, thereby transporting propagules internally.

Effects on Trees

Blue-stain fungi are thought to help their beetle vectors to overwhelm living trees. Many of the blue-stain fungi are able to independently kill host trees inoculated at densities or doses comparable to those introduced by beetle attacks. The blue-stain fungi cause long lesions in the bark and wedge-shaped occlusions in sapwood beneath the point of inoculation. It is unknown whether beetles are able to kill living trees in the absence of their associated fungi. However, a few beetle species not known to carry fungi have reportedly killed trees.

Disease Management

Many situations can influence epidemics of bark beetles and therefore enhance the spread of blue-stain fungi. Various stress situations may allow beetle populations to increase to epidemic levels. Factors such as stand age, stand density, and the composition of tree species are of significance and can be controlled by skillful forest management.

Selected References

Schowalter, T. D., and Filip, G. M. 1993. Beetle-Pathogen Interactions in Conifer Forests. Academic Press, San Diego, CA.

Wingfield, M. J., Seifert, K. A., and Webber, J. F., eds. 1993. *Ceratocystis* and *Ophiostoma:* Taxonomy, Ecology and Pathogenicity. American Phytopathological Society, St. Paul, MN.

(Prepared by H. Solheim)

Pine Wilt

Causal organism: *Bursaphelenchus xylophilus* (Steiner & Buhrer) Nickle
Hosts: Diploxylon pines, especially *Pinus thunbergiana* and *P. densiflora*
Distribution: United States and eastern Asia

Pine wilt, caused by the pinewood nematode, *Bursaphelenchus xylophilus*, is an extremely devastating disease of Japanese black (*Pinus thunbergiana*) and red (*P. densiflora*) pines in the southern half of Japan and localized areas of China, including Taiwan, and South Korea. Recent estimates in Japan indicate that more than 1 million cubic meters of timber are lost annually to the disease, which affects more than 20% of the 2.6 million hectares of pine forests. In warm areas of the United States, pine wilt kills some exotic and native pines, especially those grown as ornamentals or Christmas trees. The pathogen, but not the disease, is also found in Canada and Mexico. However, the mere presence of the pathogen in a country can result in economic hardship, since forest products destined for export must normally be free of the pinewood nematode. Thus, green lumber for export must be heat treated, and certain products whose treatment is not economically feasible (e.g., softwood chips) can no longer be exported. In Canada alone, such restrictions affect $3 billion (Canadian dollars) worth of forest product exports. A closely related nematode, *B. mucronatus* Mamiya & Enda, which has a mucronated tail compared with the rounded tail of *B. xylophilus,* and several nematodes with intermediate tail shapes are often obtained from dead and declining conifers in areas with and without pine wilt (e.g., Europe and Russia, respectively). Although many of these nematodes are pathogenic when inoculated into conifer seedlings, only the pinewood nematode is considered to be important in forest management or to warrant quarantine sanctions.

The pinewood nematode occurs in dead or dying trees and is vectored by cerambycid beetles in the genus *Monochamus* (Plate 36). Primary transmission occurs when young beetles, heavily infested with pinewood nematodes, emerge from diseased trees early in the spring and fly to young shoots of nearby healthy trees, where they feed and mature. Juvenile nematodes leave the beetle as it feeds, enter the tree through the beetle's feeding wounds, molt to adults, and migrate throughout the tree, feeding on parenchyma cells of the ray canals and on cambial tissues. At 25°C, the life cycle is completed in 4–5 days, there is a dramatic increase in the pinewood nematode populations, and the nematodes migrate throughout the tree. At these high temperatures and under conditions of moisture stress, disease symptoms, including needle chlorosis and then wilting and browning followed by tree death, develop quickly. Several months after the tree's death, dispersal and propagative third-stage and dispersal fourth-stage (dauerlarvae) juvenile nematodes appear. Dauerlarvae, which are resistant to desiccation, seek out *Monochamus* larvae, enter their tracheae, and remain there until the beetles emerge.

During secondary transmission, dauerlarvae leave the cerambycid beetles as they oviposit in dead or dying trees or in newly cut logs. There the nematode populations feed on a variety of fungi, and these trees or logs act as reservoirs of pinewood nematodes for later dispersal by *Monochamus* sp. Primary transmission occurs mainly in Japan and other countries where susceptible tree species are abundant and the hot (mean July temperatures above 20°C), dry summers stress host trees and favor both nematode buildup and multiple generations of *Monochamus* beetles. Secondary transmission occurs in areas with cooler summers and a variety of resistant host tree species (e.g., most of North America).

In countries such as Japan where pine wilt is widespread, one long-term forest management option is to gradually replace susceptible tree species with resistant species. Such species are identified by field observations and inoculations and by seedling inoculations that have established the relative susceptibility of both pines and other conifers. Another long-term solution is the development of genetically resistant pines. In China (Plate 37), where the disease affects Japanese black and red pines in discrete localities, both diseased trees within an affected area and susceptible trees in a 1- to 2-km-wide surrounding belt are cut and destroyed. Strict quarantines that prevent movement of pinewood nematode- and beetle-infested logs from those areas have also been implemented. Trees in Japanese parks and urban areas have been protected by using insecticides to kill potential vector beetles each year before they begin feeding during maturation. Sometimes, systemic insecticides or nematicides or both are employed to protect trees of extreme value. On a large scale, chemical or heat treatment of low-value commodities such as wood chips has not proved to be either efficient or economically feasible. However, gamma irradiation may be worthwhile for high-value logs. International quarantines and treatment requirements have been implemented to limit movement of both pinewood nematode- and *Monochamus*-infested products, but they are not always easy to enforce because infested lumber often appears in unsuspected products, such as dunnage.

Selected References

Sutherland, J. R., and Webster, J. M. 1993. Nematode pests of forest trees. Pages 351-380 in: Plant Parasitic Nematodes in Temperate Agriculture. K. Evans, D. L. Trudgill, and J. M. Webster, eds. CAB International, Wallingford, England.

Wingfield, M. J., ed. 1987. Pathogenicity of the Pine Wood Nematode. American Phytopathological Society, St. Paul, MN.

(Prepared by J. R. Sutherland)

Stem Decays

Stem decays are caused by fungi that utilize components of the wood cells. Decay results from the action of a relatively small number of specialized fungi, usually hymenomycetes, which possess combinations of the laccase, lignin peroxidase, and manganese-dependent lignin peroxidase enzyme systems capable of degrading lignin. In addition to these white rot species, a smaller number of fungi are capable of causing brown rot through selective removal of cellulose and other polysaccharides from the woody cell walls.

In living trees, there are several mechanisms that resist infection and development of decay. The first line of defense is the bark, a protective layer that cannot be penetrated by most decay fungi. Should this protective layer be broken (for example, by wounding), the living sapwood actively responds to the damage and invasion by decay fungi. Phenolic compounds are mobilized to the wound area, and a wound periderm (necrophylactic periderm) is laid down behind the wound in an attempt to isolate the damaged and possibly infected tissue. This response, together with the high moisture content of the sapwood, makes living sapwood very resistant to decay. Heartwood in living trees is more susceptible to decay than sapwood but is partially protected by its interior location and by the extractive content of the wood. In dead trees, however, it is the sapwood that is most susceptible to decay because the dynamic cytological response is lost and there are no extractives, such as those in the heartwood, to protect the sapwood from invasion.

Despite these protective mechanisms, both sapwood and heartwood are susceptible to infection by decay fungi, generally termed the saprots and heartrots, respectively. Decay fungi are also categorized by their mode of entry. The true heartrot fungi use natural disruptions in the bark surface to gain entry to the heartwood, whereas the wound decay fungi require a wound to access the heartwood. Because of the tremendous resistance of living sapwood to decay, saprots are usually restricted to dead trees or dead parts of trees. However, despite our attempts to group them, decay fungi do not always fit neatly into these categories. They form a continuum from parasites of living trees to strict saprophytes on dead trees. Furthermore, some decay fungi, such as those that cause the true heartrots, operate within a narrow range on this continuum, while others operate over a broader spectrum.

True heartrot fungi are confined to the heartwood and usually fruit only on living trees, although fruiting may continue on freshly cut logs. Sporocarps are most frequently produced at branch stubs, and new infections are caused by basidiospores that germinate and infect living trees through branch stubs or branchlet scars. Wound-entry decay fungi require a wound for infection. Sporocarps produced by these fungi are generally found on dead trees or parts of trees and are not frequently associated with branch stubs.

Any damage to the protective outer tissues on the surface of a tree may be considered a wound. Throughout its life span, a tree is exposed to numerous naturally occurring abiotic and biotic agents with the potential for causing wounds. Fire, wind, heavy snow, lightning, sun scorch, excessive heat or cold, drought, and insects or other animals all provide infection courts for wood decay fungi. In addition, harvesting operations and other forest-management procedures can result in wounding. The significance of any wound depends on the extent of the damage as well as the ability of the tree to seal off the wound and prevent entry or minimize the spread of decay-causing microorganisms.

Trees with heartrot do not usually show symptoms because the living, functional parts of the tree are not infected. Wounds, broken tops, and broken branches are indicators of heartrot caused by wound-entry fungi, but detection of true heartrots is difficult without the presence of sporocarps (conks). Increment cores may be used to detect signs of heartrot fungi if the heartrot occurs in the location of the core sample. The species of heartrot fungus can usually be determined by examining the type and appearance of decay (white or brown rot and texture of the decay) and the host species. Confirmation is achieved through isolation and culture of the fungus and examination of cultural characteristics and microscopic features.

Longitudinal decay in the heartwood progresses much more rapidly than radial decay and can become quite extensive and cause significant losses in wood volume. Losses to decay increase with increasing tree age and are usually insignificant in trees less than 90 years old. Trees with heartrot also provide habitat for cavity-nesting animals because the easily excavated heartwood is protected by a firm shell of sapwood. Generally, the roots of trees with heartrot are not affected unless the heartrot results from the extension of a root-rot fungus up into the bole. This occurs with several root-disease fungi, most notably *Heterobasidion annosum* on several conifer species and *Inonotus tomentosus* on spruce.

Management of damage caused by true heartrot fungi is based on rotation ages that minimize losses to heartrots. Minimizing wounds to trees during thinning, pruning, and other activities helps to reduce losses to heartrot fungi that enter through wounds but is ineffective for true heartrots. In suitable climates, pruning during the dry season, when sporulation is decreased, may reduce the incidence of some wound decays.

Selected References

Bondartsev, A. S. 1953. The Polyporaceae of the European USSR and Caucasia. Izdatel'stvo Akademii Nauk SSSR, Moskva-Leningrad. (Translation by Israel Program Scientific Translation, Jerusalem, and available from U.S. Department of Commerce, National Technical Information Service, Springfield, VA)

Gilbertson, R. L., and Ryvarden, L. 1987. North American Polypores, vols. 1 and 2. Fungiflora, Oslo.

Stalpers, J. A. 1978. Identification of wood-inhabiting Aphyllophorales in pure culture. Stud. Mycol. 16. Centraalbureau voor Schimmelcultures, Baarn, Netherlands.

(Prepared by K. Lewis and S. Woodward)

Wound Decays

Red Rot

Other names: red heart, red streak, mottled bark disease, bleeding fungus; sapin rouge (French); Blutender Schichtpilz (German); and blödskinn (Swedish)

Causal organism: *Stereum sanguinolentum* (Albertini & Schwein.:Fr.) Fr. (syns. *Haematostereum sanguinolentum* (Albertini & Schwein.:Fr.) Pouzar and *Stereum balsameum* Peck)

Hosts: wide host range including *Abies, Larix, Pinus, Picea,* and *Pseudotsuga* spp.

Distribution: Europe, North America, East Africa, Australia, New Zealand, and Japan

Symptom and Diagnosis

Stereum sanguinolentum causes a white rot. Initially, a yellow stain develops in the affected wood; further decay reduces the wood to a dry, fibrous state with yellow to orange brown streaks. The thin, leathery basidiocarps may be resupinate or develop as small brackets in large groups. They are commonly found on downed woody material or on the ends of infected logs (Plate 38). The gray hymenium is smooth, not poroid, and "bleeds" if scratched or cut, giving rise to the common name, bleeding fungus.

Disease Cycle

Basidiospores are produced under cool, moist conditions throughout the year and distributed by air currents. Infection is through fresh wounds. *S. sanguinolentum* is a relatively poor competitor and is unable to colonize wood previously infected by other species. The fungus grows rapidly in wounded xylem tissues, progressing more than 400 mm per year. Basidiocarps are produced annually from well-established infections.

Effects on Tree and Forest

In Europe, *S. sanguinolentum* is considered the most common cause of wound decay in conifers. It has been reported to colonize 24–44% of the wounds in *Picea abies* in Norway. The decay may spread up to 2.5 m on either side of the wound (Plate 39). It is also of significance in the Pacific Northwest region of North America. In temperate regions, *S. sanguinolentum* is frequently found as a primary colonizer of conifer stumps, where it may compete with the root and butt rot pathogen, *Heterobasidion annosum*. The common and apparently ubiquitous occurrence of *S. sanguinolentum* on conifer wood left in the forest suggests that the fungus is important in degrading lignified material in the coniferous forest ecosystem.

Predisposing Factors

Any fresh wound provides a suitable infection court for *S. sanguinolentum* infection, but infections are more likely in large wounds and if the sapwood is splintered during wounding. *S. sanguinolentum* has been reported to colonize wounds on *Pinus patula* and *P. radiata* made by elephants and buffalo in East Africa and Malawi. In the United Kingdom, damage to spruce bark caused by deer provides a common entry route. Increased incidence of *S. sanguinolentum* has been associated with severe infection by *Mycosphaerella pini* (anamorph *Dothistroma septospora*) in *P. radiata* plantations in New Zealand. Since the decay caused by *S. sanguinolentum* develops slowly, losses may increase with tree age and therefore may be reduced through the use of short rotations.

Selected References

Etheridge, D. E. 1973. Wound parasites causing tree decay in British Columbia. Can. For. Serv. Pac. For. Res. Cent. For. Pest Leafl. 62.

Gregory, S. C. 1984. Micro-organisms isolated from wounded stems of *Picea sitchensis*. Trans. Br. Mycol. Soc. 83:683-725.

Griffin, H. D. 1967. Further studies on *Stereum sanguinolentum* Alb. and Schw. ex Fries in Kenya forest plantations. East Afr. Agric. For. Res. Org. Mycol. Note 49.

Hallaksela, A.-M. 1984. Causal agents of butt-rot in Norway spruce in southern Finland. Silva Fenn. 18:237-243.

Ivory, M. H. 1987. Diseases and disorders of pines in the tropics: A field and laboratory manual. Overseas Dev. Admin. Oxford For. Inst. Overseas Res. Publ. 31.

(Prepared by S. Woodward)

Amylostereum Rot

Causal organisms: *Amylostereum areolatum* (Fr.) Boidin and *A. chailletii* (Pers.:Fr.) Boidin (syn. *Stereum chailletii* (Pers.:Fr.) Fr.)

Hosts: wide host range, including *Larix, Picea, Pinus,* and *Pseudotsuga* spp.

Distribution: Europe, North America, and Australasia

Symptoms and Diagnosis

Amylostereum areolatum and *A. chailletii* cause slowly expanding white rots. The color of the initial stain varies according to host; in *Picea* spp., it is usually red to red brown, whereas in *Larix* spp., it is dark brown. Although the two fungi may invade open wounds independently, it is generally accepted that their most frequent routes of entry are wounds made by wood wasps of the family Siricidae, particularly *Sirex* and *Urocerus* species. Sporophores are similar to those of the closely related *Stereum* species. However, the strain of *A. areolatum* prevalent in New Zealand, Tasmania, and Victoria has never been found to produce sporophores in nature. During severe attacks, the foliage often wilts as a result of a translocatable phytotoxin in the mucus of the wood wasp.

Disease Cycle

Individual species of *Sirex* appear to carry only one species of *Amylostereum*. *A. areolatum* is carried by *S. cyaneus, S. noctilio,* and *S. nitobie,* while *A. chailletii* is the symbiont of *S. areolatus, S. californicus, S. imperialis,* and *Urocerus* spp. These wood wasps are widely distributed throughout the northern hemisphere as far south as India, and they have been introduced on timber imports into New Zealand and Australia.

A female wood wasp carries oidia of the fungus in paired, invaginated, intersegmental sacs connected by ducts to the base of her ovipositor. How the female acquires the fungus is not certain, but it probably occurs in the tunnels of the host tree, either during or immediately after pupation. Fungal propagules remain in deep skin folds on the insects through each instar. During oviposition, the fungus is injected into the wound along with the eggs in a mucus carrier and develops in the tracheids of the host tree. After hatching, larvae of the wood wasp feed on the fungus and the degraded host tissues. Although the larvae feed predominantly on the mycelium, the fungus is absent from the larval gut, suggesting that the enzymes required for the breakdown of the mycelium are produced externally, possibly by secretion onto the mouth parts.

Effects on Tree and Forest

On *Pinus radiata* in New Zealand, severe outbreaks of the *Sirex-Amylostereum* problem occurred between 1946 and 1951, affecting more than 125,000 ha of plantations, although mortality was confined mainly to overstocked stands. The epidemic reached Tasmania during 1950 and 1951, and about half the trees in plantations near Hobart were killed. In Europe, sapwood decay caused by *A. chailletii* and *A. areolatum* is usually of secondary importance to that caused by *Stereum sanguinolentum*. Trees debilitated by other agents, however, may be more severely affected than healthy trees.

Predisposing Factors

Insect vectors of *A. areolatum* and *A. chailletii* attack trees affected by other debilitating factors such as suppression and overcrowding, waterlogging, drought stress, and fire damage. It appears that the intensity of the damage varies greatly with climatic and silvicultural factors. Wood wasps are attracted to the trees by resin volatiles. A low wood moisture content, 40–75%, has been reported as optimum for oviposition in *P. radiata*.

Disease Management

Good silvicultural practices, including correct choice of species for the site and optimum planting density, minimize wood wasp attack. During the severe outbreaks in New Zealand and Australia, various parasitic insects and nematodes were tested as biological control agents. Larvae of the insects *Ibalia leucospoides* and *Rhyssa persuasoria* attack the larvae of *Sirex* spp. inside the host tree. Two parasitic nematodes, *Deladenus wilsoni* and *D. siricidicola*, which sterilize both sexes of *Sirex* spp., were released in Tasmania and Victoria during 1970 and 1971, and by 1974, populations of wasps and nematodes had stabilized.

Selected References

Gaut, I. P. C. 1969. Identity of the fungal symbiont of *Sirex noctilio*. Aust. J. Biol. Sci. 22:905-914.
Morgan, F. D., and Stewart, N. C. 1966. The biology and behaviour of the woodwasp *Sirex noctilio* F. in New Zealand. Trans. R. Soc. N.Z. Zool. 7:195-204.
Rawlings, G. B. 1955. Epidemics in *Pinus radiata* forests in New Zealand. N.Z. J. For. 7(2):53-55.
Talbot, P. H. B. 1977. The *Sirex-Amylostereum-Pinus* association. Annu. Rev. Phytopathol. 15:41-54.

(Prepared by S. Woodward)

Yellow Brown Top Rot

Other name: brown cubical rot
Causal organism: *Fomitopsis cajanderi* (P. Karst.) Kotlaba & Pouz. (syns. *Fomes cajanderi* Weir, *Trametes subrosea* Weir, *Fomes subroseus* Weir (Overh.), and *Trametes carnea* C. G. Lloyd)

Hosts: many members of Pinaceae, particularly Douglas-fir (*Pseudotsuga* spp.) but also reported on *Prunus* spp.
Distribution: widespread in young coniferous forests of the northern hemisphere, especially in western North America

Symptoms

Fomitopsis cajanderi is a wound-invading heartrot fungus. Fruiting occurs commonly on wounds but is most often observed on fallen, infected boles where decay continues (Plate 40). It occurs infrequently on branch stubs of trees that have been infected for a long period of time.

Top breaks more than 2.5 cm in diameter within the upper host crown are indicators of potential subsequent decay (Plate 41). Trees with small breaks regenerate normal-appearing but usually infected crowns. If breaks are greater than 15 cm in diameter or below the live crown, *F. pinicola*, rather than *F. cajanderi*, colonizes the exposed wood. Bole crooks are indicators of heartrot and are visible years after breakage. The probability of infection and extent of decay increase with greater crook offset. Similarly, a whorl of branches of greater than normal diameter may indicate infection. Trees with large breaks regenerate several dense competing crowns above the infected break.

Advanced decay of living trees confined to the heartwood is typically yellowish brown and fractured cubically with minimal or no mycelial felt in shrinkage cracks. Yellowish, incipient decay may extend several feet beyond the advanced rot and show greenish streaks.

F. cajanderi is a pinkish brown to blackish, imbricate, effused-reflexed, rarely solitary fungus, not to be confused with the similarly colored, solitary, ungulate, largely sympatric *F. rosea*.

Disease Cycle

The time and amount of infection in young forests depend on the infrequent glaze storms that may break off up to 80% of the tops and also upon more frequent heavy, wet snow that breaks an occasional tree. Within the susceptible size range, nearly 100% of the breaks become infected by abundant inoculum from the common, perennial fruiting bodies. Most infected trees fall from the stand, so occurrence of the fungus in old-growth forests is rare.

Effects on Tree and Forest

Most crowns that regenerate after breakage are associated with rotten heartwood at the break and are consequently highly susceptible to later breakage by forces usually much less severe than those that caused the original breaks. These breaks, combined with crown shortening by natural pruning, so decrease crown mass and tree height that broken trees are in a vulnerable, intermediate position within the stand. Death results as growth of infected trees is suppressed by that of healthier neighbors. The consequences of second breaks may range from beneficial thinning to severe understocking.

Contributing Factors

Forests are predisposed to top rot by high stand density and uneven tree spacing, which result in asymmetrical crowns among the codominants. Such crowns load unevenly with ice and snow and break when weight becomes excessive. Tops of trees of intermediate size may be broken by falling debris. Top rot can be minimized by spacing to allow symmetrical crown development. Marketable broken trees should be removed when circumstances allow.

Selected References

Boyce, J. S., and Wagg, J. W. B. 1953. Conk rot in old-growth Douglas fir in western Oregon. Bull. 4. Oregon Forest Products Laboratory, Corvallis.

Roth, L. F. 1970. Disease in young-growth stands of Douglas fir and western hemlock. Pages 38-43 in: Management of Young-Growth Douglas Fir and Western Hemlock. A. B. Berg, ed. Oreg. State Univ. Sch. For. Pap. 666.

(Prepared by L. F. Roth)

Brown Crumbly Rot

Other names: red belt fungus, Rotrandiger Baumschwamm (German), klibbticka (Swedish)
Causal organism: *Fomitopsis pinicola* (Sw.:Fr.) P. Karst. (syn. *Fomes pinicola* (Sw.:Fr.) Cooke)
Hosts: primarily members of Pinaceae but also other conifers and several hardwoods
Distribution: North America, Europe, Asia, and Australia

Symptoms and Diagnosis

Fomitopsis pinicola is one of the most common and conspicuous saprophytic wood decayers in coniferous forests of temperate areas, where it causes a brown, cubical rot of the sapwood and heartwood of dead trees. The fungus is found most often on the wood of stumps, fallen logs, or standing trees (killed by other agents), but in some northern forests, *F. pinicola* is also the cause of an important wound decay of live trees (Plate 42). It is particularly abundant on conifer wood, even though it is reported from more genera of hardwoods (14) than of conifers (10). In the early stage of decay caused by *F. pinicola*, the wood has a faint yellowish to brownish discoloration; wood in an advanced stage of decay is brown and crumbly and has a tendency to break into cubical shapes. The decay is a typical brown rot type, in which only cellulose is degraded. Prominent, white, mycelial felts develop in cracks, which enlarge as the decayed wood shrinks. As a saprophyte of dead trees, the fungus generally attacks sapwood rapidly and then progresses into the heartwood.

Basidiocarps are highly variable in form, but in general they are hard, woody in texture, perennial, up to 60 cm or more in width, and shelf (Plate 43) or hoof (Plate 44) shaped. Flattened, appressed forms sometimes occur on the undersides of logs. The upper surfaces of *F. pinicola* basidiocarps are smooth and gray to black. A wide band, ranging in color from red to orange, occurs along the upper margin of some basidiocarps and is the source of one common name, the red belt fungus. The lower surfaces are usually creamy white with small, circular pores (approximately five to six per linear millimeter). The context of a basidiocarp is cream colored to buff, corky to woody, and trimitic in hyphal structure. Tube layers are stratified but sometimes indistinctly so. Basidiospores are cylindric-ellipsoid, hyaline, smooth, and 6–9 × 3.5–4.5 µm. No known conidial spores are formed.

Disease Cycle

F. pinicola produces abundant basidiocarps and basidiospores. Spores are generally dispersed in air currents and germinate readily upon contact with wood or many other substrates. As the fungus colonizes wood, mycelia from single basidiospores anastomose with other sexually compatible mycelia to form dikaryotic mycelia with clamp connections. Basidiocarps are usually produced abundantly to complete the fungus's life cycle.

Wood of dead conifers with adequate moisture content is quickly colonized by *F. pinicola*. Large bole wounds, broken branches or tops, or old dwarf mistletoe infections may be infection courts for heartrot of live trees. Heartrot levels are positively correlated with tree and stand age.

Ecological Significance and Disease Management

Its widespread occurrence as a dominant saprophyte of woody debris indicates that *F. pinicola* is extremely important ecologically as a recycler of nutrients. It is also a key producer of brown rot residues, which are stable soil components in coniferous forest ecosystems. *F. pinicola* causes large, old trees to die by stem breakage, making it a major disturbance agent creating canopy gaps. Forest managers who wish to maximize timber production can limit heartrot losses by reducing stem and root wounding during stand entries and by shortening stand rotations. Saprophytic losses can be reduced by timely salvage of dead trees. Regardless of whether its host is living or dead, *F. pinicola* contributes to trees becoming hazardous in some settings.

Selected References

Buckland, D. C., Foster, R. E., and Nordin, V. J. 1949. Studies in forest pathology. VII. Decay in western hemlock and fir in the Franklin River area, British Columbia. Can. J. Res. Sect. C 27:312-331.
Kimmey, J. W. 1956. Cull factors for Sitka spruce, western hemlock and western redcedar in southeast Alaska. U.S. Dep. Agric. For. Serv. Alaska For. Res. Cent. Stn. Pap. 6.
Mounce, I. 1929. Studies in forest pathology. II. The biology of *Fomes pinicola* (SW.) Cooke. Dom. Can. Dep. Agric. Bull. 111 (N.S.).

(Prepared by P. E. Hennon)

Other Wound Decays

Cylindrobasidium evolvens (Fr.:Fr.) Jülich (syns. *Corticium evolvens* Fr.:Fr. and *C. laeve* Pers.:Fr., sensu Jülich) causes a wound decay on many conifers in Europe. The resupinate fruit body is white or cream colored and is commonly found on discarded butt logs or on branches on the forest floor. The decay is a white rot; however, since it develops slowly, it is less significant than that caused by *Stereum sanguinolentum*.

Some investigators have found that *C. evolvens* is second to *S. sanguinolentum* in abundance at certain sites. For example, Solheim and Selas found that up to 39% of artificial wounds made in *Picea abies* stems were infected by *C. evolvens* after 2 years.

The time of year at which wounds occur has a marked effect on colonization by *C. evolvens*; more infections occur during the winter and spring than during the summer. Roll-Hansen and Roll-Hansen suggested that there is a relationship between infection by *C. evolvens* and *S. sanguinolentum*, although the exact nature of this relationship was not investigated.

White stringy rot, caused by *Resinicium bicolor* (Albertini & Schwein.:Fr.) Parmasto (syn. *Odontia bicolor* (Albertini & Schwein.:Fr.) Quél.), is found on many conifers (*Abies, Larix, Picea, Pinus, Pseudotsuga, Thuja,* and *Tsuga* spp.) in Europe, North America, and South Africa. While the decay is more commonly associated with wounds on roots, occasionally *R. bicolor* has been found to cause wound decay in stems.

A resupinate sporophore is formed on infected stumps and timber left in the forest. During the incipient stage, the decay is pink to red brown and becomes white, pitted, and stringy with black flecks during the advanced stage. Accurate diagnosis can be made by microscopic examination of material from the decay in which the characteristic encrusted, capitate cystidia can be seen.

R. bicolor is capable of growth through the soil via the production of mycelial cords (discrete bundles of hyphae without an apical meristem) (Plate 45). These cords are believed to be important mechanisms for spread of individual genets of the fungus, and their production is stimulated in an antagonistic environment. However, it is likely that colonization of stem wounds results from direct infection by basidiospores.

Previously, *R. bicolor* was considered to be a pathogen of conifer roots, but it is now thought to be only weakly pathogenic or saprophytic. It is frequently found colonizing stump surfaces. In areas where *R. bicolor* is abundant, the rate of infection by *Heterobasidion annosum* appears to be reduced, leading to suggestions that *R. bicolor* be used as a biological control agent for root and butt rot. In root-colonized hosts, vertical spread of the rot rarely exceeds 3 m. In some areas, *R. bicolor* occurs commonly on conifer stumps, suggesting an important role in the decomposition of lignified tissues.

The fir hydnum, or coral fungus, *Hydnum abietis* (Wier ex Hubert) (syn. *Hericium abietis* (Wier ex Hubert) K. A. Harrison) (Plate 46), causes a white pocket rot sometimes called long-pitted rot or long-pocket rot. It is found on *Abies* spp. and occasionally on *Tsuga* spp. in the forests of western North America.

Incipient decay is characterized by a yellow to yellow brown stain of the sapwood with darker spots. In advanced decay, the elongated pockets, sometimes containing white material, are interspersed with red brown, firm wood. The elegant, annual, corallike sporophores are white to cream colored, and the hymenium covers teethlike projections. Little is known of the biology of this hymenomycete. Infection of wounds and dead branches is likely to occur via basidiospores.

H. abietis can be important on *Abies* spp. in some areas, particularly the Pacific Northwest of North America. For example, of decay losses representing 3.1% of timber volume of *A. concolor* in Oregon, 34.4% of the decay was caused by *H. abietis*. Nitrogen-fixing bacteria have been detected in association with decay caused by the species in *A. concolor* in Oregon.

Selected References

Aho, P. E., Seidler, R. J., Evans, H. J., and Raju, P. N. 1974. Distribution, enumeration, and identification of nitrogen-fixing bacteria associated with decay in living white fir trees. Phytopathology 64: 1413-1420.

Filip, G. M., Kanaskie, A. M., and Frankel, S. J. 1984. Substantial decay in Pacific silver fir caused by *Hericium abietis*. Plant Dis. 68:992-993.

Kirby, J. J. H., Stenlid, J., and Holdenrieder, O. 1990. Population structure and response to disturbance of the basidiomycete *Resinicium bicolor*. Oecologia 85:178-184.

Nobles, M. K. 1953. Studies in wood-inhabiting hymenomycetes. I. *Odontia bicolor*. Can. J. Bot. 31:745-749.

Roll-Hansen, F., and Roll-Hansen, H. 1980. Microorganisms which invade *Picea abies* in seasonal stem wounds. I. General aspects. Hymenomycetes. Eur. J. For. Pathol. 10:321-339.

Solheim, H., and Selas, P. 1986. Misfarging og mikroflora i ved etter saring av gran. I. Utbredelser etter 2 ar. Rapp. Nor. Inst. Skogsfor. 7/86.

(Prepared by S. Woodward)

True Heartrots

Red Ring Rot

Other names: conk rot, white pocket rot; tramète des pins (French); Kiefern Feuerschwamm (German); la pudricion blanca (Spanish); and tallticka (Swedish)

Causal organism: *Phellinus pini* (Thore:Fr.) A. Ames (syns. *Fomes pini* (Thore:Fr.) P. Karst., *Trametes pini* (Thore:Fr.) Fr., and *Porodaedalea pini* (Thore:Fr.) Murrill)

Hosts: Most frequently found on hemlocks, spruces, larch, and Douglas-fir. Pines and red cedar are occasionally attacked. True firs appear to be less susceptible.

Distribution: widespread throughout the coniferous forests of the world

Symptoms and Diagnosis

The first symptom of the decay caused by *Phellinus pini* is a red stain, which often appears initially as a ring when an infected stem is viewed in cross section (hence the common name) (Plate 47). Next, the characteristic white, spindle-shaped pockets of decay develop parallel to the wood grain. In advanced stages, the wood has a honeycombed appearance caused by localized removal and utilization of lignin and cellulose by the fungus. Within the heartwood, incipient decay (red-stained wood) may be several meters ahead of the advanced decay, and the decay column may extend as much as 8 m above and below a sporophore. Sporophores are usually found at branch stubs, are perennial and bracketlike, and vary widely in thickness and size (Plates 48 and 49). The upper surface is brownish black and rough with concentric rings that parallel a wavy, yellowish to reddish brown margin. The lower surface is yellow brown with large, angular pores (two to three per millimeter). The context is cinnamon brown and punky.

Disease Cycle

Basidiospores are dispersed by wind, and the infection court is thought to be at small branch stubs where the heartwood is exposed. In *Pinus strobus*, infection is associated with old injuries caused by the white pine weevil. Basidiospores germinate, and the monokaryons initiate decay. Dikaryons are formed when compatible monokaryons anastomose. Once the dikaryon reaches the stem heartwood, the fungus begins to decay the heartwood in both directions. *P. pini* can infect only living trees, although in trees that have been killed or broken, the decay can continue until the resource is exhausted. In living trees, the fungal mycelium is thought to follow branch traces from the heartwood to the bark, where a sporocarp is produced. On dead and downed trees, sporocarps tend to be resupinate and frequently develop from cracks in the bole.

The Pathogen

The taxonomy and nomenclature of *P. pini* is confused. Recent research on intersterility groups and DNA polymorphisms suggests that as many as 13 different species may be involved, although most cannot be separated reliably on the basis of morphological features. In Europe, *P. pini* occurs on pines, and the fungus on spruce is commonly referred to as *P. chrysoloma*. Available evidence suggests that neither species, in the strict sense, is found in North America, but further work is needed before name changes can be justified. *P. cancriformans* is a distinctive species that causes a canker rot in *Abies* spp. in western Oregon (Plate 50).

Effects on Tree and Forest, Ecological Role

Losses caused by *P. pini* are greater than those caused by any other heartrot in the northern temperate forests, in part because of the relatively broad host range of this fungus and its common occurrence. The greatest damage occurs in older stands where the fungus has had time to cause more infections and develop in the heartwood. In British Columbia, it is estimated that volume losses to this heartrot are 5–10% of harvested volume. Because of the nature of the rot, *P. pini* contributes significantly to the provision of habitat for cavity-nesting species. It also predisposes trees to breakage and windthrow and therefore contributes to habitat for organisms that utilize downed logs. It is a decomposing fungus, one of the first in a succession of decomposers that are important for the recycling of nutrients locked within the wood structure.

Older trees are more susceptible to damage by this fungus, although even very young trees are susceptible to infection. Infections occur most frequently in dense stands where self-pruning creates infection courts for the fungus.

Disease Management

The fungus is not thought to utilize wounds for infection courts. Therefore, avoidance of wounding during silvicultural and harvesting operations does not affect *P. pini*. The most effective management for this heartrot is to use a "pathological rotation age" to determine when stands should be harvested. Selective harvesting may be effective in promoting multiple ages and a more open canopy structure, both of which reduce the rate of infection and decay development.

Selected References

Boyce, J. S., and Wagg, J. W. B. 1953. Conk rot of old-growth Douglas-fir in western Oregon. Bull. 4. Oregon Forest Products Laboratory, Corvallis.

Fischer, M. 1996. Molecular and microscopical studies in the *Phellinus pini* group. Mycologia 88:230-238.

Hunt, R. S., and Etheridge, D. E. 1995. True heart-rots of the Pacific region. Can. For. Serv. Pac. For. Res. Cent. Pest Leafl. 55.

(Prepared by K. Lewis)

Rust-Red Stringy White Rot

Other names: the Indian paint fungus
Causal organism: *Echinodontium tinctorium* (Ellis & Everh.) Ellis & Everh.
Hosts: All native species of *Abies* and *Tsuga* in western North America are major hosts; *Picea engelmannii*, *Picea glauca*, and *Pseudotsuga menziesii* are minor hosts.
Distribution: coniferous forests in western North America from Mexico to Alaska

Symptoms and Diagnosis

The early symptom of the decay caused by *Echinodontium tinctorium* is a brown to reddish orange rot of heartwood that may significantly weaken the wood, even during incipient stages. In advanced stages of decay, the wood becomes stringy and eventually takes on a whitish cast as lignin is removed (Plate 51). Occasionally, wood separations (laminations) may occur between annual growth rings as the wood dries. The decay is classified as a white rot, since both lignin and cellulosic components are removed. The bright orange coloration often associated with decay (Plate 52) may occasionally protrude from the bark in rusty punk knots, which are reliable indicators of the fungus. Decay may extend up to 7 m above and below a basidiocarp and into large branches. Two conks separated vertically by at least 1.5 m on the bole indicate extensive rot, and several conks indicate that the tree is unmarketable. The large (up to 40 cm wide × 20 cm thick), hoof-shaped, perennial, woody basidiocarps (conks) are the most visible and diagnostic signs of this disease (Plate 53). *E. tinctorium* is the only hydnaceous hymenomycete in western North America that produces a perennial basidiocarp. Basidiocarps usually develop under dead branch stubs, but they occasionally form on the undersides of dead branches close to the bole. The upper surface is dark gray brown to black and appears hairy with light margins when immature; it becomes harder and deeply cracked with concentric grooves and ridges at maturity. The lower surface is light gray to tan and irregularly poroid at the margin and bears long, thin and brittle to thick and rigid pendant teeth (spines) that turn white with buff, felty tips during active sporulation. The teeth are flattened to cylindric and may fuse in older basidiocarps. The dark orange to brick red internal contextual tissues were commonly used by western aboriginals as a source of pigment for making dyes and paints (Plate 54).

Disease Cycle

Sporulation occurs predominately during cool, wet, fall and spring months, although light sporulation can occur throughout the year when mean daily temperatures are 4–16°C. Basidiospores are dispersed primarily by wind and must be exposed to subfreezing temperatures for a period of one to several months before significant germination can occur. This may explain why the fungus has adapted to cold, high-elevation, interior mountainous habitats. Most infections probably occur during the spring. The primary infection courts of this fungus are currently believed to be small (1–2 mm in diameter), recently shade-killed twigs and exposed stubs of living secondary branchlets on branches near the bole. Basidiospores germinate to form monokaryotic hyphae that penetrate and colonize the dead wood and pith tissue of the branch. Studies indicate that monokaryons are immediately capable of causing substantial decay on their own prior to development of the dikaryon. The fungus is heterothallic, and its mating behavior is controlled by a bifactorial sexual-incompatibility system. Consequently, infections by at least two compatible mating types in relatively close proximity are required for anastomosis and formation of the dikaryon. Once it reaches the bole, the dikaryotic thallus causes rapid decay of heartwood and produces an extensive decay column before a fruiting body can form. Mycelia appear to grow along branch traces from the heartwood to reach the bark and amass growth under branch stubs to initiate basidiocarp formation. The basidiocarp may continue to develop on standing and downed trees for many years and release basidiospores annually until the nutrient and energy reserves in the heartwood are exhausted.

Effects on Tree and Forest, Ecological Role

The Indian paint fungus is the predominant cause of heartrot and volume losses in native species of living true firs and hemlocks in coniferous forests of western North America. Heartwood volume losses attributed to this fungus in conifer stems are second only to those caused by *Phellinus pini* in this region. The fungus causes the greatest damage in mature and overmature trees that have exceeded the 100- to 150-year pathological rotation age, although significant volume losses can occur in much younger trees. *E. tinctorium* contributes significantly to the predisposition of living trees to windthrow and bole break, accelerating their demise by decomposing organisms. It also serves as an early decomposer itself by initiating decomposition processes that contribute to nutrient and carbon recycling from the heartwood of mature standing trees. These trees later become nurse logs, releasing nutrients and mulch essential for the survival of conifer seedlings during regeneration.

Predisposing Factors

Trees growing at sites that reduce vigor and in dense stands in which shade-suppressed trees have the small, dead twigs and exposed branch stubs that serve as infection courts are predisposed to attack. Infections of trees in the vicinity of infected trees bearing basidiocarps also are more numerous on densely shaded sites. This is particularly true in cold, humid habitats that favor sporulation and basidiospore germination. Large wounds on boles or branches do not appear to be utilized by the fungus to enter its hosts, but they may serve as sites for basidiocarp formation. Stands in which the trees are of various ages tend to have more decay than those in which trees are of similar age. Older trees are at greater risk of attack and are more susceptible to windthrow because of their larger volume of heartwood.

Disease Management

Planting should be avoided on sites that do not promote vigorous growth of fir and hemlock species. Sites on cold, shaded slopes that favor sporulation and basidiospore germination are most likely to harbor diseased trees. Nonhost species should be planted on high-hazard sites with high incidence of the fungus. Forest stands containing fir or hemlock as dominant species should be thinned to prevent shade suppression and reduce relative humidity under the canopy, thus avoiding conditions that promote sporulation and formation of infection courts. Trees bearing basidiocarps should be removed as a sanitation measure during precommercial thinning cuts and after logging, since decay incidence and volume loss increase in stands containing actively sporulating basidiocarps. Wounding of merchantable timber during thinning should be avoided, since wounds may provide exit points for basidiocarp formation and entry points for other decay fungi. Management strategies should favor stands with trees of similar age to avoid the higher incidence of the fungus and the greater damage it causes in stands with trees of different ages. Commercial harvests should be completed before the dominant, merchantable species have reached their predetermined pathological rotation ages specific for each site.

Selected References

Etheridge, D. E., and Craig, H. M. 1976. Factors influencing infection and initiation of decay by the Indian paint fungus (*Echinodontium tinctorium*) in western hemlock. Can. J. For. Res. 6:299-318.

Maloy, O. C. 1991. Review of *Echinodontium tinctorium*, 1895-1990. Wash. State Univ. Coop. Ext. Bull. 1592.

Thomas, G. P. 1958. Studies in forest pathology. XVIII. The occurrence of the Indian paint fungus, *Echinodontium tinctorium* E. & E., in British Columbia. Can. Dep. Agric. For. Biol. Div. Publ. 1041.

Wilson, A. D. 1990. The genetics of sexual incompatibility in the Indian paint fungus, *Echinodontium tinctorium*. Mycologia 82:332-341.

(Prepared by A. D. Wilson)

Stem Rusts

All the stem rusts of pines are in the closely related genera *Cronartium*, *Peridermium*, or *Endocronartium*, except pine twist rust, caused by *Melampsora pinitorqua*. The alternate hosts of these rusts are a diverse array of angiosperms. Indeed, the host association is often the best aid for identification, since morphological distinctions are subtle or lacking in many species.

Cronartium telia produce teliospores packed together in columns that resemble hairs. They are usually formed on the undersides of leaves. Aecia are peridermioid, and yellow, echinulate aeciospores form beneath the peridium, which ruptures at maturity.

Most *Cronartium* rust fungi are heteroecious (host-alternating) and macrocyclic, forming three infectious spore stages during their life cycles. Aeciospores are disseminated by the wind during the spring and are capable of long-distance dispersal to alternate hosts. Uredia and urediospores are formed first on the alternate host and function as conidia, causing new infections on the same plant species and allowing an increase in inoculum during the growing season. Telia typically form in the uredial pustules, often during periods of cool weather. When leaves are wet, teliospores germinate to form basidia, and basidiospores are released.

Basidiospores are also dispersed by wind, but they are more delicate than aeciospores or urediospores. Effective dispersal is generally limited to distances of 500 m or less. Basidiospores germinate on needles or young shoots, and germ tubes either enter through stomata or penetrate the epidermis directly. Pycnia (spermagonia) appear the following spring or later, and pycniospores are carried by insects from one pycnium to another, effecting fertilization. Aecia are formed on fertilized infections a year later. *Cronartium* rust fungi form a perennial mycelium in the stems of their pine hosts, producing pycnia along the margins of expanding cankers and aecia in tissues that bore pycnia the previous year. In most cases, the rusts are ephemeral on the angiosperm host.

Autoecious, pine-to-pine rusts have a simpler life cycle. Pycnia, uredia, telia, and basidiospores are not formed. Spores morphologically identical to aeciospores are formed on pine, are dispersed by the wind, and initiate new infections on pines. These "imperfect" rusts are placed in the form genus *Peridermium* on the basis of aecial morphology. Alternatively, they may be grouped in *Endocronartium* on the basis of cytological evidence that meiosis occurs in the germinating "aeciospores." *Peridermium* (*Endocronartium*) species are closely allied with individual *Cronartium* species on the basis of hosts, morphology, and most recently, molecular evidence.

Rusts of Soft Pines

White Pine Blister Rust

Other names: rouille vèsiculeuse des pin a cinq feuilles (French); Blasenroste (German); moho ampoua del pino blanco (Spanish)

Causal organism: *Cronartium ribicola* J. C. Fisch.

Hosts: Telial hosts include *Pedicularis* spp. (Asia only) and all *Ribes* spp.; aecial hosts include all *Pinus* spp. within the foxtail, stone, and white pine groups of the *Haploxylon* subgenus.

Distribution: throughout the range of the hosts in Asia, Europe, and North America

Symptoms and Diagnosis

On pines, conspicuous, orange aeciospores are produced during the spring on slightly swollen fusiform cankers that encircle branches and near the margins of diamond-shaped cankers on stems (Plate 55). Cankers on smooth stems and branches have orange margins, which are readily seen when moistened. Old cankers often have roughened bark where aeciospores have ruptured the periderm. They also may have resinous constrictions and evidence of secondary organisms, which cause necrotic patches, dead branches, and dead cankers. Since most stem cankers originate from branch cankers, they frequently have a branch or branch stub at the center.

Disease Cycle

Germ tubes from basidiospores enter pines through stomata of needles during late summer (Plate 56). The first symptom is a yellow or red needle spot that appears the following spring. By fall, the first signs of orange hyphae are clearly visible in the bark at the bases of infected needles. Within 2 years, dark lesions containing pycnia develop in the infected area. Flies attracted to the sweet pycniospore droplets facilitate dikaryon formation between compatible mating types. Genetically diverse, dikaryotic aeciospores are produced during the following spring. The fungal tissue in the canker margin remains haploid, and pycniospores are produced annually from spring to fall along the periphery. Germ tubes from aeciospores enter alternate hosts (usually *Ribes* spp.) via stomata. Urediospores are produced throughout the summer and infect additional *Ribes* leaves by entering through stomata. Occasionally, aeciospore infection leads directly to production of telial columns, but more often telia are produced during the fall (Plate 57).

Effects on Tree and Forest, Ecological Role

C. ribicola is native to Asia and was introduced into Europe and North America. American five-needle pines have little resistance to the fungus, and it has been highly destructive in some regions (Plate 58). It causes cankers, dieback, dead tops, and death and predisposes trees to attack by bark beetles. As susceptible species are eliminated from the forest, they are replaced by other, less commercially desirable native conifers. In some areas, western white pine has become locally extinct. In addition to causing loss of a valuable timber species, white pine blister rust also has had a significant impact on some forest ecosystems. For example, stone pine seeds are important foods for nutcrackers, and white-bark pine cones and seeds are important foods for grizzly bears. White pines are tall trees used as perches and nesting sites for birds of prey, and the rough bark provides a preferred foothold as an escape route for young black bears. White pines are also tolerant of some root diseases (e.g., laminated root rot), but blister rust limits use of white pine as a root-disease management tool.

Disease Management

Alternate hosts. Seeds of *Ribes* spp. are persistent and germinate readily when exposed to light, making eradication of the alternate host difficult. However, eradication is practiced in high-value sites, such as nurseries and watersheds. Often *Ribes* spp. are concentrated along streams, and nonhost barrier trees can be used there in an effort to impede spore movement between hosts. *Ribes* spp. that grow with pines are usually more important sources of inoculum than those concentrated along streams. The unit density of *Ribes* spp. bushes can be used as a factor in determining hazard zones; usually the more bushes per hectare, the greater the incidence of disease.

Hazard zones. Blister rust hazard zones are delineated by climate and/or topography. Blister rust is more prevalent in cool, moist habitats than in warm, dry areas and in sites where there is air movement to transport basidiospores. The hazard is greater in open stands where spores may blow in and in areas where a luxuriant shrub layer may increase dew formation.

Fungicides. Several preventative fungicides have been developed and are sometimes used in nursery situations. There are no published reports of success with therapeutic fungicides.

Quarantines. Quarantines are common between countries but rare within countries. They usually prevent the movement of pine hosts other than seeds and restrict the movement of *Ribes* spp. to times of the year when they are leafless and dormant.

Pruning. There are usually more infections in the lower part of the tree because conditions there favor infection. Pruning the lower branches before cankers reach the stems is a widely practiced control method. In order to be effective, pruning must be done as soon as it is practical. Frequently, trees are pruned to a height of 1–2 m and then pruned a second time to 3–4 m. These treatments are frequently combined with precommercial thinning operations, which allow preferential removal of stem-cankered trees.

Resistance. Several selection programs for blister rust resistance exist for eastern white, western white, and sugar pines. There is also a modest one for white-bark pine. Seed orchards of cloned seedlings or their parents have been established from families that display resistant reactions to artificial inoculation. There is a significant reduction in rust incidence in plantations established from the earliest of these seed orchards. Only a few plantations, usually those with large populations of *Ribes* spp., have inadequate stocking levels. New generations of seed orchards are expected to possess superior rust resistance.

Selected References

Hagle, S. K., McDonald, G. I., and Norby, E. A. 1989. White pine blister rust in northern Idaho and western Montana: Alternatives for integrated management. U.S. Dep. Agric. For. Serv. Gen. Tech. Rep. INT-261.

Hodge, J. C., Gross, H. L., and Retnakaran, A. 1989. White pine blister rust and white pine weevil management guidelines for white pine in Ontario. Ont. Minist. Nat. Resour. Pest Manage. Sec. Rep. PM-36.

Hunt, R. S., comp. 1988. Western white pine management symposium. Nakusp, B.C., Canada.

Kaneko, S., Katsaya, K., Kakishima, M. and Ono, Y., eds. 1995. Proc. IUFRO Rusts Pines Work. Party Conf., 4th. Forestry and Forest Products Research Institute, Ibaraki, Japan.

(Prepared by R. Hunt)

Other Soft Pine Stem Rusts

Pinyon Blister Rust

Cronartium occidentale Hedgc., Bethel, & N. Hunt infects pinyon pines (*Pinus edulis* and *P. monophylla*) and currant (*Ribes* sp., Grossulariaceae). On pinyons, the rust is reported from Arizona, California, Colorado, Nevada, New Mexico, and Utah; on *Ribes*, the rust is also known from Idaho, Kansas, Montana, Nebraska, and Washington. Although its life cycle is similar to that of *C. ribicola*, *C. occidentale* has different resistant and favored host species for production of telia. Spores of the two rusts on *Ribes* are morphologically and biochemically distinguishable. This rust, native to North America, is never as damaging as the nonindigenous white pine blister rust, presumably because pinyons grow in drier areas that are less conducive to disease development and have more natural resistance. Although the rust readily kills seedlings, old trunk infections display little necrosis. Cankers on *P. monophylla* from an outbreak during the 1920s were still obvious in 1960. It has never threatened whole stands, and control is not practiced.

Pine-to-Pine Rusts (Soft Pines)

Two autoecious species have been described from *Pinus pumila* in Japan. *Peridermium yamabense* Saho & Takahashi and *Endocronartium sahoanum* Imazu & Kakishima cause geographically separated, morphologically distinct rusts on native pine at high elevations in Japan. Both have the aeciospore germ tube cytology typical of *Endocronartium* spp.

Selected References

Imazu, M., Kakishima, M., and Katsuya, K. 1991. Morphology, cytology, and taxonomy of stem rusts on five-needle pines in Japan. Pages 76-91 in: Rusts of Pine. Proc. IUFRO Rusts Pine Work. Party Conf., 3rd. Y. Hiratsuka, J. K. Samoil, P. V. Blenis, P. E. Crane, and B. L. Laishley, eds. For. Can. Northw. Reg. North. For. Cent. Inf. Rep. NOR-X-317.

Kimmey, J. W. 1946. Notes on visual differentiation of white pine blister rust from pinyon rust in the telial stage. Plant Dis. Rep. 30:59-61.

(Prepared by R. Peterson and B. Geils)

Rusts of Hard Pines

Fusiform Rust

Causal organism: *Cronartium quercuum* (Berk.) Miyabe ex Shirai f. sp. *fusiforme* (Hedgc. and N. Hunt) Burdsall and G.

Snow (syns. *C. fusiforme* Hedgc. & N. Hunt, *C. cerebrum* (Peck) Hedgc. & W. H. Long, *C. quercuum* (Berk.) Miyabe ex Sharai, and *Peridermium cerebrum* Peck; anamorph, *Peridermium fusiforme* Arth. & Kern)

Hosts: Primary pine hosts are loblolly (*Pinus taeda*) and slash pines (*P. elliottii*). Telial hosts are members of the red oak group, especially water (*Quercus nigra*) and willow (*Q. phellos*) oaks.

Distribution: Fusiform rust is found only in the United States from Maryland south to Florida and west to Texas and Arkansas.

Fusiform rust is a limiting factor in the commercial production of loblolly and slash pines in a corridor from central South Carolina to southern Louisiana. Pond pine (*Pinus serotina*) and South Florida slash pine (*P. elliottii* var. *densa*) are also infected. Longleaf (*P. palustris*) and shortleaf (*P. echinata*) pines are tolerant or resistant. Hosts from outside the natural range include pitch pine (*P. rigida*) and the highly susceptible western species Monterey (*P. radiata*), Jeffrey's (*P. jeffreyi*), and ponderosa (*P. ponderosa*) pines.

The most susceptible telial hosts are members of the red oak group. White oaks are rarely infected. Some other members of the family Fagaceae, including species of chestnut and chinquapin, are also moderately susceptible as is the western species, California black oak (*Quercus kelloggii*).

Symptoms and Diagnosis

On the pine hosts, the pathogen induces formation of perennial, spindle-shaped swellings or galls on stems or branches (Plate 59). The galls, particularly on loblolly pine, may be irregular in shape (Plate 60). Infections on nursery seedlings are usually at the ground line or slightly above and result in early death. On slash pine, multiple infections on young seedlings sometimes result in witches'-brooms (Cronartium bushes). Slash pine is generally considered more susceptible than loblolly, and death caused by the rust is common (Fig. 5).

On oaks, the fungus produces symptoms that are minor and not considered injurious. The earliest symptoms are chlorotic spots. Then minute, orange uredial pustules and finally the brown, hairlike telial columns develop.

Disease Cycle

Cronartium quercuum is macrocyclic and heteroecious. Pycnia are produced on galls during September to February, and aecia are formed a year later from February to April, depending on the geographic area. Aeciospores are very abundant and infect young, immature leaves of susceptible oaks, and within 10–14 days, uredia begin to appear. After an additional 7–10 days, telia develop either in old uredial sori or separately and remain viable on oak leaves for several weeks.

Pines may be infected by basidiospores from early April through June, depending on the latitude. Only young, succulent shoots and needles of pines are susceptible. When a basidiospore germinates on susceptible pine tissue, the germ tube penetrates the epidermis directly. The first symptoms of pine infection are reddish brown to purple spots. In 6 months to a year, hyperplasia and hypertrophy of infected cells result in formation of a gall on the pine stem or branch. These galls are perennial and can persist for decades. Pycniospores ooze from the bark in a yellowish orange exudate. It is presumed that these spores function in the diploidization of the organism, but this has not been demonstrated.

Effects on Tree and Forest, Ecological Role

Galls on older trees are often attacked by insects or other fungi and subsequently develop into resinous cankers. Stem galls and cankers lead to wind breakage and early death; how-

Fig. 5. Damage to 14-year-old slash pine in coastal South Carolina caused by fusiform rust.

ever, the major economic loss results from degradation of the tree's butt log.

The pine forests of the southern United States originally contained a significant proportion of both long- and shortleaf pine stands with trees of various ages. Both of these species are resistant or at least tolerant to fusiform rust. However, difficulty in planting and slow growth have made both species less attractive for the intensive, plantation-type forestry practiced today. Rapid, early growth encouraged foresters to plant rust-susceptible slash and loblolly pines so extensively that approximately two-thirds of the total inventory of slash pine is now in plantations and almost half of the loblolly in the five southeastern states was planted. Unfortunately, during the 1940s and 1950s, seeds of these species were often collected from diseased parents, which amplified the susceptibility and made the disease more prevalent.

Contributing Factors

The use of susceptible loblolly and slash pines in widescale planting operations, with the resulting rapid growth, leads to a higher incidence of rust. Improved fire protection has also increased the hazard of disease by allowing oaks to flourish.

Intensive site preparation increases the early growth of pines. With slash pine, the increase in growth results in a higher incidence of fusiform rust. With loblolly pine, however, the relationship between intensive site preparation and increased levels of rust is less clear. There is an increase in rust on high-hazard sites but not on low-hazard sites.

Disease Management

The intensive forest management currently practiced in the southern United States has greatly influenced the current fusiform rust epidemic. Many silvicultural options are available to forest land managers for reducing rust losses, even where rust hazard is quite high. Selection of planting stock, intensity of site preparation, spacing, fertilization, and the type of product that is to be produced should all be considered in determining how to deal with fusiform rust. Perhaps the most important factor has been the development during the 1980s and 1990s of highly rust-resistant planting stock for use in areas of high rust hazard. This single achievement has probably been the greatest contribution made to forest pest management in this area in recent times. The use of resistant seedlings plus the other silvicultural manipulations now available have helped tremendously in reducing damage from rust.

When stems reach merchantable size, selective cutting of trees that are severely infected with fusiform rust decreases potential mortality and improves stand quality. In a recent study, the selective cutting of high-risk trees in 15- to 20-year-old slash and loblolly pine plantations lowered the incidence of stem infections, reduced aeciospore production, and improved the quality of thinned stands significantly.

Selected References

Powers, H. R., Jr., and Kraus, J. F. 1983. Developing fusiform rust-resistant loblolly and slash pines. Plant Dis. 67:187-189.
Powers, H. R., Jr., Schmidt, R. A., and Snow, G. A. 1981. Current status and management of fusiform rust on southern pines. Annu. Rev. Phytopathol. 19:353-371.
Wells, O. O., and Wakely, P. C. 1966. Geographic variation in survival, growth, and fusiform-rust infection of planted loblolly pine. For. Sci. Monogr. 11. Society of American Foresters, Washington, DC.
Wilcox, P. L., Amerson, H. V., Kuhlman, E. G., Liu, E. H., O'Malley, D. M., and Sederoff, R. R. 1996. Detection of a major gene for resistance to fusiform rust disease in loblolly pine by genomic mapping. Proc. Natl. Acad. Sci. USA 93:3859-3864.

(Prepared by H. Powers and E. G. Kuhlman)

Eastern Gall Rust

Other names: pine-oak gall rust
Causal organism: *Cronartium quercuum* (Berk.) Miyabe ex Shirai f. sp. *banksianae*
Hosts: Jack pine (*Pinus banksiana*) is the principal host. Uredinial and telial hosts are various species of oak (*Quercus*) and chinquapin (*Castanopsis*).
Distribution: eastern North America from the northern limit of *Quercus* spp. to Florida and Texas

Several host-specialized forms of *Cronartium quercuum* are recognized: f. sp. *banksianae* (the cause of eastern gall rust), f. sp. *fusiforme* (the cause of fusiform rust), f. sp. *densiflorae* (an Asian form), f. sp. *echinatae* (found on shortleaf pine), and f. sp. *virginianae* (found on Virginia pine).

Symptoms and Diagnosis

Conspicuous, perennial, globose galls are produced on the stems of hard pines. Pycnia form during the spring and are followed in May to July by powdery, orange yellow aeciospores (Plate 61). In some areas of North America, this rust overlaps with another gall-forming rust, western gall rust (*Endocronartium harknessii*), and examination of germinating spores with a microscope is necessary to differentiate the two.

Effects on Tree and Forest

Unless galls are produced on the main stems, this disease does not affect growth of trees significantly. Trees with galls on main stems tend to be deformed and are easily broken at the galls. Therefore, they are unsuitable for utilization. Nursery infections and infections on cultivated Christmas trees or ornamental trees are economically important.

Disease Management

Nursery infections can be reduced by using protective fungicides during the time that basidiospores are produced or by removing alternate hosts in and around the site.

Selected References

Anderson, G. W., and French, D. W. 1965. Differentiation of *Cronartium quercuum* and *Cronartium coleosporioides* on the basis of aeciospore germ tubes. Phytopathology 55:171-173.
Burdsall, H. H., Jr., and Snow, G. A. 1977. Taxonomy of *Cronartium quercuum* and *C. fusiforme*. Mycologia 69:503-508.
Gross, H. L. 1983. Negligible cull and growth loss of jack pine associated with globose gall rust. For. Chron. 59:308-311.
Nighswander, J. E., and Patton, R. F. 1965. The epidemiology of the jack pine-oak gall rust (*Cronartium quercuum*) in Wisconsin. Can. J. Bot. 43:1561-1581.

(Prepared by Y. Hiratsuka)

Western Gall Rust

Other names: pine-pine gall rust, globose gall rust
Causal organism: *Endocronartium harknessii* (J. P. Moore) Y. Hiratsuka (anamorph *Peridermium harknessii* J. P. Moore)
Hosts: Major hosts include jack pine (*Pinus banksiana*), lodgepole pine (*P. contorta* var. *latifolia*), and ponderosa pine (*P. ponderosa*). Bishop's pine (*P. muricata*), Monterey pine (*P. radiata*), and introduced species such as mugo pine (*P. mugo*), Austrian pine (*P. nigra*), maritime pine (*P. pinaster*), and Scot's pine (*P. sylvestris*) are also affected.
Distribution: across northern North America and south to Virginia in the east and south to Arizona and northern Mexico in the west

Western gall rust has often been confused with host-alternating eastern gall rust (pine-oak gall rust), caused by *Cronartium quercuum* f. sp. *banksianae*, which also produces globose galls. However *C. quercuum* f. sp. *banksianae* is rare in Canada. In areas where both rusts occur, microscopic examination of germinating spores is necessary to differentiate the two.

Symptoms and Diagnosis

Conspicuous, perennial, globose galls are produced on the stems of hard pines (Plate 62). From May to July, powdery, orange yellow spores are produced on the surfaces of the galls. Very young galls are sometimes spindle shaped rather than spherical and can be confused with the spindle-shaped swellings produced by comandra blister rust.

Disease Cycle

Western gall rust has an autoecious life cycle; i.e., it is capable of infecting pine directly without going to an alternate host. Spores are produced on the galls from the end of May to July, become airborne, and infect the green tissue of young shoots. Small galls appear a few months after infection but do not produce spores until the following year. Galls grow each year and produce spores every spring until the gall tissue dies with the stem or sori are inactivated by mycoparasites.

Effects on Tree and Forest

Main-stem galls often kill small trees (Plate 63), but on larger stems, galls usually increase in size for many years without killing the trees. Trees with galls on the main stems tend to be deformed and are easily broken at the gall. Therefore, they are unsuitable for utilization (Plate 64). Branch galls on large trees do not affect the vigor of trees significantly.

This disease tends to intensify in highly managed, young pine forests probably because of 1) the pine-to-pine life cycle, 2) the high susceptibility of vigorously growing shoots, and 3) the perennial nature of active galls, which serve as inoculum sources. A significant amount of western gall rust infection of nursery origin has been found in some plantations, emphasizing the importance of producing disease-free planting stock.

Contributing Factors

Fast-growing seedlings in nurseries and plantations are often more susceptible to infection because of their longer, susceptible candles.

Disease Management

One should avoid planting susceptible pine species near heavily infected stands. Young, naturally regenerated or planted seedlings with main-stem galls should be removed. Eradication of galls during thinning operations may be effective in intensive management situations. Timely protective fungicide applications are effective and economically feasible in tree nurseries or highly managed tree farms where high-value ornamental pines are grown. Eradication of infected trees or pruning of infected branches in and around tree nurseries should minimize the level of infection.

Selected References

Hiratsuka, Y., and Maruyama, P. J. 1985. Western gall rust. Can. For. Serv. North. For. Res. Cent. For. Pest Leafl. 27.

Merrill, W., and Kistler, B. R. 1976. Phenology and control of *Endocronartium harknessii* in Pennsylvania. Phytopathology 66:1246-1248.

Powell, J. M., and Hiratsuka, Y. 1973. Serious damage caused by stalactiform blister rust and western gall rust to a lodgepole pine plantation in central Alberta. Can. Plant Dis. Surv. 53:67-71.

(Prepared by Y. Hiratsuka)

Resin Top Disease

Other names: rouille vésiculeuse des pins à deux feuilles (French), Kiefernrinden-Blasenrost (German), törskate (Swedish)

Causal organisms: full cyclic rust, *Cronartium flaccidum* (Alb. & Schwein.) Winter (syn. *C. asclepiadeum* (Willd.) Fr. var. *thesii* Berk.); monocyclic rust, *Peridermium pini* (Pers.) Lev. (sometimes known as *Endocronartium pini*)

Hosts: *Pinus halepensis, P. mugo, P. nigra, P. pinaster, P. pinea,* and *P. sylvestris*

Distribution: The disease is found throughout Europe; *C. flaccidum* is predominant in the southern part of the continent and *P. pini* in the north.

Symptoms and Diagnosis

The first noticeable symptom of resin top disease is the production in early summer of pink aecia, containing abundant orange spores, over a 5- to 15-cm band of tissue on 2- to 3-year-old shoots. With each successive year, new bands of aecia form, both distally and proximally. The foliage distal to the lesion can remain alive and healthy for several years. Eventually, however, death of tissue leads to girdling of the shoot and the appearance of foliar symptoms. Downward growth of the parasite can lead to progressive invasion of older stems, and in this way, infections originating on lateral branches can reach and eventually girdle the main stem. The sudden death of the crown above this point is the most conspicuous symptom of the disease (Plate 65). It has been noted that production of aecia on *Peridermium pini* lesions is less conspicuous on large stems than on small ones, while resin exudation becomes more and more abundant with decreasing stem size. When the trunk of a mature tree is girdled, very heavy resin flow occurs; hence, the name resin top (Plate 66).

Disease Cycle

Two closely related rust fungi cause resin top disease. *Cronartium flaccidum* has a long cycle, and aeciospores infect several herbaceous alternate hosts. The chief of these is *Vincetoxicum hirundinaria*, although other plants such as *Paeonia, Pedicularis* and *Tropaeolum* spp. can also become diseased.

Uredinia are formed within 8–14 days, and then teliospores and basidiospores are formed from mid- to late summer. Basidiospores infect pine needles, which show yellow spots 1–2 months later. The fungus invades the shoot, and spermagonia and aecia are produced after 2–3 years (Plate 67).

Peridermium pini has a short cycle, and the aeciospores can reinfect pine directly. Infection is presumed to take place through the needles, although there is little direct evidence for this. *P. pini* exists in two forms that differ in spore germination characteristics and cytology. Both forms are capable of sustained growth in pure culture.

Effects on Tree and Forest, Ecological Role

In general, neither *C. flaccidum* nor *P. pini* has a major impact on forest management. However, in Italy, *C. flaccidum* has destroyed whole plantations of *P. pinea, P. pinaster,* and *P. nigra*; trees up to 20 years old are particularly susceptible. The most notable damage caused by *P. pini* in recent years has been to plantations of 50- to 60-year-old *P. sylvestris* in eastern England. Trees less than 40 years of age were rarely attacked.

Disease Management

In areas subject to severe *C. flaccidum* attack, plantations of pure pine should be avoided where *V. hirundinaria* is present. Management of disease caused by *P. pini* is normally restricted to the selective removal of obviously diseased trees during thinning. No sustained breeding programs have been developed, but with certain pines, such as *P. pinaster*, great variation between provenances in susceptibility to *C. flaccidum* has been found.

Selected References

Gibbs, J. N., England, N., and Wolstenholme, R. 1988. Variation in the pine stem rust fungus *Peridermium pini*, in the United Kingdom. Plant Pathol. 37:45-53.

Pei, M. H., and Pawsey, R. G. 1991. Axenic culture of *Peridermium pini*. Mycol. Res. 95:108-115.

Ragazzi, A. 1988. *Cronartium flaccidum* (Alb. & Schwein.) Winter. Pages 475-476 in: European Handbook of Plant Diseases. I. M. Smith, J. Dunez, R. A. Lelliott, D. H. Phillips, and S. A. Archer, eds. Blackwell Scientific, Oxford.

(Prepared by J. Gibbs)

Comandra Blister Rust

Causal organism: *Cronartium comandrae* Peck (syn. *C. pyriforme* Hedgc. & W. H. Long)

Hosts: principal aecial hosts, *Pinus banksiana*, *P. contorta*, and *P. ponderosa*; telial hosts, *Comandra* and *Geocaulon* spp. (Santalaceae), bastard toadflax or comandra; perennial herbs of grasslands and shrublands (*Comandra* spp.) or wetlands (*Geocaulon* spp.)

Distribution: North America, coast to coast from the northernmost pines across Canada southward to California, Arizona, New Mexico, Arkansas, and northern Alabama

Comandra blister rust is the most common of the canker rusts in the western United States; in several forest districts, half the lodgepole pine are cankered by this rust. Comandra rust causes slight fusiform swellings on young pine stems, branch flagging, and characteristic long cankers on older trunks. Length-to-width ratios of well-developed cankers are 2:1 to 3:1. (Plate 68). Aeciospores of *Cronartium comandrae* are unique among the *Cronartium* rust fungi; they are tapered, long (38–65 μm), and two or three times longer than they are wide.

Diseased seedlings die a few months to a few years after infection. The average number of years for a canker to girdle the trunk equals the diameter of the trunk in centimeters. Trees with large cankers or severe top kill grow slowly (Fig. 6). The stemwood of treetops killed by comandra blister rust (which may be half the tree height) is usually checked, but resin impregnation and loss to decay are slight. A high percentage of cankers are gnawed to the wood by porcupines and other rodents seeking both the fungus and the infected inner bark (Plate 69).

In young, dense stands, mortality can provide favorable release thinning. However, infection in older, less dense stands is generally undesirable; mortality and topkill are more frequent in the dominant and codominant species, thereby increasing rotation age and reducing effective stocking.

In regions where the telial host of comandra blister rust (Plate 70) is restricted to shrublands adjacent to the forest, disease incidence and damage vary with distance up to 8 km between aecial and telial host populations. The coincidence of susceptible hosts, viable basidiospores, and suitable environmental conditions occurs so infrequently that in most years, few infections are established. Large numbers of infections may develop only after large-scale, warm, frontal rains.

Selected References

Bergdahl, D. R., and French, D. W. 1976. Epidemiology of comandra rust on jack pine and comandra in Minnesota. Can. J. For. Res. 6:326-334.

Geils, B. W., and Jacobi, W. R. 1993. Effects of comandra blister rust on growth and survival of lodgepole pine. Phytopathology 83:638-644.

Krebill, R. G. 1968. Histology of canker rusts in pine. Phytopathology 58:155-164.

Peterson, R. S., and Jewell, F. F. 1968. Status of American stem rusts of pine. Annu. Rev. Phytopathol. 6:23-40.

Vogler, D. R., Cobb, F. W., Jr., Geils, B. W., and Nelson, D. L. 1996. Isozyme diversity among hard pine stem rust in the western United States. Can. J. Bot. 74:1058-1070.

(Prepared by R. Peterson and B. Geils)

Stalactiform Blister Rust

Causal organism: *Cronartium coleosporioides* Arth. (anamorph *Peridermium stalactiforme* Arth. & F. Kern; *C. stalactiforme* nomen nudum)

Hosts: principally *Pinus contorta* in western North America and *P. banksiana* in eastern North America; telial hosts include several species of Scrophulariaceae

Distribution: North America from southeastern Alaska to the Maritime Provinces, southward to California, Utah, Colorado, and the lake states; not known at the northern limit of pines in Canada nor near the Atlantic coast

Well-developed stem cankers can be up to 10 m long with a length-to-width ratio greater than 10:1 (Plate 71). Trunk cankers that arise by invasion from branches may exist for more than a century before the host is girdled. Although stalactiform blister rust quickly girdles saplings, its lateral expansion is so slow that larger trees are rarely girdled. Although stalactiform blister

Fig. 6. Spike top of ponderosa pine caused by comandra blister rust in Oregon. (Courtesy R. Hunt)

rust cankers increase the resin content of wood in a large portion of the bole, the effect on wood quality is modest. A high percentage of cankers are gnawed by squirrels, porcupines, and other rodents, which can arrest development of the rust or speed girdling and death of the host. However, the more immediate effect of the gnawing is that it makes diagnosis difficult, because rust cankers are often described as "mammal damage."

Stalactiform blister rust is particularly common in lodgepole pine forests of central Idaho, the Sierra Nevada, and western Canada. The telial hosts of stalactiform blister rust are usually widely and sporadically distributed across and within the forest; therefore, quantifying the effect is difficult. However, this rust, at least in Idaho, is common on cool, dry, midelevation sites. Although some infection may be possible during most years, especially suitable conditions occur about once per decade in certain "wave years."

Selected References

Anderson, N. A., French, D. W., and Anderson, R. L. 1967. The stalactiform rust on jack pine. J. For. 65:398-402.

Krebill, R. G. 1968. Histology of canker rusts in pine. Phytopathology 58:155-164.

Peterson, R. S. 1967. The *Peridermium* species on pine stems. Bull. Torrey Bot. Club 94:511-542.

Peterson, R. S., and Jewell, F. F. 1968. Status of American stem rusts of pine. Annu. Rev. Phytopathol. 6:23-40.

Vogler, D. R., Cobb, F. W., Jr., Geils, B. W., and Nelson, D. L. 1996. Isozyme diversity among hard pine stem rust in the western United States. Can. J. Bot. 74:1058-1070.

(Prepared by R. Peterson and B. Geils)

Sweetfern Blister Rust

Causal organism: *Cronartium comptoniae* Arth. (anamorph *Peridermium comptoniae* Orton & J. F. Adams)

Hosts: Principally *Pinus banksiana* in the eastern part of North America and, less frequently, *P. contorta* in the west. *C. comptoniae* is the only known stem rust on *P. resinosa*. Telial hosts include *Comptonia* and *Myrica* spp. (Myricaceae), sweetfern and sweetgale, shrubs of sandy sites or wetlands, respectively.

Distribution: North America from southeastern Alaska across Canada and southward to northern Oregon, the lake states, Indiana, Ohio, through the northeastern states, and south to Georgia

Sweetfern blister rust induces more xylem hyperplasia than the other pine canker rusts. The hyperplasia appears as a ridge or as clusters of small swellings that become thick, ellipsoid galls. Length-to-width ratios of well-developed cankers are 5:1 to 10:1 (Plate 72). Young cankers and those too close to the ground to permit symmetrical development are less elongate. Ruptured aecia of *Cronartium comptoniae* display sharp peridial "teeth" and more projecting filaments than the other *Cronartium* spp., but even microscopic distinctions may fail to differentiate specimens of *C. coleosporioides* from those of *C. comptoniae*.

Most cankers of *C. comptoniae* occur near the ground. Decay, especially that caused by *Phellinus pini*, is commonly associated with sweetfern blister rust. The disease is most serious in eastern North America, where it occurs sporadically on jack pine, commonly on loblolly pine, and so frequently on ponderosa and lodgepole pine that attempts to cultivate these western species beyond their native ranges have been abandoned.

The principal factor in rating the hazard in pine stands is the distance from telial host populations. Sweetfern blister rust is a concern where aecial and telial hosts are closely associated (less than 30 m). Effective dispersal of *C. comptoniae* is limited to less than 100 m.

Selected References

Gross, H. L., Patton, R. F., and Ek, A. R. 1978. Reduced growth, cull, and mortality of jack pine associated with sweetfern rust cankers. Can. J. For. Res. 8:47-53.

Krebill, R. G. 1968. Histology of canker rusts in pine. Phytopathology 58:155-164.

Peterson, R. S. 1967. The *Peridermium* species on pine stems. Bull. Torrey Bot. Club 94:511-542.

Peterson, R. S., and Jewell, F. F. 1968. Status of American stem rusts of pine. Annu. Rev. Phytopathol. 6:23-40.

Vogler, D. R., Cobb, F. W., Jr., Geils, B. W., and Nelson, D. L. 1996. Isozyme diversity among hard pine stem rust in the western United States. Can. J. Bot. 74:1058-1070.

(Prepared by R. Peterson and B. Geils)

Other Hard Pine Stem Rusts

Appalachian Blister Rust

Cronartium appalachianum Hepting (*Peridermium appalachianum* Hepting & Cummins) infects *Pinus virginiana* and *Buckleya distichophylla* (Santalaceae) in Virginia, North Carolina, and Tennessee. *B. distichophylla,* a rare shrub of steep, rocky slopes, is obligately parasitic on gymnosperm roots, especially those of *Tsuga canadensis*. The rust is, therefore, restricted to sites where *P. virginiana*, *T. canadensis*, and *B. distichophylla* grow together. *C. appalachianum* resembles both *C. comptoniae* and *C. arizonicum* but may be distinguished by its lack of aecial filaments, its long aeciospores, and the lack of hyperplasia in infected pines.

Mistletoe Blister Rust

Peridermium bethelii Hedgc. & Long infects *Pinus contorta* and rarely *P. ponderosa* in the Rocky Mountains from central Colorado north to southern Alberta and in the central Sierra Nevada of California. Infections are associated with dwarf mistletoes (*Arceuthobium*, usually *A. americanum*) parasitic on the pines; and in fact, rust mycelium invades mistletoe as well as pine tissues. Aecia, however, are formed only on pine; no sporulation of the rust on mistletoe has been found. Morphologically and biochemically, *P. bethelii* resembles *Cronartium comandrae*, which is often found nearby. Autoecious reinfection of pine is suspected; however, aeciospore germ tubes do not resemble those of other pine-to-pine rusts. Rust-infected pine branches and the mistletoe usually die within a few years of becoming infected. The rust is relatively uncommon, and it may be even beneficial as a biological control of mistletoe.

Conigenum Rust

Piña de pino, caused by *Cronartium conigenum* Hedgc. & N. Hunt, infects most hard pine species in Mexico and Central America and produces uredinia and telia on oaks. It is known mainly as a cone rust (see Southwestern Pine Cone Rust), but it also causes conspicuous galls, formerly ascribed to *Caeoma* (*Peridermium*) *mexicanum* Pat., on half the pine species infected. By modifying apical meristems, the rust gives rise to a great variety of gall shapes (often lobed) and causes most infected tissue to become pithlike. Although conspicuous, the galls have not been associated with cankers or large stems and do not seem to cause much damage to trees, perhaps because they kill small shoots quickly. Heavy infection of seedlings (which are killed) could be damaging.

Selected References

Hawksworth, F. G., Dixon, C. S., and Krebill, R. G. 1983. *Peridermium bethelii*: A rust associated with lodgepole pine dwarf mistletoe. Plant Dis. 67:729-733.

Hepting, G. H. 1957. A rust on Virginia pine and *Buckleya*. Mycologia 49:896-899.

Peterson, R. S., and Salinas-Quinard, R. 1967. *Cronartium conigenum*: Distribución y efectos en los pinos. Inst. Nac. Invest. For. Bol. Téc. 19.

(Prepared by R. Peterson and B. Geils)

Limb Rusts

Causal Organisms

Although the taxonomy of the limb rust fungi is very confused and more work remains to determine their phylogenetic relations, the limb rusts are easily distinguished from canker rusts by the fact that sporulation is restricted to branches and the fungi spread systemically within the xylem (Plate 73). Taxonomically, the limb rusts can be sorted into five groups on the basis of details of morphology and life cycle.

1) *Cronartium arizonicum* Cummins (*Peridermium filamentosum* Peck) is a full-cycle species formerly included in the "*Cronartium coleosporioides* complex." The name *Peridermium filamentosum* is also applied to the two following taxa, so to avoid confusion, the name Coronado Peridermium is used to denote this host-alternating fungus.

2) The Powell Peridermium differs from *C. arizonicum* slightly in morphology and phenology and greatly in the germination behavior of the aeciospores.

3) The Inyo Peridermium differs from the Powell Peridermium in morphological details.

4) *Cronartium coleosporioides* Arth. (sensu lato, as a limb rust referred to as Washoe Peridermium) is a full-cycle species that is markedly distinct from the preceding three taxa in aecial morphology and somewhat distinct from *C. arizonicum* in the uredinial and telial stages. There are few morphological differences between the Washoe Peridermium and stalactiform blister rust (described above as a canker rust). It is not yet clear whether the Washoe Peridermium differs specifically or racially from stalactiform blister rust. There are notable differences in host range and macromolecular biochemistry (DNA) but few useful field characteristics to differentiate these taxa.

5) Finally, there are several additional taxa in Mexico and Guatemala that behave as limb rusts, including a rust on *Pinus lawsonii* in Oaxaca that resembles *Cronartium flaccidum* (*Peridermium pini*).

Hosts

The limb rust fungi infect pines of the subsection Ponderosae: in the United States, *Pinus jeffreyi*, *P. ponderosa*, and *P. arizonica*; and to the south also *P. cooperi*, *P. durangensis*, *P. engelmannii*, and *P. michoacana*. Both *C. arizonicum* and *C. coleosporioides* infect *Castilleja*, *Orthocarpus*, and *Pedicularis* spp.; *Cordylanthus* and *Lamourouxia* spp. are probably also infected by limb rust. All five are closely related genera of Scrophulariaceae and root parasites of many angiosperms.

Distribution

C. arizonicum is known from the Black Hills of South Dakota westward and southward through the Rocky Mountains and Mexico to western Guatemala. Distribution of the Powell Peridermium is within that of *C. arizonicum* in Colorado, Utah, and northern Arizona, but few localities are known where both occur. The Inyo Peridermium is found in the Sierra Nevada of California and Nevada. *C. coleosporioides* as a limb rust (Washoe Peridermium) is known definitely only on *P. jeffreyi* in California and Nevada.

Symptoms and Diagnosis

Infection by limb rusts is systemic, and spread occurs in the wood. After reaching the trunk, mycelium grows upward and downward (about 18–21 cm per year in each direction) and then out into branches where it sporulates. Infected branches die within a few years, and the crown is progressively killed (Plate 74). Trunks show no external sign of infection, and their bark is not invaded. A single mycelium may be hundreds of years old and hundreds of feet long, more than 99% of it in the host xylem (tracheids). Aecial structure and aeciospore characteristics clearly distinguish the Washoe Peridermium from the other limb rusts. This species also usually causes rough bark ("cankers") and a little hyperplasia on branches where it sporulates; these branches may be larger in diameter than the twigs on which the other fungi sporulate (without causing roughening or hyperplasia). The Washoe Peridermium attacks trees of any age, and infected seedlings and saplings die before the limb rust syndrome is evident. The fungi may be distinguished from one another by life cycle and by the long aeciospore germ tubes of *C. arizonicum* versus the short tubes of Powell and Inyo Peridermia. Aecia of the Powell and Inyo Peridermia are tough, persisting more or less intact for months, whereas those of the two host-alternating rusts usually shatter during late spring.

Effects

Long before they complete the slow, steady destruction of a tree's crown, the limb rusts have so weakened their hosts that the trees succumb to bark beetles or other biotic agents or abiotic conditions. Before death, however, progressive branch necrosis leads to loss of stem growth and typical crown form. Endemic bark beetle populations commonly persist in limb rust-infected trees. There appears to be less limb rust in the southwestern United States and the Sierra Nevada now than there was 30 years ago as a result of reduced infection rates and silvicultural removal of the conspicuously damaged trees. If future outbreaks are identified early, serious damage can be avoided by selectively cutting diseased trees.

Selected References

Christenson, J. A. 1968. A cytological comparison of germinating aeciospores in the *Cronartium coleosporioides* complex. Mycologia 60:1169-1177.

Peterson, R. S. 1972. Pine limb rust fungi in Mexico. Plant Dis. Rep. 56:896-898.

Peterson, R. S. 1973. Studies of *Cronartium* (Uredinales). Rep. Tottori Mycol. Inst. 10:203-223.

Peterson, R. S., and Shurtleff, R. G., Jr. 1965. Mycelium of limb rust fungi. Am. J. Bot. 52:519-525.

Ziller, W. G. 1970. Studies of western tree rusts. VII. Inoculation experiments with pine stem rusts (*Cronartium* and *Endocronartium*). Can. J. Bot. 48:1313-1319.

(Prepared by R. Peterson and B. Geils)

Pine Twist Rust

Other names: rouille courbeuse, (French), Kieferndreher or Drehrost (German), ruggine curvatrice (Spanish), knäckesjuka (Swedish)

Causal organism: *Melampsora pinitorqua* Rostr., regarded by some as a forma specialis of *Melampsora populnea* (Pers.) P. Karst. or of *M. tremulae* Tul. (syn. *Caeoma pinitorquum* A. Br.)

Hosts: *Pinus sylvestris, P. pinaster,* and *P. pinea;* also *P. halepensis, P. montana,* and *P. nigra* and nonnative *P. contorta* and *P. strobus.* Telial hosts, species of *Populus,* section Leuce; other than near the Mediterranean, mainly *P. tremula;* in Mediterranean areas, also *P. alba* and *P. canescens* (syn. *P. alba* × *tremula*)

Distribution: throughout Europe and eastward across northern Siberia to the Far East

Like some other *Melampsora* species, *M. pinitorqua* infects a wide range of young conifer seedlings, including many additional Eurasian and American hard pines, most importantly the highly susceptible *P. ponderosa,* white pine (section Strobus), three *Larix* species, and *Pseudotsuga menziesii.* Azbukina reported an outbreak in a *Larix decidua* stand in Magadanskaiâ Oblast, far east of Siberia's hard pines. Since 1950, twist rust has become epidemic in Italy within the natural range of *P. pinaster.*

Symptoms and Diagnosis

During the spring, narrow, yellow patches up to 3 cm long appear on current-year pine shoots. Within these patches are darker yellow spermagonia, which are 130 μm wide. Normally dormant buds are activated, but there is little or no swelling. The rust's common names refer to the C- or S-shaped curvature of pine shoots resulting from the necrosis along one side. Aecia (2–3 cm long) lack obvious peridia and are borne on current-year shoots or occasionally on needles during the spring. Uredinia and telia on poplars are distinguishable from those of other *Melampsora* spp. by morphology and period of germination.

Disease Cycle

Basidiospores originate from telia on fallen, overwintering poplar leaves during February through June (northern Italy) or beginning in April (Bulgaria, Russia, and France) to early June (Lithuania, Sweden, and Finland). The basidiospores become windborne and infect current-year tissue of pine seedlings and shoots through the cuticle and cell walls. Aeciospores are produced in June after 9–16 days of development and are carried by the wind to poplar leaves, where urediniospores are produced during June and July. Teliospores form later in the summer but do not germinate until spring. Variations in this sequence can occur: 1) urediniospores can develop during the spring from mycelium that has overwintered in poplar buds; and 2) although rare, aeciospores can be produced from mycelium that has overwintered in pine stems and appear earlier than those from infections of the current year. Mycelia in pine are short lived. Small (less than 5 mm in diameter), infected shoots usually shrivel and die quickly, and larger shoots may wall off the infection and survive rust free.

Effects

Heavy infection levels destroy huge numbers of pine shoot tips, reducing growth, deforming trees, and killing seedlings. *Botrytis cinerea* may follow infection by the rust, thereby increasing the damage to developing shoots. Although a single year of infection does not destroy a stand that is older than 2 or 3 years, damage can be significant. Successive years of infection, however, may destroy stands over large areas.

Contributing Factors

A coincidence of basidiospore production, expanding pine shoots, and prolonged high humidity (with free moisture on the surface of pine tissue) is required but is lacking at any given place in most years. The most important predictive factor for abundant infection is the proximity of poplar hosts to infected pines. Basidiospores are ephemeral, and abundant infection of pine occurs only within 150 m of infected aspen.

Disease Management

Susceptible pines should be planted at least 250 m from aspen, resistant pines can be planted (e.g., *P. nigra* rather than *P. sylvestris*), or aspen can be eliminated. Some countries have enacted statutes that forbid growing alternate hosts of *M. pinitorqua* together. Long-term, site-specific disease forecasts aid silvicultural management. Seedlings are sometimes treated with fungicides. There is much variation in susceptibility between and within pine species, and breeding programs in Europe are working toward the development of resistant planting stock.

Selected References

Azbukina, Z. M. 1984. Opredelitel' Rzhavchinnykh Gribov Sovetskogo Dal'nego Vostoka. Akademia Nauka SSSR, Moscow.

Kurkela, T. 1973. Epiphytology of Melampsora rusts of Scot's pine (*Pinus sylvestris* L.) and aspen (*Populus tremula* L.). Commun. Inst. For. Fenn. 79(4):1-68.

Naldini, B., Longo, N., Drovandi, F., Gonnelli, T., and Moriondo, F. 1992. Observations on some Italian provenances of *Melampsora populnea.* III. Presence of *Melampsora pinitorqua* in possible outbreak areas. Eur. J. For. Pathol. 22:188-191.

Pinon, J. 1973. Les rouilles du peuplier en France: Systematique et repartition du stade uredien. Eur. J. For. Pathol. 3:221-228.

Siwecki, R. 1974. A review of studies on the occurrence of *Melampsora pinitorqua* in Central and Eastern Europe. Eur. J. For. Pathol. 4:148-155. (Includes important results from A. Yu. Rimkus, 1968 and 1969, in Lithuanian publications not readily available.)

Vanin, S. I. 1955. Lesnaya Fitopatologiia. Suppl. 4th ed. Goslesbumezdat, Moscow.

(Prepared by R. Peterson and B. Geils)

Gymnosporangium Stem Rusts

Causal Fungi, Hosts, and Distributions

Approximately 51 species of *Gymnosporangium* (including a few doubtful ones) and three related *Roestelia* and *Uredo* species infect trees and shrubs of Cupressaceae (Plates 75–77). Fifteen are limited to "foliage" (scale leaves and green shoots), leaving 39 stem rusts. These 39 are listed in Table 1 with principal telial and aecial hosts, symptom types, and distributions. Some species are host specific (e.g., *G. nidus-avis* Thaxt. and *G. kernianum* Bethel grow side by side on different members of the subgenus *Sabina* without infecting the other's host), whereas *G. clavipes* (Cooke & Peck) Cooke & Peck in Peck and *G. sabinae* infect species of *Sabina, Oxycedrus,* and *Cupressus.*

Symptoms, Diagnosis, and Effects

Nearly all species of *Gymnosporangium* produce brown to orange telia that gelatinize during spring rains. Effects on conifer hosts are an array of swellings (annual or living for centuries), distortions, and witches'-brooms. A few species canker or deform large trunks and therefore can be important in forestry. Others, such as *G. libocedri* (Henn.) F. Kern, reduce growth by heavily infecting the foliage. On their aecial hosts, some species destroy fruits and leaves; a few species cause cankers and witches'-brooms.

TABLE 1. *Gymnosporangium* spp. and Related Stem Rusts of Cupressaceae

Causal Fungus	Telial Host[a]	Symptom Type[b]	Aecial Host	Distribution
Gymnosporangium				
atlanticum	Sabina	F	Unknown	Northwestern Africa
bermudianum	Sabina	G	Sabina	Southeastern United States, Bahamas, Bermuda
bethelii	Sabina	G	Crataegus	Western Canada, United States, Mexico
biseptatum	Chamaecyparis	CFL	Amelanchier	East coast of United States
clavariiforme	Oxycedrus	FW	Crataegus, Amelanchier	North America, Eurasia, northwestern Africa
clavipes	Cupressus	F	Crataegus	Mexico, Guatemala
	Oxycedrus	FL	Cydonia, Malus, Crataegus	Canada, United States
	Sabina	F	Cydonia, Malus, Crataegus	Canada, United States
confusum	Oxycedrus, Sabina	F	Mespilus, Crataegus	Eurasia, northwestern Africa, western United States
connersii	Sabina	G	Crataegus	Canada
cornutum	Oxycedrus	F	Sorbus	Eurasia, northwestern Africa, North America
cunninghamianum	Cupressus	CFG	Amelanchier, Pyrus	India, United States, Mexico, Guatemala
dobozrakovii	Oxycedrus, Sabina	F	Pyrus	Ukraine
effusum	Sabina	CF	Unknown	Eastern United States
ellisii	Chamaecyparis	CFW	Myrica, Comptonia	Eastern United States
exterum	Sabina	F	Gillenia	Eastern United States
floriforme	Sabina	GL	Crataegus	Southeastern United States
globosum	Sabina	G	Crataegus, Pyrus	Eastern North America
gracile	Oxycedrus, Sabina	FW	Amelanchier, Crataegus	Southern Europe, northwestern Africa, southwestern North America
hyalinum	Chamaecyparis	FL	Crataegus	East coast of the United States
inconspicuum	Sabina	FL	Amelanchier	Western North America
japonicum	Sabina	F	Photinia	Eastern Asia
juniperi-virginianae	Sabina	G	Malus	Eastern North America
kernianum	Sabina	WL	Amelanchier, Pyrus	Western United States, northwestern Mexico
libocedri	Calocedrus	WL	Amelanchier, Malus, Pyrus	West coast of North America
miyabei	Chamaecyparis	FG	Malus, Sorbus	Japan
nelsonii	Sabina	G	Amelanchier	North America
nidus-avis	Sabina	WL	Amelanchier, Chaenomeles	Canada, United States, China
padmarense	Sabina	F	Unknown	India
sabinae	Cupressus, Oxycedrus, Sabina	F	Pyrus	Southern Europe, Eurasia, northwestern Africa, North America[c]
speciosum	Sabina	CFGW	Fendlera, Philadelphus	Western United States, Mexico, Guatemala
taianum	Cupressus	F	Unknown	Yunnan, China
trachysorum	Sabina	G	Crataegus	Southeastern United States
tremelloides	Oxycedrus	FG	Malus, Amelanchier	Eurasia, northwestern Africa, western North America
tsingchensis	Cupressus	FW	Unknown	Szechuan, China
turkestanicum	Sabina	G	Sorbus	Central Asia
vauqueliniae	Sabina	FWL	Lindleya, Vauquelinia	Southwestern United States, Mexico
yamadae	Sabina	GL	Malus	Eastern Asia
Roestelia brucensis	Sabina	GL	Sabina	Ontario, Canada
Uredo				
apacheca	Sabina	F	Unknown, none?	Southwestern United States, Mexico
cupressicola	Cupressus	CF	Unknown, none?	Western United States to Guatemala

[a] *Oxycedrus* and *Sabina* are subgenera of *Juniperus*.
[b] C = canker; F = fusiform gall; G = globose gall; W = witches'-broom; and L = leaf rust.
[c] Introduced.

Disease Cycles

The telium-producing *Gymnosporangium* species produce basidiospores during spring or summer rains. The species differ greatly in effective distance for infection of aecial hosts. Spermagonia and aecia develop during the summer, and aeciospores infect the telial host. *G. bermudianum* Earle in Seym. & Earle, *Roestelia brucensis*, and probably *Uredo cupressicola* R. S. Peterson and *U. (Caeoma) apacheca* R. S. Peterson, however, infect conifers without host alternation. Mycelium overwinters in conifer hosts and in some areas also in angiosperm hosts.

Disease Management

The primary aim of management strategies is to protect aecial hosts, especially apple and pear, which may be severely damaged. Keeping alternate hosts apart is the best preventive measure and in some places legally required. Planting resistant species and cultivars is part of this program. Fungicidal sprays of aecial hosts are more effective than those of telial hosts.

Selected References

Bernaux, P. 1956. Contribution a l'etude de la biologie des *Gymnosporangium*. Ann. Epiphyt. Ser. C 7:1-210.

Kern, F. D. 1973. A Revised Taxonomic Account of *Gymnosporangium*. Pennsylvania State University Press, University Park.

Parmelee, J. A. 1965. The genus *Gymnosporangium* in eastern Canada. Can. J. Bot. 43:239-267.

Parmelee, J. A. 1971. The genus *Gymnosporangium* in western Canada. Can. J. Bot. 49:903-926.

Peterson, R. S. 1982. Rust fungi (Uredinales) on Cupressaceae. Mycologia 74:903-910.

(Prepared by R. Peterson and B. Geils)

Parasitic Plants

The mistletoes are the most common and widespread parasitic seed plants that infect broad-leaved and conifer trees. Since the 1970s, a large body of scientific and technical literature has accumulated on the mistletoes, particularly those infecting conifers in North America, Europe, and Asia. At least a dozen new species of mistletoes and dwarf mistletoes have been named, and there is a much better understanding of the distribution, biology, and host associations of these parasites. Suppressing the damage caused by some mistletoes is an important component of forest management in some countries, while other mistletoes are protected for their pharmaceutical, cultural, and ecological values.

By far, most species parasitize tropical and temperate broad-leaved trees and shrubs, but in North America, many conifers are hosts of mistletoes. North America is home to many different conifers and the center of the greatest diversity of pines (*Pinus* spp.); therefore, it is not surprising that the greatest collection of mistletoe species on conifers occurs on this continent.

The mistletoes belong to two families, Loranthaceae and Viscaceae. The members of the two families are closely related and similar in many ways, but they are easily distinguished by differences in floral structure. Flowers of members of Viscaceae are usually small and indistinct, whereas those of Loranthaceae are conspicuous and colorful. Several genera in both families are parasites of conifers.

Several root parasites, mostly in the figwort family (Scrophulariaceae), infect conifers; but with few exceptions, they are uncommon and not considered to be a problem in conifer forests.

Dwarf Mistletoes

Other names: muerdago enano (Spanish)
Causal organisms: *Arceuthobium* spp.; 44 taxa, mostly specific to a single genus or even species of conifer
Hosts: most genera of Pinaceae and *Juniperus* spp. in the Cupressaceae
Distribution: northern hemisphere, especially western North America, southern Europe, and the Himalayas

The dwarf mistletoes, *Arceuthobium* spp. (Viscaceae), are the most common, widespread, and damaging parasitic seed plants of conifers in the world. In addition, they make up one of the most complex and difficult groups of plants to describe and classify. Besides the visual morphological characters used to determine species, certain cytological, palynological, physiological, and chemical features and host affinities and host responses have been used in the classification of *Arceuthobium* spp. A major monograph on the genus (Hawksworth and Wiens, 1996) details the biology, formal taxonomy, criteria for classification, host relationships, biogeography, and evolutionary trends and is the source of much of the information presented here.

Causal Organisms

Dwarf mistletoes are for the most part small, dicotyledonous, dioecious plants characterized by several unusual features (Plates 78 and 79), including bicolored fruit (Plate 80), an explosive seed-dispersal mechanism, an endosperm that contains chlorophyll, a ringlike archesporium, a germinating radicle that contains stomata, and shoots that contain no central vascular cylinders or sieve tube elements.

Forty-four species or subspecies of dwarf mistletoe occur in both the Old and New Worlds, most of them in North America. Distribution in the New World extends from Alaska and central Canada as far south as Honduras and Hispaniola (Plate 81). The greatest collection of species occurs in the western United States and Mexico (Tables 2 and 3). Only members of Pinaceae are parasitized by the New World species.

In the Old World, eight known species or subspecies of dwarf mistletoes parasitize genera in Pinaceae and Cupressaceae. They range from the Azores through northern Africa, southern Europe, the Middle East, the Himalayas, and into China (Table 4 and Plate 82).

The geological history of *Arceuthobium* spp. is limited, but the oldest fossil pollen similar to that of the present-day *A. oxycedri* was found in eastern Germany in a stratum of the late tertiary period of the Cenozoic era. This indicates that dwarf mistletoes have been on earth for at least 30–50 million years.

Infection and Development

Unlike most mistletoe seeds, which are spread by birds, those of dwarf mistletoe are dispersed mainly by an explosive

TABLE 2. Distribution and Principal Hosts of Dwarf Mistletoes in Canada, the United States, and Northern Mexico

Arceuthobium sp.	Distribution	Principal Hosts
A. abietinum		
f. sp. *concoloris*	Western United States	*Abies* spp.
f. sp. *magnificae*	Western United States	*Abies magnifica*; *A. procera*
A. americanum	Western United States, central and western Canada	*Pinus contorta*
A. apachecum	Southwestern United States, northern Mexico	*Pinus strobiformis*
A. blumeri	Arizona, northern Mexico	*Pinus strobiformis*
A. californicum	California, Oregon	*Pinus* (white pines)
A. campylopodum	Western Canada, western United States, Baja California, Mexico	*Pinus ponderosa*
A. cyanocarpum	Western United States	*Pinus* (white pines)
A. divaricatum	Western United States, Baja California, Mexico	*Pinus* (pinyons)
A. douglasii	Western Canada, western United States, northern Mexico	*Pseudotsuga* spp.
A. gillii	Arizona, Northern Mexico	*Pinus* (*leiophyllae*)
A. laricis	Northwestern United States, southwestern Canada	*Larix* spp.
A. microcarpum	Arizona, New Mexico	*Picea engelmannii, P. pungens*
A. monticola	Southern Oregon, northwestern California	*Pinus monticola*
A. occidentale	California	*Pinus* (*sabinianae, oocarpae*)
A. pusillum	Central and eastern Canada, eastern United States	*Picea* spp.
A. siskiyouense	Southern Oregon, northwestern California	*Pinus attenuata*
A. tsugense		
subsp. *mertensianae*	Western British Columbia south into California	*Tsuga mertensiana*
subsp. *tsugense*	British Columbia, Alaska, Washington, Oregon, California	*Tsuga heterophylla*
A. vaginatum subsp. *cryptopodum*	Southwestern United States, northern Mexico	*Pinus ponderosa*

fruit mechanism. Only occasionally has the spread of dwarf mistletoe seeds by birds been observed. Mature seeds, usually one per fruit, are forcibly discharged up to 10 m or farther to surrounding trees, where they initially stick to needles. The seeds become slippery when it rains and slide onto twigs, where they adhere, germinate, and infect the host. Subsequent production of aerial shoots takes place entirely from the rootlike system growing within the host. It is usually 1 year or longer after infection before any external evidence of a dwarf mistletoe plant is seen. Flowering and fruit production usually take place within 1–3 years after shoot production. Therefore, an average of about 4–6 years, depending on the species, is required for dwarf mistletoe to complete its life cycle. Because dwarf mistletoes are perennial plants, shoots, flowers, and fruit can be produced each year thereafter.

Damage

Dwarf mistletoes are widely distributed and are natural components of the conifer ecosystem in western North America. Forest surveys show that in California, Oregon, and Washington, about 23% (nearly 5.5 million hectares) of the major conifer forests are infested with one or more species of dwarf mistletoe. In addition, dwarf mistletoes occur on more than a third of the ponderosa pine forests in Arizona and New Mexico.

TABLE 3. Distribution and Principal Hosts of Dwarf Mistletoes in Central America

Arceuthobium sp.	Distribution	Principal Hosts
A. abietinum	Chihuahua, Mexico	Abies durangensis
A. abietis-religiosae	Central and southern Mexico	Abies religiosa, A. vejarii
A. aureum		
subsp. aureum	Guatemala, Belize	Pinus spp.
subsp. petersonii	Chiapas, Mexico	Pinus spp.
A. bicarnatum	Hispaniola	Pinus occidentalis
A. durangense	Northern Mexico	Pinus montezumae, P. durangesis
A. globosum		
subsp. globosum	Northern Mexico	Pinus spp.
subsp. grandicaule	Southern Mexico, Guatemala	Pinus spp.
A. guatemalense	Guatemala	Pinus avacahuite
A. hawksworthii	Belize	Pinus caribaea var. hondurensis
A. hondurense	Honduras	Pinus oocarpa
A. nigrum	Central and southern Mexico	Pinus spp.
A. oaxacanum	Oaxaca, Mexico	Pinus spp.
A. pendans	Central and southern Mexico	Pinus discolor, P. cembroides subsp. orizabensis
A. rubrum	Northern Mexico	Pinus spp.
A. strictum	Durango, Mexico	Pinus spp.
A. vaginatum		
subsp. vaginatum	Mexico	Pinus spp.
A. verticilliflorum	Durango, Mexico	Pinus spp.
A. vecorense	Northern Mexico	Pinus spp.

TABLE 4. Distribution and Principal Hosts of Old World Dwarf Mistletoes

Arceuthobium sp.	Distribution	Principal Hosts
A. azoricum	Azores	Juniperus brevifolia
A. chinense	Yunnan and Szechwan, China	Abies spp., Keteleeria spp.
A. juniperi-procerae	Ethiopia, Kenya	Juniperus spp.
A. minutissimum	Western Himalayas	Pinus griffithii
A. oxycedri	Mediterranean, Europe	Juniperus spp.
A. pini	Yunnan, Szechwan, southwestern China and Tibet	Pinus tabulaeformis
A. sichuanense	Southwestern China	Picea spp.
A. tibetense	Southwestern China	Abies spp.

Dwarf mistletoes occur in many other forests in the west as well and in some conifer forests in the north central and eastern United States. Therefore, the amount of forest land in the United States occupied by the parasite is enormous. Dwarf mistletoes are also widespread in the forests of Canada and Mexico.

Although it is difficult to put an economic value on the total losses caused by dwarf mistletoes, the losses in wood volume alone are fairly well known. Drummond estimated that loss of wood fiber in the United States from growth reduction and mortality amounted to more than 11.5 million cubic meters per year. The forest stand dynamics and ecological factors in relation to spread, impact, and control are also reasonably well understood. In addition to tree mortality and the impact on growth, poor lumber quality, lowered reproduction rates, changes in species composition, tree deformation (Plate 83), and predisposition of infected trees to fire and attack by insects and disease are all factors in the conifer damage and changes in stand dynamics attributable to dwarf mistletoes (Fig. 7).

Disease Management

Damage to conifer forests in western North America caused by dwarf mistletoes has been recognized for many years, and numerous attempts have been made to control them. A rating system of 1 to 6, in which 1 is the least amount of infection and 6 the most, is commonly used to evaluate the amount of dwarf mistletoe in trees. In general, trees with ratings of 3 or less suffer little or no damage from dwarf mistletoe.

Silvicultural approaches have been the most widely and successfully applied methods of control. Removal of infected trees, thinning to promote growth of residual trees, favoring less susceptible tree species, pruning infected branches, and even prescribed fire have provided some measure of control. One of the earliest methods of controlling dwarf mistletoes was clear-cutting infected forests. It was thought that the only

Fig. 7. Large, deformed branches of *Pseudotsuga menziesii* in southern Oregon caused by *Arceuthobium douglasii*.

effective way to control the disease was to eradicate the parasite. Current control methods recognize the limitations of clearcutting as a management practice and are aimed not at eradicating the parasite, but at reducing or holding dwarf mistletoes to acceptable levels in the forest so that infected trees suffer little or no disease impact.

Several other silvicultural methods have been used to keep dwarf mistletoes in check. In the mixed-species forests of the western United States, usually not all species are infected. Removing infected trees and favoring the growth and reproduction of nonsusceptible conifer species keeps dwarf mistletoes at tolerable levels. Because the distance the parasite can spread is limited, buffers of nonsusceptible trees or natural openings have also reduced infection.

In other forests, proper tree spacing and vigorous growth are enough to keep dwarf mistletoes under control. In fir (*Abies* spp.) forests in California, for example, young, infected trees on sites conducive to growth can grow in height faster than the mistletoe can spread upward, assuming that adjacent, larger, infected trees are removed. As such, young fir forests are less severely infected by the parasite over time.

Fire is being recognized as a natural part of the western forest ecosystem. Historically, wildfire has played a part in the natural control and distribution of dwarf mistletoe. Managers have used prescribed fire in some cases to reduce dwarf mistletoe populations or to eliminate infected trees and stands when the stands are too severely damaged to be managed by other methods.

On high-value forested sites, such as campgrounds and park and recreation areas, intensive methods to control dwarf mistletoes have been used. In lightly infected trees, pruning individual infected branches has reduced population levels and prevented spread of the parasite. In more heavily infected trees, for example, the older pines in southern California, pruning only the large, old, infected, and deformed branches increased tree vigor and growth and better enabled the trees to survive periods of drought and insect attack. In addition, simple horticultural methods, such as irrigation, fertilization, weed control, and prevention of injury, that increase vigor of infected trees have been recommended to lessen the impact of dwarf mistletoes.

Attempts have also been made to control dwarf mistletoes with various herbicides and other chemicals. To date, no chemical has been found that will selectively kill the mistletoe without killing the host, although chemical defoliants have been shown to kill shoots and provide some short-term control of spread. There are several native pathogens of dwarf mistletoes (Plate 84), but they apparently have little effect on mistletoe populations.

Resistance to dwarf mistletoe in ponderosa pine has been recognized for many years and has been found recently in Jeffrey pine and western hemlock. The U.S. Forest Service is currently selecting, testing, and propagating ponderosa pine for resistance to dwarf mistletoe in California and Oregon.

Selected References

Drummond, D. B. 1982. Timber loss estimates for the coniferous forests in the United States due to dwarf mistletoes. U.S. Dep. Agric. For. Serv. For. Pest Manage. Rep. 83-2.

Hawksworth, F. G. 1983. Mistletoes as forest parasites. Pages 317-333 in: The Biology of Mistletoes. M. Calder and P. Bernhardt, eds. Academic Press, New York.

Hawksworth, F. G., and Wiens, D. 1996. Dwarf mistletoes: Biology, pathology, and systematics. U.S. Dep. Agric. For. Serv. Agric. Handb. 709.

Kuijt, J. 1987. Novelties in Mesoamerican mistletoes (Loranthaceae and Viscaceae). Ann. Mo. Bot. Gard. 74:511-532.

Scharpf, R. F., and Parmeter, J. R., Jr., tech. coords. 1978. Proceedings of the symposium on dwarf mistletoe control through forest management. U.S. Dep. Agric. For. Serv. Gen. Tech. Rep. PSW-31.

Scharpf, R. F., and Roth, L. F. 1992. Resistance of ponderosa pine to dwarf mistletoe in central Oregon. U.S. Dep. Agric. For. Serv. Res. Pap. PSW-208.

Scharpf, R. F., Smith, R. S., and Vogler, D. 1988. Management of western dwarf mistletoe in ponderosa and Jeffery pines in forest recreation areas. U.S. Dep. Agric. For. Serv. Gen. Tech. Rep. PSW-103.

Smith, R. B., Wass, E. F., and Meagher, M. D. 1993. Evidence of resistance to hemlock dwarf mistletoe (*Arceuthobium tsugense*) in western hemlock (*Tsuga heterophylla*) clones. Eur. J. For. Pathol. 23:163-170.

(Prepared by R. F. Scharpf, B. Geils, D. Wiens,
C. Parker, and W. Forstreuter)

Leafy Mistletoes and Other Parasitic Plants

Other names: Mistel (German), le gui (French), muerdago verdadero (Spanish)
Causal organisms: *Phoradendron*, *Viscum*, and *Psittacanthus* spp.
Hosts: Pinaceae, especially *Abies* spp., and Cupressaceae
Distribution: northern hemisphere

Phoradendron spp.

Except for dwarf mistletoes, the various species of *Phoradendron* are the most widespread and damaging mistletoes of forest trees in western North America. Unlike the dwarf mistletoes, several species of *Phoradendron* infect a wide range of hardwoods in both the eastern and western United States as well as in Mexico. Nearly all of the species that parasitize conifers

TABLE 5. *Phoradendron* spp. on Conifers in the United States, Mexico, and Central America

Phoradendron sp.	Distribution	Conifer Hosts
P. abietinum	Chihuahua to Jalisco, Mexico	*Abies durangensis*
P. acuminatum	Guatemala	*Cupressus lusitanica*
P. adamsii	Northeastern Mexico	*Juniperus* spp.
P. bolleanum	Mexico	*Juniperus* and *Cupressus* spp.
P. capitellatum	Arizona; New Mexico, northernern Mexico	*Juniperus* spp.
P. densum	California, southern Oregon, central Arizona, Baja California, Mexico	*Cupressus* and *Juniperus* spp.
P. hawksworthii	Southern New Mexico, western Texas, northeastern Mexico	*Juniperus* spp.
P. juniperinum	Western United States, northern Mexico	*Cupressus* and *Juniperus* spp.
P. libocedri	California, southern Oregon, Baja California, Mexico	*Libocedrus decurrens*
P. minutifolium	Northern Mexico and the Sierra del Carmen	*Juniperus* spp.
P. olivae	Coahuila and Jalisco, Mexico	*Cupressus lusitanica*
P. pauciflorum	California, southern Arizona, Baja California, Mexico	*Abies concolor*
P. rufescens	San Luis Potosi, Queretaro, and Hidalgo, Mexico	*Juniperus* spp.
P. saltillense	Eastern and northeastern Mexico	*Juniperus* spp.
P. sedifolium	Chiapas and Hidalgo, Mexico	*Cupressus lusitanica*

are found in western North America (Table 5). Only one genus of Pinaceae (*Abies*) and three genera of Cupressaceae (*Cupressus*, *Juniperus*, and *Libocedrus*) are infected by *Phoradendron* spp.

Phoradendron spp. are dioecious, green, flowering plants that are spread by birds (Plate 85). Birds seek out and feed on the mistletoe berries, especially during the winter when other sources of food are limited. Mistletoe is often more common in the tops of large trees because many birds like to perch there (Fig. 8). A sticky substance and hairlike threads on the seeds enable them to adhere to branches after they are carried on a bird's feet, wiped from its beak on a branch, or excreted. A germinating seed produces a hypocotyl and a structure commonly called a "holdfast," with which it adheres to the branch. A structure then develops from the holdfast and penetrates the bark to the living host tissues. A root system develops in the host, and aerial shoots are produced from the growing hypocotyl. Over time, additional shoots can arise along the branch from buds produced on the growing root system. Some *Phoradendron* spp. are leafless and closely resemble dwarf mistletoes, whereas others bear conspicuous leaves. These mistletoes contains chlorophyll and manufacture most of their own food. However, in the absence of shoots, the root system can live inside the host for years. Heavily infected trees are weakened, predisposed to insect attack, reduced in growth rate, and sometimes killed, particularly during periods of environmental stress such as a drought (Fig. 9).

Control of *Phoradendron* spp. in the forest is not considered economically practical. Homeowners or public agencies with high-value trees can remove infected branches to control the parasite if the trees are not heavily infected. Periodic defoliation every 2–3 years will limit fruit production and localized spread of the parasite but will not prevent continued reintroduction by birds.

Viscum spp.

Three species of *Viscum* parasitize trees in Europe. The most well known, *V. album*, infects many different hardwoods. The other two species infect only conifers. The Tannen-Mistel (*V. abietis* Fritsch) infects primarily *Abies* spp. but also occurs on *Larix* and *Picea* spp. (Plate 86). The Kiefern-mistel (*V. laxum* Boiss. & Reut.) occurs in southern Europe and is considered a minor parasite on pines (Plate 87), whereas *V. abietis* causes substantial damage to native stands of *A. alba* in Spain, France, Greece, and Romania.

Because of its historic role in legend and folk medicine, renewed interest and some measure of protection is offered the various species of *Viscum*. The plant is still collected and used in parts of Europe during the winter holiday season, and a large body of research information has been compiled on the pharmaceutical uses of *Viscum* spp. and their derivatives.

In spite of its protected status in some areas, control of the parasite has been attempted, mostly of *V. abietis* on *A. alba* in France and Spain where damage is very severe. Tests of various chemicals by both foliar application and trunk injection for control of *V. abietis* show some promise.

Psittacanthus spp.

The genus *Psittacanthus* (Loranthaceae) is another mistletoe in North America that parasitizes conifers. Of the 14 known species, only four (*P. americanus* Mart., *P. calyculatus*, *P. macranthera* Torr. ex Benth., and *P. schiedeanus*) infect *Pinus*, *Abies*, or *Cupressus* spp. in Mexico and Central America. The remaining species infect a variety of other hosts from Mexico to as far south as Argentina.

The genus includes some of the most beautiful and colorful mistletoes that grow on conifers. The plants are usually large and bear conspicuous leaves and colorful flowers (Plate 88). Like those of *Phoradendron* spp., the seeds are dispersed by

Fig. 8. *Phoradendron pauciflorum* in the top of *Abies concolor* in central California.

Fig. 9. *Libocedrus decurrens* in northern California heavily infected by *Phoradendron libocedri*.

TABLE 6. Other Mistletoes That Infect Conifers

Mistletoe	Distribution	Conifer Host
Cladocolea cupulata	Jalisco, Mexico	*Pinus* spp.
Dendrophthoe falcata	India	*Pinus kesiya*
Ileostylus micracanthus	New Zealand	*Pinus muricata*
Korthalsella dacrydii	Indonesia	*Podocarpus imbricata* and *Picea* spp.
Pseudixus japonica	Japan	*Pinus* spp.
Psittacanthus angustifolius	Nicaragua	*Pinus* spp.
Scurrula parasitica	Bhutan	*Pinus* spp.
Struthanthus		
deppeanus	Vera Cruz, Mexico	*Pinus patula*
palmeri	Sonora, Mexico	*Taxodium mucronatum*
quercicola	Vera Cruz, Mexico	*Pinus* spp.
Taxillus		
kaempferi	Himalayas; Japan	*Pinus, Abies,* and *Tsuga* spp.
matsudai	Taiwan	*Pinus taiwanensis*
Tripodanthus acutifolius	Brazil	*Pinus* and *Cupressus* spp.

birds. *Psittacanthus* spp. are widespread on conifers through much of Mexico but are common on pines only in the central and southernmost regions where the hosts occur. Several species of *Psittacanthus* cause considerable damage to their hosts. Pines infected by the parasite have been found to suffer from reduction in growth, poor reproduction, lowered cone and seed production, and death. Some attempts have been made to control *Psittacanthus* spp. on pines with chemicals. Preliminary results indicate that 2,4-D and perhaps several other herbicides show some promise for controlling the parasite with little or no damage to the host.

Other species of mistletoes that infect conifers are listed in Table 6.

Parasitic Plants Other than Mistletoes

A number of root parasites have been shown to parasitize conifers in field and greenhouse tests in the southern United States, but the extent to which they infect and damage trees in the forest is not known. Only one species of root parasite, *Seymeria elliottii*, was reported to damage conifers when it was found killing young *Pinus elliottii* in a plantation in Florida. Most of the root parasites are herbaceous annual or perennial plants, but some are woody shrubs. Most are in the figwort family, including members of the genera *Agalinis, Areolaria, Buchnera, Dasistoma, Macranthera,* and *Schwalbia*. At least one member of the parasitic family Orobanchaceae, *Orobanche minor*, has been reported to parasitize roots of *Juniperus* spp.

More research is needed on the extent to which these parasites occur in forests and damage conifers.

Selected References

Basanez, M. C. 1989. Los muerdagos (Loranthaceae) del Estado de Jalisco. Page 45 in: Memoria de Resumes. Symp. Natl. Sobre Parasitol. For., 5th.

Bello Gonzales, M. A., and Gutierrez, G. M. 1985. Clave para la identificacion de la familia Loranthaceae en la porcion del eje neovolcanico localizado dentro del estado de Michoacan. Cienc. For. 10(54):3-33.

Cházaro, B. M., and Olivia, R. H. 1988. Loranthaceas del centro de Veracruz y zona limitrofe de Puebla. III. Cactaceas y Succulentas Mexicanas 33:14-19, 71-75.

Delabraze, P., and Lanier, L. 1972. Contribution a la lutte chimique contre le gui (*Viscum album* L.). Eur. J. For. Pathol. 2:95-103.

Farah, A. F. 1991. The parasitic angiosperms of Saudi Arabia—A review. Pages 68-75 in: Int. Symp. Parasit. Weeds, 5th.

Forstreuter, W. 1988. Zur Morphologie, Anatomie und Okologie von *Tripodanthus acutifolius* (Ruiz et Pav.) Tiegh. (Loranthaceae). Ph.D. diss. Philipps-Universitat, Marburg, Germany.

Hawksworth, F. G. 1983. Mistletoes as forest parasites. Pages 317-333 in: The Biology of Mistletoes. M. Calder and P. Bernhardt, eds. Academic Press, New York.

Kuijt, J. 1975. The identity of *Struthanthus haenkei* (*Spirostylis haenkei*) (Loranthaceae). Can. J. Bot. 53:249-255.

Kuijt, J. 1987. Novelties in Mesoamerican mistletoes (Loranthaceae and Viscaceae). Ann. Mo. Bot. Gard. 74:511-532.

Musselman, L. J., and Mann, W. F., Jr. 1978. Root parasites in southern forests. U.S. Dep. Agric. For. Serv. Gen. Tech. Rep. SO-20.

Oti, K., and Satomi, N. 1953. The hosts of *Pseudodipus japonica* Hayata. Hokuriku J. Bot. 2:49-51.

Parker, C. 1992. Weeds of Bhutan. National Plant Protection Centre, Thimphu, Bhutan.

Rawlins, G. B. 1950. Native mistletoe on exotic pine. N.Z. For. Serv. For. Res. Inst. For. Res. Pathol. Notes 1(2), 18.

Scharpf, R. F., and Hawksworth, F. G. 1974. Mistletoes on hardwoods in the United States. U.S. Dep. Agric. For. Serv. For. Pest Leafl. 147.

Shaw, C. G., III. 1993. First report of *Dendrophthoe falcata* on *Pinus kesiya*. Plant Dis. 77:847.

Vasquez-Collazo, I., Perez-Chavez, R., and Perez-Chavez, R. 1987. Control quimico del Muerdago (*Psittacanthus* sp.) en la Sierra Purepecha (Meseta Tarasca). Cienc. For. II 59:106-126.

von Hartmann, T. 1990. Die kiefernmistel im Raum Schwabach/Mittelfranken. Allg. Forst Z. 45:914-916.

Wiens, D. 1964. Revision of the acataphyllous species of *Phoradendron*. Brittonia 16:11-54.

(Prepared by R. F. Scharpf, B. Geils, D. Wiens, C. Parker, and W. Forstreuter)

Cankers and Twig Blights

A canker is a sunken, necrotic lesion caused by localized death of the cambium on a tree's stem or branch. Cankers on main stems may eventually result in death of the tree if they expand and girdle the stem. They also greatly increase the chance of stem breakage by causing irregular radial growth or by opening the wood to decay organisms. The value of lumber from cankered trees may be reduced by discoloration, abnormally high proportions of nonwoody cells, and resin-soaking, which affects the wood's appearance and strength and interferes with the penetration of preservatives. Pulp value is also lowered because it is more difficult to bleach the discolored wood in cankers and there are debarking problems and high levels of wood fibers and pitch in chips made from malformed wood.

The extent and severity of a canker is dependent upon variables such as environmental conditions, the species of the pathogen, and the host response as well as the size and age of the host tissue infected and the time of year. Many types of cankers are caused by facultatively parasitic fungi, which require a wound or other host stress to become established. The fungi are generally ascomycetes or their deuteromycete associates. Other cankers are caused by obligate parasites such as rusts or by environmental factors such as freezing or drought.

Most cankers can be placed into one of three categories.

Perennial or target cankers. These generally circular to lens-shaped cankers persist for years and slowly expand at about the same rate as the radial growth of the affected tree. Perennial cankers gradually take on a sunken appearance because tissues under the dead cambium do not grow along with the surrounding wood. The cankers are often surrounded by concentric ridges of host callus tissue formed during each growing season in response to the infection, resulting in a pattern reminiscent of a target. In rapidly growing trees, the callus may eventually grow over and cover infected areas.

Diffuse cankers. The expansion of diffuse cankers is more rapid than the radial growth of the tree, and thus they include little or no callus. Because of their rapid growth, diffuse cankers often girdle a tree within a few years. A diffuse canker is likely to be encountered in young tissue invaded by a fungal pathogen.

Annual cankers and twig blights. These cankers also contain little or no callus but usually develop rapidly and for one season only. They frequently occur on young stems or twigs that have been predisposed to opportunistic fungal infections by frost, drought, other stresses, or wounding. The dead bark may be overgrown by callus during the following growing season, resulting in complete healing if the external stresses are no longer present.

(Prepared by B. Callan)

Larch Canker

Other names: European larch canker; chancre du milhze (French), Lärchenkreb (German), chancro del alerce (Spanish), lärkkräfta (Swedish)
Causal organism: *Lachnellula willkommii* (R. Hartig) Dennis (syns. *Dasyscyphus willkommii* (R. Hartig) Rehm and *Trichoscyphella willkommii* (R. Hartig) Nannf.)
Hosts: *Larix* spp.
Distribution: Europe, Japan, and northeastern North America

Larch canker has been known for a long time; the early studies were reviewed by Yde-Andersen in 1979 and 1980. The disease occurs on trees of all ages. Cankers more than 10 cm in diameter may be present on the stems and branches of mature trees (Plate 89). Such cankers generally are quite old, do not exude resin, and produce few, if any, fructifications of the fungus. In younger parts of trees or on young trees, incipient infections and young cankers can be seen on axial or lateral shoots as young as 1–2 years. The most typical feature of infection is canker development from a short shoot (Plate 90). The needles of a short shoot at first shrivel up and then yellow and fall prematurely; the bark around the short shoot dies and shrinks, and callus grows around it; resin exudes; and fructifications of the fungus develop. Resinosis and fructifications are persistent and renewed for many years as the canker increases in size. Infection may result in the death of the distal part of the shoot, sometimes within a few months if the shoot is small in diameter.

Diagnosis

Lachnellula willkommii is a discomycete (Ascomycetes, Helotiales) that produces two types of fructifications. Hyaline spermatia (2–8 × 1–2 μm) gathered in whitish globules up to 1 mm in diameter break through the bark. Mature apothecia are cup-shaped, 2–6 mm in diameter, and white and densely piliferous outside and bright orange inside. A short stalk not longer than 1 mm is present. Ascospores (13.5–28 × 4.5–11 μm) are elliptic, unicellular, hyaline, and smooth. Apothecia may produce ascospores for several months. On standard agar medium, mycelium is white with velvety to felty aerial growth. The colony is thicker at the center and has a dense margin. Optimal temperature for growth is about 16–18°C, and radial growth is slow (about 0.6 mm per day). Only spermatia are produced in pure culture.

Disease Cycle

Ascospores are considered the only infectious propagules. They are produced year-round, depending upon humidity and temperature. Active spores are released after rainfalls and only if the temperature is greater than 0°C. They germinate easily (the rate is usually more than 80% after 24–48 hr) within a broad range of temperatures (0–30°C). Natural means of infection are not fully known. Wounds in the bark, for example, feeding wounds made by weevils (i.e., *Hylobius abietis*), may be infection courts. Since cankers are most often located around short shoots, it is strongly suspected that infection starts in either needle scars or the feeding sites of aphids (i.e., *Adelges larici*). Host susceptibility is not uniform throughout the year. Artificial inoculation of wounds with ascospores showed a period of higher susceptibility during the spring when flushing and shoot elongation occur.

Effects on Tree and Forest

Cankers may be numerous on branches and twigs without causing any marked effect on tree growth and tree health; nevertheless, they are the main sources of local inoculum. When young plantations are heavily infected, their future is very much in question. When cankers are present on the trunks, trees can usually live as long as uninfected trees, but the value of the boles will be greatly reduced. Cankered stems can also break in the wind. In Europe, where larch canker is an endemic disease, damage is variable. In the natural area of *L. decidua* (the Carpathians and the Alps of central Europe), damage is usually minimal. Where larch is planted outside this area (at lower altitudes and in an oceanic climate), infection levels often exceed 40–50%; sometimes 100% of the trees are diseased. Because of this threat, European larch has not been used very often in plantations, in spite of its fast growth and very valuable wood. At the beginning of the twentieth century, the disease was introduced into North America (Massachusetts) where eradication measures were applied. In New Brunswick and Nova Scotia, Canada, numerous infections were first observed in 1981 on young natural regeneration of eastern larch (*L. laricina*); in 1992, infections were seen in *L. laricina* regeneration and *L. decidua* plantations on Prince Edward Island. Disease incidence often exceeded 80% (Plate 91).

Predisposing Factors

Many environmental factors reportedly predispose trees and stands to larch canker development, including high tree density, poor growth, cool and humid climate, stagnant cold air, and frequent fog. Results and opinions on this matter are conflicting, however. Edaphic conditions, altitude, and exposure have never been clearly associated with the disease. Stand composition could be a factor: pure larch stands appear to have more cankers than stands mixed with beech. However, mixture with Norway spruce may increase disease, which has led to the speculation that the adelgid *A. larici*, which spends a part of its life cycle on *Picea abies,* may play a role in spread of the disease.

The most important factor affecting infection is the susceptibility of the host itself. Knowledge in this field comes chiefly from observations made during many years in Europe in comparative plantations and experiments. The species considered moderately to very resistant are *L. leptolepis* (= *L. kaempferi*), *L. siberica*, *L. gmelini,* and the hybrid *L.* × *eurolepis*. *L. laricina* is considered not resistant on the basis of observations made in Canada. *L. decidua* is by far the most susceptible species, but its susceptibility varies greatly. Field observations suggest that provenances from central Europe are generally resistant to very resistant, i.e., those from Sudeten (Czech Republic, "Sudetica" race) and the center of Poland ("Polonica" race) and some from northeast of the Alps (Austria). Provenances from south and west of the Alps are very susceptible. The high susceptibility of *L. decidua* from the French Alps and the resistance of trees from Poland and the Sudeten (which was equal to that of *L. leptolepis* and *L.* × *eurolepis*) were confirmed through artificial inoculations.

Disease Management

Because of the low incidence of larch canker in natural forests, the small number of existing European larch plantations, and the sporadic incidence of the disease, no specific management strategies have been defined for European larch. The only method considered very effective in controlling the disease in *L. decidua* is the use of resistant provenances in new plantations. *L.* × *eurolepis*, which has good vigor and wood quality, is also recommended. Additional precautions that might be considered to lower the epidemic hazard, especially in young stands, include 1) minimizing adelgid populations on larch by not planting them with spruce and using insecticides where practical; 2) controlling weevil populations (e.g., by trapping mature insects and using insecticides) to reduce feeding wounds on receptive young shoots; and 3) eliminating infected trees and pruning as early and as often as possible to reduce local inoculum.

Selected References

Simpson, R. A., and Harrison, K. J. 1993. First report of European larch canker on Prince Edward Island, Canada. Plant Dis. 77:1264.

Sylvestre-Guinot, G. 1986. Etude des sites d'infection du *Lachnellula willkommii* (Hartig) Dennis chez le *Larix decidua* Miller. Ann. Sci. For. 43:199-206.

Sylvestre-Guinot, G., Paques, L., and Delatour, C. 1994. Une méthode d'inoculation pour l'évaluation précoce du comportement du Mélèze vis a vis du *Lachnellula willkommii*. Eur. J. For. Pathol. 24:160-170.

Yde-Andersen, A. 1979. Host spectrum, host morphology and geographic distribution of larch canker, *Lachnellula willkommii*. Eur. J. For. Pathol. 9:211-219.

Yde-Andersen, A. 1979. Disease symptoms, taxonomy and morphology of *Lachnellula willkommii*. Eur. J. For. Pathol. 9:220-228.

Yde-Andersen, A. 1980. Infection process and the influence of frost damage in *Lachnellula willkommii*: A literature review. Eur. J. For. Pathol. 10:28-36.

(Prepared by C. Delatour)

Sphaeropsis Shoot Blight and Canker

Other names: Diplodia blight, tip blight, collar rot, sap stain; Triebsterben der Kiefer (German)

Causal organism: *Sphaeropsis sapinea* (Fr.:Fr.) Dyko & Sutton in Sutton (syns. *Macrophoma sapinea* (Fr.:Fr.) Petr., *Sphaeropsis ellisii* Sacc., *Diplodia pinea* (Desmaz.) J. Kickx. fil., and *Macrophoma pinea* (Desmaz.) Syd. & Petr.)

Hosts: many *Pinus* spp. (especially two- and three-needled pines); also reported on species of *Abies*, *Araucaria*, *Cedrus*, *Chamaecyparis*, *Cupressus*, *Larix*, *Picea*, *Pseudotsuga*, and *Thuja* and possibly other conifers

Distribution: worldwide where hosts are native or planted as exotic species

"Shoot blight and canker" incompletely describes the variety of damage caused by this disease to different tree parts at different stages of tree development. Seed cones colonized by the fungus may be shrunken, shriveled, and partially or entirely necrotic. Exudation of resin from cones and other colonized tree parts is common. Symptoms resulting from seed-associated *Sphaeropsis sapinea* include reduced germination, seed rot, decay of emerging radicles, and late damping-off of young seedlings.

Shoot blight of seedlings or older trees is characterized by death of new shoots during elongation and needle expansion (Plate 92). Resin droplets may appear on infected tissues before the development of other symptoms. Individually infected needles initially darken and cease expanding. Small, water-soaked lesions that become purplish brown and necrotic develop at points of stem infection. The rapidly invaded shoots typically are killed before needles fully expand, sometimes even before needles extend beyond fascicle sheaths. Dead, pale gray or yellow to red brown needles often remain attached to bent, curled or twisted, stunted, and desiccated shoots that may be encrusted with dried resin. Shoot death also may result from seedling collar rot, which produces discolored, necrotic bark tissue and black streaks in the wood of stems and root collars.

Shoot blight or cone infections on established trees may result in cankers. Elongated areas of depressed, brown, necrotic bark are indicative of cankers on young branches with smooth bark. The underlying wood is discolored green or brown and soaked with resin, and there may be a distinct demarcation between healthy and diseased tissue in both bark and wood (Plate 93). If the pathogen reaches a node, the entire whorl may be killed. However, disease development may be halted and another bud or branch may become dominant; thus, repeated infection may result in excessively branched, malformed crowns.

Older cankers on both branches and stems may be bounded by callus tissue and become quite large and misshapen. Faces of older cankers are sunken, and sloughing of dead, rough bark reveals underlying darkly stained wood. Girdling of branches or stems causes death of all distal parts, resulting in striking discoloration of foliage on recently killed branches and tops. Persistence of dead tops results in a "stag-head" appearance.

Rapid wilting of foliage on branches or entire crowns also occurs without the external stem symptoms of cankers. Shoots become limp and foliage chlorotic, and then needles quickly brown as branches die. Wilt results from colonization of the sapwood, indicated by wedge-shaped, gray to blue black stains. Sap stain also develops in standing trees wounded by pruning or logging operations and in recently felled timber.

Diagnosis of Sphaeropsis shoot blight and canker is aided by recognition of fruiting bodies and cultural characteristics of the pathogen. Immersed, dark brown or black pycnidia are up to 250 μm in diameter and globose or have only a slight neck. Pycnidia may be solitary or aggregated and erupt through surfaces of colonized needles (Plate 94) (especially below the fascicle sheath), young shoots, bark, and umbos of cone scales (Plate 95). Conidiogenous cells (without conidiophores) develop from cells lining the inner wall of the pycnidium. Oblong conidia (approximately 30–45 × 10–16 μm) are thick walled and truncate at the base. They vary in color from hyaline or yellow to dark brown when mature and may be aseptate or have one, or rarely more than one, septum. Hyaline microconidia or spermatia (2.5 × 1.0 μm) also may be produced in the pycnidia. No sexual stage is known.

Two groups within *S. sapinea* are recognized on the basis of cultural morphology. Isolates of the more widespread type (referred to as "A") grow more quickly on agar media and produce abundant, white to gray green aerial mycelium. Isolates of the second type ("B") are currently known only from the north central United States. These isolates grow more slowly and have dark gray mycelium that is closely appressed to the agar surface. The two morphotypes also are distinguished by analyses of isozymes and random amplified polymorphic DNA.

Disease Cycle

Conidia of *S. sapinea* are released during wet weather throughout the growing season and disseminated by splashed and wind-driven rain. Spores are most abundant, however, during the spring and early summer when new, expanding shoots are susceptible to infection. Spores germinate rapidly when moist, and germ tubes penetrate stomata on young needles. The fungus directly penetrates beneath hyphal aggregations formed on the surface of young, expanding stem tissue and can also infect second-year cones early in the growing season. Fresh wounds, e.g., those created by insects, hail, or pruning, on older branches and stems also are infection courts. Fully elongated needles and shoots and bark on the previous year's shoots apparently are not susceptible to infection. However, *S. sapinea* has been found to persist on or in asymptomatic needles, cones, and stems, and it can proliferate to cause disease in stems subjected to water stress. Pycnidia with mature conidia may form within a few weeks on any colonized tree part, and secondary cycles of infection may occur if susceptible tissue is present or exposed during conducive weather. The pathogen overwinters as spores in pycnidia or as hyphae in colonized host material.

Predisposing Factors

The frequency and severity of Sphaeropsis shoot blight and canker are at least in part attributable to inherent host characteristics. Experience and experimentation have established that variation in susceptibility exists both between and within some host species. For example, in the northeastern and central United States, ornamental Austrian pines (*P. nigra*) are more often both chronically and severely affected than native pines. Variation in susceptibility among host species also has been established for several pines grown in plantations in South Africa as well as between selections of *P. radiata*.

Severe damage is often associated with climatic and site conditions. Disease of established trees historically has been associated with lack of adequate precipitation exacerbated by droughty soils and competing vegetation. Enhancement of colonization in moisture-stressed trees has been demonstrated for several hosts. Severe disease outbreaks also have been associated with altered host nutrition. Epidemics have been attributed to excessive nitrogen from atmospheric sources in the Netherlands and from paper mill wastes spread in plantations in Wisconsin. Severe symptoms also often develop rapidly on trees wounded and stressed by excessive pruning and on those wounded by hail, especially when they have been predisposed by other factors.

Disease Management

Practices for minimizing the impact of Sphaeropsis shoot blight and canker vary with the stage of host development, the situation in which it is grown, and the intended product. In nurseries, seed orchards, Christmas tree plantations, and ornamental plantings where there is a history of disease, use of protective or systemic fungicides has become routine. Fungicides also have been applied to prevent severe damage to conifer windbreaks in agricultural areas. In these situations, reducing inoculum by removing nearby diseased trees, roguing affected trees from a planting, or pruning affected branches might reduce disease incidence.

In commercial timber stands, both the incidence and severity of symptoms and associated losses should be monitored. If disease progresses to seriously affect large numbers of trees or the entire stand, early harvest may be desirable. Likewise, in precommercial plantings, if cumulative mortality reduces stocking below a minimum desired level, reestablishment of the plantation might be considered.

Effective, long-term disease management in forests depends on site selection and implementation of management procedures that minimize stress on appropriately selected host material. Sites with soil nutrient imbalances or those prone to drought should be avoided. Nursery stock should be carefully handled and the need for management of competing vegetation considered. Pruning should not be excessive and should be done during the winter when inoculum is less likely to be abundant or, as in the shearing of Christmas trees, only during dry weather. On sites with a history of losses from Sphaeropsis shoot blight and canker, planting of less susceptible species or of selected stock with useful levels of resistance should be considered.

Selected References

Smith, D. R., and Stanosz, G. R. 1995. Confirmation of genetic basis for two distinct populations of *Sphaeropsis sapinea* in the north central United States using RAPDs. Phytopathology 85:699-704.

Stanosz, G. R., Prey, A. J., and Cummings Carlson, J. 1994. Biology and control of *Sphaeropsis sapinea* in nurseries and plantations in Wisconsin, USA. Pages 13-26 in: Disease and Insects in Forest Nurseries. Proc. Meet. IUFRO Working Party S2.07.09, 2nd. R. Perrin and J. R. Sutherland, eds. Institut National de la Recherche Agronomique, Paris.

Swart, W. J., and Wingfield, M. J. 1991. Biology and control of *Sphaeropsis sapinea* on *Pinus* species in South Africa. Plant Dis. 75:761-766.

(Prepared by G. Stanosz)

Scleroderris Canker

Other names: Brunchorstia dieback, Scleroderris Krankheit (German), knoppoch grentorka (Swedish)

Causal organism: *Gremmeniella abietina* (Lagerberg) Morelet; (syns. *Scleroderris lagerbergii* Gremmen and *Ascocalyx abietina* (Lagerberg) Schläpfer; anamorph *Brunchorstia pinea* (P. Karst.) Höhn.)

Hosts: pines, especially *Pinus sylvestris*, *P. nigra*, *P. contorta*, and *P. resinosa*; *Picea abies*; and *Abies* spp.

Distribution: widely spread in the northern hemisphere

Gremmeniella abietina was first found on *Pinus sylvestris* during the late nineteenth century. It has since been reported as a pathogen on conifers in many genera throughout the northern hemisphere. Susceptibility and the type of damage caused vary greatly among host species, but throughout its range, the fungus seems to be most damaging to pines, especially in plantations of exotic species or in off-site areas.

In Europe during the first decades of the twentieth century, the pathogen was known to attack mainly *P. nigra*, which was extensively used in afforestations in Europe far outside the natural range of the tree species. Reports of severe disease in native tree species were rare during the first half of the century. This situation changed during the 1950s, when the disease killed millions of *P. sylvestris* seedlings in nurseries. Extensive outbreaks have subsequently occurred; and during the late 1970s and 1980s, disease intensity increased and plantations and forests of both native and introduced species all over Europe were devastated (Plate 96). The number of outbreaks peaked during about 1987–1988, and the disease is now declining and is regularly found only in the north of Fennoscandia.

In North America, *G. abietina* was first reported during the early 1950s in Michigan. The disease pattern was characterized by cankers and shoot blight on seedlings and within about 2 m of ground level on larger trees. During the mid-1970s, a more virulent outbreak began in New York on pole-sized *P. resinosa* and *P. sylvestris* and led to the introduction of quarantine regulations. However, these regulations did not stop the spread of the disease in the northeastern United States and into eastern Canada. The fungus has also been found on *Picea mariana* and *P. glauca* and in Newfoundland on *Abies balsamea*. In western North America, *G. abietina* can occasionally be found on *Pinus contorta*, *P. ponderosa*, and *P. monticola*. In Japan, the disease occurs on *A. sachalinensis*.

The Pathogen

G. abietina is an ascomycete of the order Helotiales. It fruits on dead host tissue and forms apothecia that are dark brown to black on the outside and have a hyaline to greenish hymenium that is 1–2 mm across when wet. Ascospores (14–20 × 3.3–5 µm) are colorless and consist of three to four cells. Pycnidia emerge from short shoot scars or directly through the bark. They are spherical, black, and 0.5–2 mm across. When pycnidia are ripe, pink tendrils of conidia are exuded. Conidia (usually 20–55 × 2.5–3 μm) are sickle shaped and contain two to seven septa. Cryptopycnidia, formed immersed in the bark, and also sometimes ordinary pycnidia contain microconidia that are similar to normal conidia but smaller.

Disease Cycle

The current year's shoots are typically infected during wet periods of the growing season. After germination, the fungal hyphae grow through stomata or penetrate the host at bud scars or needle bases. Wounds may be infected but are not required for colonization. Once established, the fungus enters a latent phase until the host becomes dormant. Depending on the number of days during the winter period conducive to disease development (temperatures of −5 to +5°C), the fungus may penetrate and colonize the periderm.

Infections by *G. abietina* may be detected during late winter or early spring when small necrotic areas start to develop under infected short shoots or in buds. The disease becomes more evident as the needles turn red or brown, starting at the shoot axis and progressing outward for about 1 year after infection. This rapidly expansive phase of the disease takes place during the spring, and infected shoots usually die before the new flush. When a shoot, branch, or main stem is encircled by dead tissue, the distal parts of the tree die. Cankers are formed when the infection does not manage to encircle the shoot, branch, or stem during one season. The cankers can persist for many years, producing fruiting bodies and functioning as perennial infection sources.

Under some conditions, the first pycnidia begin to appear on the shoot late in the summer or autumn of the year in which the shoot died, but the peak of conidial release is during the period of shoot elongation the next spring. Apothecia develop somewhat later than pycnidia and ripen from summer until late autumn. Little is known about the relative extent to which conidia and ascospores contribute to infections in new shoots. Since the two types of spores have slightly different dispersal periods, they may have different methods of infection. Conidial dispersal coincides with shoot elongation; the shoots are more mature when ascospores are released.

Disease in pole-sized trees. In most of continental Europe and Scandinavia at low elevations (and in the recent outbreak in North America), trees 10–40 years old are most frequently attacked (Plate 96). A typical shoot blight ensues, starting in the lower crown. Trees suffer reduced growth but typically recover after a few years, unless they are repeatedly infected. Trees exposed to repeated, severe attacks usually die from the disease or from attack by secondary pathogens or insects. Depending on the tree species, cankers are more or less common. Cankers occur on branches and twigs of *P. sylvestris* but rarely on the main stem; large and deep cankers are common on *P. resinosa*, and attacks frequently result in death of the tree. Large cankers weaken the stem, and the tree will become more vulnerable to snapping in the wind. In southern Sweden, up to about 30% of the trees in severely attacked stands were killed during a 3-year outbreak of the disease. Norway spruce growing beneath infected pine can also be infected and suffer top dieback (Plate 97).

Disease in young trees in harsh climates. A second type of disease occurs in plantations of *P. resinosa* and *P. banksiana* in northern North America, on introduced *P. contorta* in northern Scandinavia, and on *P. cembra* in regions of the Alps and Apennines. This disease pattern is confined to areas with harsh climate marked by short growing seasons, cold winters, and a long period of deep snow cover. The snow cover is probably a prerequisite for this type of disease, since it occurs on seedlings and parts of trees covered by snow during the winter. The temperature in deep snow during the winter is close to 0°C, which has been shown to promote disease development. The temperature above the snow is usually too low for the fungus to grow. In this type of disease, whole branches or several successive shoots can be killed at a time. Cankers on branches and on the main stem are common. Heavily attacked young trees are killed, and plantations can be devastated within a few years.

Population Structure

The two major patterns of disease caused by *G. abietina*, one affecting the canopy of pole-sized trees and the other occurring on stems and shoots of saplings growing in harsh climates, are determined by the taxonomic variety and race of *G. abietina* as well as by host and climate. The population structure of this pathogen is complicated (Table 7).

Three varieties have been distinguished by differences in spore shape, number of septa, and size and confirmed by protein electrophoresis and comparative DNA studies. *G. abietina* var. *balsamea* and var. *cembrae* have narrower host ranges than *G. abietina* var. *abietina*, and they attack their hosts in natural populations within their natural range of occurrence. The varieties of the fungus could have differentiated from an ancestral population as a result of geographic isolation.

Within *G. abietina* var. *abietina* are three races or strains. In North America, the European and North American races differ in disease pattern type, colony morphology in culture, serological reaction, protein electrophoresis banding pattern, RAPD (random amplified polymorphic DNA) banding pattern, and DNA restriction sites. No difference has been found between the European race in North America and *G. abietina* var. *abietina* from central Europe, suggesting recent introduction into North America. Although the European race has been present in North America for two decades, no hybrids with the North American race have been found. The differences and reproductive isolation between the two races of *G. abietina* indicate that the two in fact could be considered separate biological species. A third race within *G. abietina* var. *abietina* was found in Japan, where it causes severe damage on *A. sachalinensis*.

Populations of *G. abietina* occurring on the European continent are not uniform. The first indication of these differences was found in populations in the Alps that have unusually long conidia and that were later recognized as *G. abietina* var. *cembrae*. In Scandinavia, the typical European race of *G. abietina* var. *abietina* (with short conidia) is found in pole-sized stands

TABLE 7. Taxonomy and Population Structure of *Gremmeniella abietina*

Taxon / Race	Hosts	Distribution	Disease type
var. *abietina*	*Pinus*, *Picea*, and *Abies* spp.	Northern hemisphere	Various
North American	*Pinus* spp., especially *P. resinosa*	United States and Canada	Shoot blight and cankers below 2 m
European	*Pinus sylvestris*, *P. nigra*, and *P. resinosa*; *Picea abies*	Europe, northeastern United States, and Canada	Shoot blight at all heights
North European	*Pinus contorta* and *P. sylvestris*	Scandinavia	Shoot blight and cankers below 2 m
Asian	*Abies sachalinensis*	Japan	Shoot blight and cankers below 2 m
var. *cembrae*	*Pinus cembra* and *P. mugo*	Alps and Appeninnes	Cankers below 2 m
var. *balsamea*	*Abies balsamea*; *Picea glauca* and *P. mariana*	Eastern Canada	Shoot blight and cankers

of *P. sylvestris* mostly in the south but also occasionally in the north. In northern Scandinavia, especially on *P. contorta*, disease is commonly expressed as cankers near the ground, as is that caused by *G. abietina* var. *cembrae* in the Alps and the North American and Asian races of *G. abietina* var. *abietina*. Like the conidia of *G. abietina* var. *cembrae*, those from northern Scandinavian isolates are long and multiseptate. Inoculation experiments with the two types of isolates (from northern and southern Scandinavia) showed differences in colonization rates in *P. sylvestris* and *P. contorta*. Clear differences were detected in the DNA of the types, and no hybrids were detected with RAPD markers. These differences suggest that there are two types of *G. abietina* in northern Europe. Protein electrophoresis has indicated that *G. abietina* var. *cembrae* and the northern Finnish isolates are similar but not identical, despite morphological and ecological resemblance. This could be a result of convergent evolution occurring in at least three places around the world. Alternatively, there may be a common ancestor of these populations in arctic regions, and they were later separated after periods of glaciation.

Selected References

Bernier, L., Hamelin, R. C., and Ouellette, G. B. 1994. Comparison of ribosomal DNA length and restriction site polymorphisms in *Gremmeniella* and *Ascocalyx* isolates. Appl. Environ. Microbiol. 60:1279-1286.

Dorworth, C. E., and Krywienczyk, J. 1975. Comparisons among isolates of *Gremmeniella abietina* by means of growth rate, conidia measurement, and immunogenic reaction. Can. J. Bot. 53:2506-2525.

Gibbs, J. N. 1984. Brunchorstia dieback in Europe. Pages 32-41 in: Scleroderris Canker of Conifers. P. D. Manion, ed. Martinus Nijhoff/Dr. W. Junk, The Hague.

Hellgren, M., and Högberg, N. 1995. Ecotypic variation of *Gremmeniella abietina* in northern Europe—Disease patterns reflected by DNA variation. Can. J. Bot. 73:1531-1539.

Patton, R. F., Spear, R. N., and Blenis, P. V. 1984. The mode of infection and early stages of colonization of pines by *Gremmeniella abietina*. Eur. J. For. Pathol. 14:193-202.

Petrini, O., Toti, L., Petrini, L. E., and Heiniger, U. 1990. *Gremmeniella abietina* and *G. laricina* in Europe: Characterization and identification of isolates and laboratory strains by soluble protein electrophoresis. Can. J. Bot. 68:2629-2635.

(Prepared by M. Hellgren and J. Stenlid)

Pitch Canker

Other names: el cancro resinoso de los pinos (Spanish)
Causal organism: *Fusarium subglutinans* (Wollenweb. & Reinking) P. E. Nelson, T. A. Toussoun, and Marasas (syns. *F. lateritium* Nees:Fr. f. sp. *pini* Hepting in W. C. Snyder, Toole, & Hepting and *F. moniliforme* J. Scheld. var. *subglutinans* Wollenweb. and Reinking)
Hosts: *Pinus* species, especially slash (*P. elliottii* var. *elliottii*), loblolly (*P. taeda*), and Monterey (*P. radiata*) pines
Distribution: southeastern United States and coastal central California; Japan, Mexico, and South Africa

Pitch canker is a serious disease of planted pines. It was first described in 1946, and interest was rekindled in 1976 when the disease became epidemic in the southeastern United States on slash and loblolly pines (Plate 98). Subsequently, it has been determined that most, if not all, southern pines are hosts of the pitch canker fungus. Since 1976, pitch canker has evolved from a regional disease to one of national and international importance. In 1986, pitch canker was found seriously damaging planted Monterey pine in coastal central California (Plate 99), and it has also been reported on pines in Japan and Mexico. In South Africa, the fungus induces a root disease of *Pinus patula* seedlings.

Pathogen

The causal fungus has gone through a number of name changes. In 1946, it was designated *Fusarium lateritium* f. sp. *pini*. Fusaria assigned to the section Lateritium typically have no microconidia or chlamydospores. During the 1970s, the most common isolates of *Fusarium* from pitch canker tissue that satisfied Koch's postulates had abundant microconidia in heads and no chlamydospores and were assigned to *F. moniliforme* var. *subglutinans* in the section Liseola. In 1983, the variety was raised to species level as *F. subglutinans*. Since not all isolates of *F. subglutinans* are pathogenic to pines, some workers have designated the pitch canker pathotype as a forma specialis (*F. subglutinans* f. sp. *pini*).

Symptoms

Two primary symptoms are associated with pitch canker on trees. The classic symptom is a bleeding, resinous canker of the trunk, terminals, large branches, and exposed roots (Plate 100). The canker is usually sunken and the bark retained, and the wood beneath the canker is deeply pitch soaked (Plate 101). The pitch-soaked wood is a diagnostic character useful in separating pitch canker from other canker diseases of pines. The other symptom is dieback of the upper crowns resulting from the formation of cankers on branches. As the branches or shoots are girdled by the fungus, the needles turn yellow to reddish brown. Dead shoots may remain in the crown for several years and serve as indicators of the disease.

Either symptom may occur on several pine species. For example, stem and branch cankers appear on Monterey (*P. radiata*), Virginia (*P. virginiana*), longleaf (*P. palustris*), and eastern white (*P. strobus*) pines. Dieback occurs primarily on planted slash, loblolly, shortleaf (*P. echinata*), and Monterey pines. In Mexico, main-stem cankers occur on *P. estevezi* and dieback on *P. arizonica* var. *stormiae*. Luchu pines (*P. luchuensis*) on Amamiooshima and the Okinawa islands of Japan have dieback and main-stem cankers.

Infection

Although the major damage from this disease typically is the infection of vegetative structures, the pitch canker fungus also infects reproductive structures, causing death of female flowers and mature cones and deterioration of seeds in several pine species. Any fresh wound, regardless of cause or location, provides an infection court for the pathogen. Insects create wounds that can be infected by airborne spores, and some insects may serve as vectors of the pitch canker fungus. In slash pine seed orchards, main-stem cankers often develop after trees are injured by mechanical cone harvesters. Weather-related injuries, e.g., those caused by wind and hail, also serve as entry points, and hurricanes and tornadoes, in particular, have contributed to the intensification of the disease in some seed orchards.

The pitch canker fungus frequently colonizes fusiform rust galls, caused by *Cronartium quercuum* f. sp. *fusiforme*. Some of the greatest damage results when the rust is followed by infection with *Dioryctria* sp. and the pitch canker fungus. Infection of rust galls by *F. subglutinans* further weakens stems of mature trees and increases the chances of breakage and tree death.

Damage

From 1945 to 1973, limited outbreaks of pitch canker were noted in the southeastern United States, but the disease was not considered to be economically important. In 1974, however, a shoot dieback, identified as resulting from pitch canker, reached epidemic proportions on slash pine in Florida and on loblolly pine in North Carolina. These outbreaks were of great concern to forest resource mangers. Since the 1970s, pitch canker outbreaks in the south have been sporadic. In 1986, pitch canker was recognized as a major component of a crown dieback of Monterey pine in California. Until the nearly simultaneous disease outbreaks at three different locations, pitch canker was not known in California. Subsequent surveys have found the disease on several pine species from San Diego to San Francisco.

Damage to pines caused by *F. subglutinans* includes death, growth suppression, stem deformation, seed and cone losses, and loss of seedlings in seedbeds and subsequent outplantings. Pitch canker is primarily a problem on planted pines in plantations, seed orchards, parks, golf courses, and the urban landscape. Although the disease may be found in natural stands, it is rarely a problem in pine ecosystems. In the southeastern United States, however, pitch canker (dieback) is currently causing some damage on shortleaf pine in natural stands and in plantations, seed orchards, and cities. The disease occurs in natural pine stands in Mexico.

Disease Management

Because each outbreak in each specific location has its own unique case history (or sequence of events), no explicit management strategy has been developed to reduce or eliminate the threat of pitch canker disease. An integrated management approach including chemical control, biological control, genetic selection for resistance, and altered cultural practices should be considered.

Chemical control with insecticides may be feasible when the wounding agent is an insect. Biological control has not been promising. Variation in the incidence of pitch canker is very common among clones within seed orchards, so genetic selection for resistance is possible. Because of the importance of wounds as infection courts in pitch canker, understanding the cause or causes of the wounding is very important in managing this disease. Any alterations in cultural control practices that reduce host susceptibility or wounding would be beneficial.

Selected References

Correll, J. C., Gordon, T. R., McCain, A. H., Fox, J. W., Koehler, C. S., Wood, D. L., and Schultz, M. E. 1991. Pitch canker disease in California: Pathogenicity, distribution, and canker development on Monterey pine (*Pinus radiata*). Plant Dis. 75:676-682.

Dwinell, L. D., Barrows-Broaddus, J. B., and Kuhlman, E. G. 1985. Pitch canker: A disease complex of southern pines. Plant Dis. 69: 270-276.

Muramoto, M., and Dwinell, L. D. 1990. Pitch canker of *Pinus luchuensis* in Japan. Plant Dis. 74:530.

Viljoen, A., Wingfield, M. J., and Marasas, W. F. O. 1994. First report of *Fusarium subglutinans* f. sp. *pini* on pine seedlings in South Africa. Plant Dis. 78:309-312.

(Prepared by L. D. Dwinell and E. G. Kuhlman)

Atropellis Cankers

Causal organisms: *Atropellis piniphila* (Weir) Lohman & Cash (Ascomycetes, Helotiales), (syn. *Cenangium pinipihilum* Weir), *A. arizonica* Lohman & Cash, and other *Atropellis* species
Hosts: *Pinus contorta* and *P. ponderosa*
Distribution: western North America

Atropellis species occur on pine and typically cause long, dark, resinous, perennial lesions on branches or trunks. Where Atropellis cankers are prevalent, most or all host trees may become infected and often develop multiple stem and branch infections. Death from stem girdling occurs only in dense stands, but wood volumes may be greatly reduced. *A. piniphila* cankers of *Pinus contorta* in Alberta reduced the volume of infected trees by up to 56.5%, and 30% of chips were rejected because of high resin content. Tree volume losses from the same disease in a 45- to 65-year-old stand in British Columbia were 7.9%, and there was a total reduction in wood value of 33.4%.

Most stem cankers start as infections on undamaged bark in the vicinity of branch whorls. The first symptom usually is a resin drop on the bark from which the underlying infection extends into the wood and blue black streaks develop in the direction of the long axes of the wood fibers. Wood at the center of a canker is resin soaked and blue black, while at canker tips, sapwood often develops a reddish discoloration before turning blue black (Plate 102). Invasion of sapwood is relatively rapid but is slowed in the heartwood. Bark around the edges of cankers becomes cracked. Cankers in *P. contorta* caused by *A. piniphila* expand radially at a rate of approximately 0.6 cm per year, while longitudinal expansion is nearly 5 cm per year, resulting in long, narrow cankers. Trunk cankers may become very large (fast-growing *A. piniphila* cankers on ponderosa pine in the southwestern United States can rapidly exceed 3 m in length) and persist for 40–50 years in vigorous trees. Large numbers of small (0.5 to several millimeters in diameter), black, stalked, saucerlike apothecia fruit in the center of the cankers and are a key diagnostic feature (Plate 103). When placed in a solution of 5% KOH, the apothecia produce a characteristic green blue stain.

Ascospores, which are forcibly ejected into the air from the apothecia, travel up to 100 m. Ascospore release, the source of primary inoculum for the disease, occurs during moist periods from early spring to midfall and even later in moist, temperate regions. Apothecia become shrunken and curled during dry periods but expand and begin to release spores again within a few hours after they are rehydrated. In dry, sunny areas, spore production on fallen trees ceases after a few weeks, but if standing dead or fallen trees remain moist or shaded, spore production may continue for a year or two.

Prior to apothecial development on cankers, which may take several years after the initial infection, globose pycnidia develop and split to expose creamy masses of small, single-celled conidia. Conidia are thought to act as spermatia, fertilizing receptive hyphae and initiating the formation of apothecia. New pycnidia continue to develop after the onset of apothecial production and may be found on old cankers among apothecia.

Disease Management

There are three types of resistance to Atropellis cankers. 1) All trees are resistant until they are about 15 years old. 2) Young tissues in the upper crowns of trees are resistant. Most infections in older trees begin in tissues 10–14 years old, and many cankers occur in tissues 15–19 years old. In older trees, tissues 5–9 years old and more than 29 years old have few infections. 3) Although it is rare, cankers in vigorous trees can be overgrown.

Wild fires historically have naturally managed pine stands for Atropellis canker. Losses can also be reduced by removing susceptible trees (more than 15 years old), even if they appear healthy, because small and incipient infections are hard to detect. Keeping a minimum of 100 m clear between old trees and susceptible young plantations and planting mixed species also reduce inoculum.

Other Atropellis Cankers

Four species of *Atropellis* have been described, all from pines in North America. *A. piniphila* is the most damaging, but *A. pinicola* is locally common and causes branch cankers on white pine in the west. Eastern species include *A. tingens* and *A. apiculata*. The apothecia of *Atropellis* species produce a characteristic green blue stain when placed in a solution of 5% KOH, although one species (*A. tingens*) produces a chocolate brown stain. Species of *Atropellis* are separated by apothecial morphology, ascospore size, KOH reaction, and host preference.

Selected References

Hopkins, J. C. 1963. Atropellis canker of lodgepole pine: Etiology, symptoms, and canker development rates. Can. J. Bot. 41:1535-1545.

Hopkins, J. C., and Callan, B. E. 1991. Atropellis canker. Can. For. Serv. Pac. For. Res. Cent. Pest Leafl. 25.

Nevill, R. J., Merler, H., and Borden, J. H. 1989. Reduced volume, grade and value of lodgepole pine lumber caused by Atropellis canker and stalactiform blister rust. For. Chron. 65:36-41.

Reid, J., and Funk, A. 1966. The genus *Atropellis*, and a new genus of the Helotiales associated with branch cankers of western hemlock. Mycologia 58:417-439.

(Prepared by B. Callan)

Other Canker Diseases

Cytospora Cankers and Dieback

Cytospora is a large and loosely characterized conidial genus. Sexual states, where known, are *Valsa* spp. (Ascomycetes, Diaporthales). Most species are associated with cankers on hardwoods, but several are relatively common on conifers throughout the northern hemisphere. Cytospora cankers, most notably those caused by *C. abietis* Sacc. (teleomorph *Valsa abietis* (Fr.:Fr.) Fr.) and *C. kunzei* Sacc. (teleomorph *Leucostoma kunzei* (Fr.:Fr.) Munk, anamorph *Leucocytospora kunzei* (Sacc.) Z. Urban), are associated with flagged branches and dead tops of affected trees. Tops of seedlings and branches may be rapidly girdled and killed by diffuse cankers following wounding or drought stress. The cankers have sunken, dead, frequently discolored bark with underlying dead cambium and resin flow at the perimeter of the dead tissue. Cankers occasionally occur on larger stems if the bark is damaged. *Abies* and *Pseudotsuga* spp. are most often attacked by *C. abietis* in North America (Plate 104), and *Tsuga* spp. and *Thuja plicata* are also occasionally cankered. In the western United States, *C. abietis* is regularly associated with dwarf mistletoe infections on *Abies grandis* and *A. concolor*. The combination causes branch death on affected trees. *Picea* spp., especially *P. pungens* where it is planted as an ornamental east of its range in North America, are attacked by *C. kunzei*.

Orange conidial tendrils are often seen at the canker margins where pycnidia form underneath the bark. The single-celled conidia are allantoid and small (3–5 × 1–1.5 µm) and are released from convoluted chambers in the flattened pycnidia from spring to autumn. The sexual states are found in dead tissue around the edges of cankers and appear as small, flat-topped pimples pierced with darker ostioles. The ascospores, released during the spring and summer, are similar to the conidia but twice as long.

Control of cankers of ornamental trees is aided by pruning, provided it is done during dry weather and the pruning tool is disinfested between cuts.

Lachnellula Cankers

Most species of *Lachnellula* (Ascomycetes, Leotiales; many formerly described as *Dasyscyphus* spp.) on conifers are associated with cankers that develop from wounds or winter damage and are thus considered opportunistic (Plate 105). However, some have proven pathogenic associations. By far the most damage is caused by *L. willkommii* (R. Hartig) Dennis (syn. *Trichoscyphella willkommii* (R. Hartig) Nannf.), which causes perennial cankers on larch in Europe and North America (European larch canker).

L. agassizii (Berk. & M. A. Curtis) Dennis has been shown to cause cankers on *Pinus* spp., *Abies balsamea,* and *A. fraseri*. *L. calyciformis* (Batsch) Dharne is pathogenic to *Abies* and *Pinus* spp. in Japan, and *L. abietis* (P. Karst.) Dennis and *L. pini* (Brunch.) Dennis also cause cankers on Japanese pine species. *L. pini* also affects five-needle pines in Europe and western North America. In North America, *L. arida* (W. Phillips) Dennis and *L. gallica* (Karst. & Har.) Dennis affect *A. balsamea*, and *L. laricis* (Cooke) Dharne and *L. occidentalis* (Hahn & Ayers) Dharne cause cankers on *Larix* spp. *L. pseudotsugae* (Hahn) Dennis causes perennial target cankers on *Pseudotsuga menziesii* on dry sites in British Columbia.

Lachnellula cankers are controlled primarily by establishing provenances suited to the environmental conditions of the growing site to ensure vigorous growth of the host. Stands previously infected with European larch canker should be monitored to ensure that remaining low levels of infection do not spread.

Phacidium Cankers

Phacidium cankers occur on most conifers; the most damaging is caused by *Phacidium coniferarum* (Hahn) DiCosmo, Nag Raj, & Kendrick (Ascomycetes, Leotiales) (syn. *Potebniamyces coniferarum* (Hahn) Smerlis). The conidial state, now considered to belong in the genus *Apostrasseria*, has been previously described in a number of different genera, including *Phacidiopycnis* and *Allantophomopsis*, but most earlier publications refer to it as *Phacidiopycnis pseudotsugae* (M. Wilson) Hahn. The confusion first arose when conidial states of a number of the *Phacidium* spp. on conifers were described as *Phomopsis* spp., even though they do not produce the beta conidia characteristic of this genus.

P. coniferarum is worldwide in distribution and is especially damaging in Douglas-fir and *Pinus* spp. plantations. In North America, it is more prevalent in the east than in the west. Cankers appear as areas of sunken bark pimpled first with black pycnidia, which split to release hyaline, irregularly rhombic conidia. Apothecia are also produced on cankers and are erumpent in black stromata, which open with irregular fissures to expose the cream-colored hymenium beneath. The apical pore of the ascus stains blue in Melzer's reagent, and ascospores are single celled (rarely with one or two septa), elliptic, hyaline, and 10–20 µm long.

Infections, which originate at wounds, are initiated during the autumn and late spring but not during the summer. In *P. sylvestris* plantations in Finland, infections are more likely if pruning has taken place from October to December. The safe pruning period was found to end in autumn when the 5-day mean temperature was below −7°C. In North America, stem and branch cankers occur on young *P. menziesii*, and infected *Larix* spp. are susceptible to top dieback. Another species, *P. balsamicola* (Smerlis) DiCosmo, Nag Raj, & Kendrick, occurs in western North America and causes cankers on *Abies* spp.

Seiridium Cankers of Cupressaceae

Several *Seiridium* species (Coelomycetes, some formerly placed in *Coryneum*, *Pestalotia*, and other genera) cause sunken, resinous cankers, branch dieback, and eventual death of susceptible members of Cupressaceae. Seiridium canker diseases occur in western North America, South America, Europe, the Commonwealth of Independent States, Asia, Australia, and New Zealand. The most common and damaging host-fungus associations are between *S. cardinale* (W. Wagener) Sutton & I. Gibson and *Cupressus macrocarpa* and *C. sempervirens*. *Chamaecyparis*, *Cryptomeria*, *Cupressocyparis*, *Juniperus*, and *Thuja* species are also potentially susceptible to Seiridium cankers.

Heavy losses to cypress plantations in Europe have been attributed to *S. cardinale* (Plate 106). Another pathogenic species, *S. unicorne* (Cooke & Ellis) Sutton (syn. *Monochaetia unicornis* (Cooke & Ellis) Sacc.) infects *Chamaecyparis* spp., *Juniperus virginiana*, and *Cupressus* spp. and is cosmopolitan in distribution. The sexual state of *S. unicorne*, *Lepteutypa cupressi* (Nattrass, C. Booth, & Sutton) H. J. Swart, produces inconspicuous, submerged perithecia on long-dead bark.

Smaller trees and trees recently pruned or otherwise wounded are most susceptible. Conidia are produced in shallow pycnidial or acervular structures on the infected periderm 5–8 weeks after the bark dies and are extruded in slimy droplets. Conidia, which are produced from annellides, are fusiform and vary in size depending on the species but range in length from 20–30 µm and have a small, pointed, unbranched hyaline appendage on each end, although the basal appendage may be missing. Conidia have four brown central cells between paler apical and basal cells. Over short distances, conidia are dispersed in water droplets, while long-range spread is by insects. Spread of the pathogen slows or stops during extended periods of dry weather.

Sirococcus Tip Blight

The coelomycete *Sirococcus conigenus* (DC.) P. Cannon & Minter (syn. *S. strobilinus* G. Preuss) causes seedling blight and shoot dieback. It occurs in Europe, North America, and northern Africa. Natural stands of *Pinus resinosa* and plantations of *Picea pungens* in midwestern and eastern North America are susceptible to shoot dieback. In coastal North America, *Tsuga heterophylla* (Plate 107) and *Picea sitchensis* are affected in natural stands, while *Pinus contorta, Picea* spp., and occasionally *Pseudotsuga menziesii* seedlings are blighted in nurseries. New and year-old shoots and seedlings are killed, and the small, brown pycnidia develop on the dead tissues (Plate 108) and contain hyaline, fusiform, uniseptate conidia, which are 12–15 μm long and spread by splashing water. Lower branches of pine trees can become browned after succeeding years of blight when damp conditions spread the conidia from the previous year's dead shoots and cone scales.

S. conigenus is also spread to new areas via contaminated seed, causing seedling blight and mortality under cool, moist conditions and low light intensity, particularly in container-grown plants in nurseries. Keeping records of contaminated seed lots, dustings seeds with fungicide, or adding fungicide to stratification water soaks will help to prevent spread of the pathogen in nursery situations. New plantations should not be established under older trees, which act as inoculum sources.

Other Cankers of Cupressaceae

Phomopsis juniperovora Hahn (Coelomycetes) is associated with nursery, seedling, and twig blight of junipers and other conifers, including *Cedrus, Chamaecyparis, Cupressus, Sequoiadendron, Taxus,* and *Tsuga* spp. It occurs in eastern North America, where it causes one of the most damaging diseases of juniper. Branch tips with dieback produce black pycnidia embedded in host tissue. The pycnidia exude cream-colored droplets of conidia during damp weather or if the diseased tissue is incubated in a damp chamber for a few days. Two types of single-celled conidia are generally emitted from the pycnidia: ellipsoid alpha conidia and filiform, curved, beta conidia, although depending on the age of the pycnidia, one or the other conidial type might be far more abundant than the other.

Spring infections are caused by conidia splashed by water from last year's dead shoots. Conidia are capable of infecting immature foliage during the growing season, resulting in yellowish spots after only 3–5 days. Diseased areas may grow to the point that the fungus invades xylem tissue and causes shoot dieback. Some juniper cultivars are more resistant to Phomopsis blight.

Kabatina juniperi R. Schneider & Arx occurs in North America and Europe on *Juniperus* and *Chamaecyparis* spp. and also affects *Thuja* spp. It is associated with branch tip dieback, and the curled, brown branchlets are covered with numerous small, dark brown acervuli containing hyaline, ellipsoid conidia. A tip blight caused by *K. thujae* R. Schneider & Arx affects *Chamaecyparis* spp. in plantations in the Pacific Northwest (Plate 109). The severity of both *Phomopsis*- and *Kabatina*-caused blights is exacerbated by high humidity, especially in dense ornamental plantings watered at night by overhead sprinklers. Incidence of cankering is also increased by wounding and host stress.

Other Douglas-Fir Cankers

Phomopsis lokoyae Hahn (Coelomycetes) causes dieback of new growth of Douglas-fir in western North America, usually after drought or stress. The teleomorph, *Diaporthe lokoyae* Funk (Ascomycetes, Diaporthales), occurs in old, dead tissue. A second rare species, *P. porteri* Funk, causes similar symptoms but differs in conidial morphology.

Coccomyces pseudotsugae Funk (Ascomycetes, Rhytismatales) is associated with leader dieback of young saplings and produces black, crustose apothecia with brown hymenia on dead tissues. *Dermea pseudotsugae* Funk (Ascomycetes, Leotiales) causes bark necrosis, cankering, and topkill of young plantation trees and nursery seedlings predisposed by frost and drought. Cankers on live trees have a reddish margin, which fades after the tree dies. Small, black apothecia appear on cankered tissue a year after its death (Plate 110). *Durandiella pseudotsugae* Funk (Ascomycetes, Leotiales) produces small, strikingly unusual cankers: cork proliferation under the cankered area causes coin-sized disks of bark to be pushed up, and small, black apothecia fruit in the centers.

Other Hemlock Cankers

Discocainia treleasei (Sacc. in Sacc., Peck, & Trel.) J. Reid & Funk (Ascomycetes, Rhytismatales) was first described as the cause of Atropellis canker of *Tsuga* and *Picea* spp. (*Atropellis treleasei* (Sacc.) Zeller & Goodd.). It causes perennial stem and branch cankers often accompanied by fusiform swelling and may colonize the entire stem after death of the tree, invading the wood and causing limited decay. *D. treleasei* produces apothecia that are light yellow when open and approximately 3 mm in diameter. The fungus is spread only by ascospores.

Botryosphaeria tsugae Funk (Ascomycetes, Dothideales) in western Canada is associated with a serious dieback of *Tsuga heterophylla* predisposed by drought. *Therrya tsugae* Funk (Ascomycetes, Rhytismatales) is associated with small, annual cankers and twig blights of suppressed *T. heterophylla* in western Canada.

Other Larch Cankers

Gremmeniella laricina (Ettlinger) Petrini et al (Acomycetes, Helotiales), previously also placed in *Encoeliopsis, Scleroderris,* and *Ascocalyx,* is associated with shoot blight and stem canker of *Larix* spp. in North America. One- to 2-year-old branches of young trees are affected. The small, black apothecia fruit in bark fissures on dead tissues and are preceded by a black pycnidial (*Brunchorstia* sp.) stage.

Other Pine Cankers

Cenangium ferruginosum Fr.:Fr. (Ascomycetes, Leotiales) occurs throughout the northern hemisphere on *Pinus* spp. predisposed by winter or insect damage. Cankers often affect only the lower, shade-weakened branches, which brown from the bases towards the tips and then lose their needles. Dead bark of cankered branches and stems bears numerous dark brown apothecia (2–5 mm in diameter) with yellowish hymenia. The apothecia roll inward when dry and have a mealy exterior texture (Plate 111).

Crumenulopsis sororia (P. Karst.) Groves (Ascomycetes, Leotiales) is associated with cankering and resinosis of *Pinus* spp. in Europe. Records of its occurrence in North America are questionable. It is reported as a damaging, and possibly primary, pathogen in plantations of western Canadian provenances of *P. contorta* in Scotland and Sweden.

Ramichloridium pini de Hoog & Rahman (Hyphomycetes) causes a dieback of *P. contorta* in England and Scotland (Plate 112). *Tympanis hyphopodia* Nyl. (Ascomycetes, Leotiales) produces clusters of small, black apothecia on living stems of *P. contorta*.

Other Spruce Cankers

Botryosphaeria piceae Funk causes perennial bark swellings of spruce of all ages in northwestern North America. The swellings may eventually girdle branches and stems. The disease is spread via ascospores, which are ejected from the small, black perithecia covering the cankers (Plate 113). *Discocainia treleasei* also infects spruce and is discussed in the section Other Hemlock Cankers.

1. Phloem lesion on Port Orford cedar caused by *Phytophthora lateralis*. (Courtesy E. Hansen)

2. Washing a log truck before it enters a "clean" area to prevent spread of *Phytophthora* sp. (Courtesy E. Hansen)

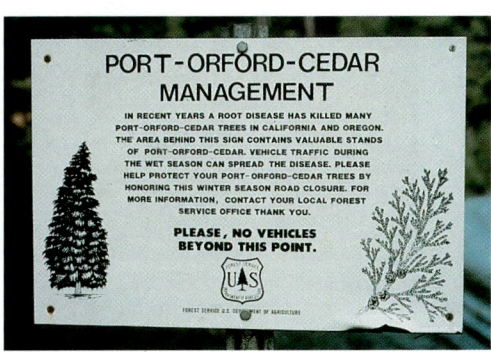
3. Road closure to stop spread of *Phytophthora lateralis*. (Courtesy E. Hansen)

4. Small, flat, young fruit bodies of *Rhizina undulata* with cream-colored rims and deep chestnut brown centers. (Courtesy K. Von Weissenburg)

5. Apothecia of *Rhizina undulata*. Note the dead *Pinus sylvestris* seedling (right) and key for scale. (Courtesy K. Von Weissenburg)

6. Mature, sporulating fruit body of *Rhizina undulata* approximately 1 month old and 7 cm in diameter. Note the strands of hyphae connecting the crust to the substrate. (Courtesy K. Von Weissenburg)

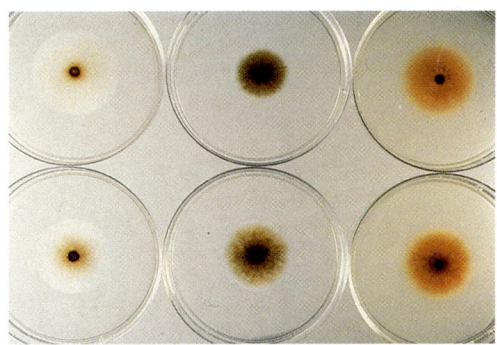
7. *Leptographium wageneri* var. *pseudotsugae* (left), var. *wageneri* (center), and var. *ponderosum* (right). (Courtesy T. Harrington)

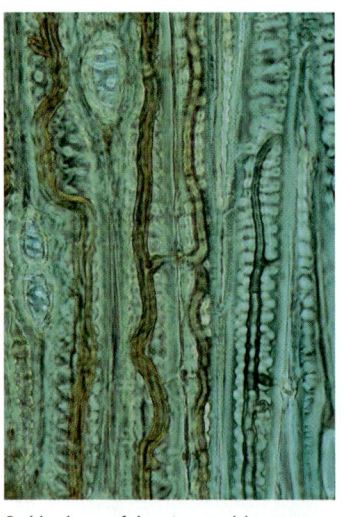
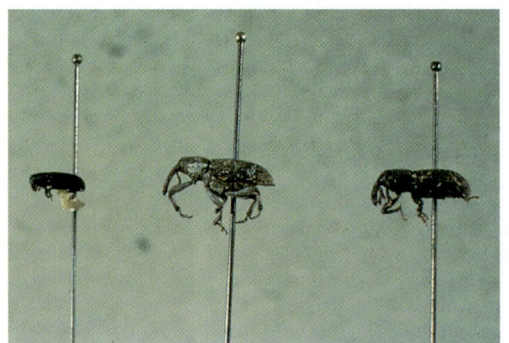

8. Black-stain root disease, caused by *Leptographium wageneri,* in the outer xylem of Douglas-fir. (Courtesy E. Hansen)

9. Hyphae of *Leptographium wageneri* in tracheids of Douglas-fir. (Courtesy E. Hansen)

10. Insect vectors of black-stain root disease: *Hylastes nigrinus* (left), *Pissodes fasciatus* (center), and *Steremnius carinatus* (right). (Courtesy E. Hansen)

11. Stain and resinosis associated with infection by *Leptographium procerum*. (Courtesy K. Lewis)

12. *Heterobasidion annosum* butt rot in Norway spruce in Sweden. (Courtesy K. Korhonen)

13. Annosum root disease center in Scot's pine, caused by *Heterobasidion annosum,* P type, on former agricultural land in Belarus. (Courtesy K. Korhonen)

14. Fruit bodies of *Heterobasidion annosum*, S type, on a Norway spruce stump in Finland. (Courtesy K. Korhonen)

15. Fruit body of *Heterobasidion annosum*, P type, at the base of a birch tree in Estonia. (Courtesy K. Korhonen)

16. Fruit body of *Heterobasidion annosum,* F type, on an *Abies alba* stump in Italy. (Courtesy K. Korhonen)

17. Resinosis caused by *Armillaria ostoyae* infection of Douglas-fir in British Columbia. (Courtesy E. Hansen)

18. Mycelial fan of *Armillaria ostoyae* on ponderosa pine in Washington State. (Courtesy E. Hansen)

19. *Armillaria ostoyae* basidiomes from Sitka spruce in England. (Courtesy E. Hansen)

20. Rootball symptom caused by laminated root rot of Douglas-fir. (Courtesy W. G. Thies)

21. Crown symptoms caused by laminated root rot of Douglas-fir. (Courtesy W. G. Thies)

22. Light-colored ectotrophic mycelium of *Phellinus weirii* enveloping roots of Douglas-fir. (Courtesy E. Hansen)

23. Collapse of a 20-year-old Sitka spruce decayed by *Phaeolus schweinitzii* in England. (Courtesy D. Barrett)

24. Fruit body of *Phaeolus schweinitzii* on a Sitka spruce in England. (Courtesy D. Barrett)

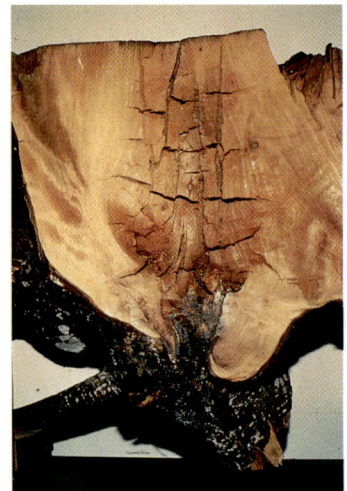
25. Red-brown rot that has invaded the butt of a Sitka spruce via the taproot in England. (Courtesy D. Barrett)

26. Tomentosus root rot of spruce in British Columbia. (Courtesy K. Lewis)

27. Reddish brown, incipient decay caused by *Inonotus tomentosus* in a spruce root in British Columbia. (Courtesy K. Lewis)

28. Collecting basidiospores from a sporocarp of *Inonotus tomentosus* in British Columbia. (Courtesy K. Lewis)

29. Young hoop pine killed by *Phellinus noxius* in Queensland. (Courtesy I. Hood)

30. Rain forest tree (*Argyrodendron trifoliatum*) colonized by *Phellinus noxius* in Queensland. Note external "stocking" of brown mycelium. (Courtesy I. Hood)

31. Brown root rot of hoop pine in Queensland. (Courtesy E. Hansen)

32. Decay of western red cedar caused by *Oligoporus sericeomollis* in British Columbia. (Courtesy E. Hansen)

33. Light-colored, resupinate fruit bodies of *Rigidoporus vinctus* on hoop pine in Queensland. (Courtesy E. Hansen)

34. *Sparassis radicata* sporocarp with a thick stalk growing from a decayed Douglas-fir root in Oregon. (Courtesy E. Hansen)

35. Blue stain associated with mountain pine beetle attack of lodgepole pine in Utah. (Courtesy E. Hansen)

36. The Oregon fir sawyer, *Monochamus scutellatus oregonensis*, a potential vector of the pinewood nematode in North America. (Courtesy J. Sutherland)

37. *Pinus massoniana* killed by the pinewood nematode in China. (Courtesy J. Sutherland)

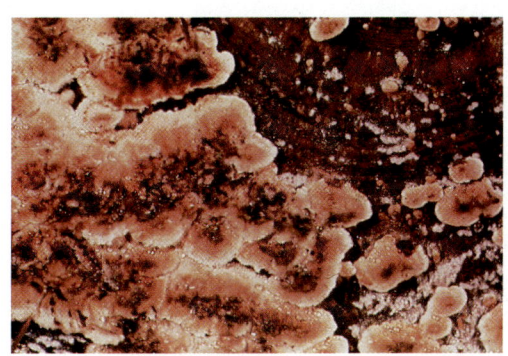

38. Resupinate fruit bodies of *Stereum sanguinolentum* on a log end of Norway spruce in Norway. (Courtesy S. Woodward)

39. Stain associated with incipient decay of wounded Norway spruce caused by *Stereum sanguinolentum* in Norway. (Courtesy S. Woodward)

40. *Fomitopsis cajanderi* fruiting on wounded Douglas-fir in Oregon. (Courtesy E. Hansen)

COLOR PLATES

41. Douglas-fir in Oregon with tops broken during an ice storm and now infected with yellow brown top rot. (Courtesy L. F. Roth)

42. *Fomitopsis pinicola* sporulating on wounded Sitka spruce in Alaska. (Courtesy E. Hansen)

43. Applanate conk of *Fomitopsis pinicola* in Alaska. (Courtesy P. Hennon)

44. Ungulate conk of *Fomitopsis pinicola* in Alaska. (Courtesy P. Hennon)

45. Cord formation by *Resinicium bicolor* on stump of Sitka spruce in Scotland. (Courtesy S. Woodward)

46. Fir hydnum sporulating on grand fir in Oregon. (Courtesy E. Hansen)

47. Red ring rot of Douglas-fir in Oregon, caused by *Phellinus pini*. (Courtesy E. Hansen)

48. *Phellinus pini* on Scot's pine in Sweden. (Courtesy E. Hansen)

49. *Phellinus pini* on spruce in British Columbia. (Courtesy K. Lewis)

50. Clustered fruit bodies of *Phellinus cancriformans* on *Abies concolor* in Oregon. (Courtesy E. Hansen)

51. White rot caused by *Echinodontium tinctorium* in grand fir. (Courtesy A. D. Wilson)

52. Red brown stringy rot, caused by *Echinodontium tinctorium,* in mountain hemlock. (Courtesy A. D. Wilson)

53. *Echinodontium tinctorium* on western hemlock. (Courtesy A. D. Wilson)

54. Sectioned conk of *Echinodontium tinctorium* on young grand fir in Oregon. (Courtesy E. Hansen)

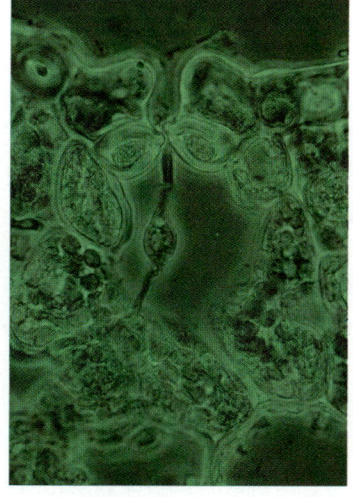

55. Aecia of *Cronartium ribicola* on western white pine in Oregon. (Courtesy E. Hansen)

56. Basidiospore germ tube of *Cronartium ribicola* penetrating stoma of white pine needle (cross section). (Courtesy R. Patton)

57. Dense cover of telia of *Cronartium ribicola* on *Ribes nigrum* resulting from artificial inoculation. (Courtesy E. Hansen)

COLOR PLATES

58. White pine blister rust on western white pine in Oregon. (Courtesy E. Hansen)

59. Stem canker caused by fusiform rust. (Courtesy H. Powers)

60. Stem and branch canker caused by fusiform rust. (Courtesy H. Powers)

61. Pycnial stage of eastern gall rust on jack pine. (Courtesy E. Hansen)

62. Western gall rust on lodgepole pine in Alberta, Canada. (Courtesy Y. Hiratsuka)

63. Stem swellings on jack pine in a nursery resulting from western gall rust. (Courtesy Y. Hiratsuka)

64. Broken stem of lodgepole pine caused by infection of the bole by western gall rust. (Courtesy Y. Hiratsuka)

65. Resin top disease of Scot's pine in England caused by *Peridermium pini*. (Courtesy E. Hansen)

66. Resin-soaked wood behind a *Peridermium pini* canker on Scot's pine in England. (Courtesy E. Hansen)

67. *Peridermium pini* aecia on a limb of Corsican pine in Scotland. (Courtesy E. Hansen)

68. Comandra blister rust on lodgepole pine in Wyoming. (Courtesy B. Geils)

69. Comandra blister rust canker on lodgepole pine in Colorado chewed by squirrels. (Courtesy B. Geils)

70. Telia of *Cronartium comandrae* on *Comandra* spp. (Courtesy Y. Hiratsuka)

71. Stalactiform blister rust on lodgepole pine in Alberta, Canada. (Courtesy Y. Hiratsuka)

72. Sweetfern blister rust on jack pine in Ontario, Canada. (Courtesy E. Hansen)

73. Aecia of limb rust on ponderosa pine in Colorado. (Courtesy B. Geils)

74. Limb rust of ponderosa pine in Arizona. (Courtesy B. Geils)

75. *Gymnosporangium* sp. on *Juniperus standleyi* in Guatemala. (Courtesy B. Geils)

COLOR PLATES

76. *Gymnosporangium clavariiforme* on *Juniperus communis.* (Courtesy B. Geils)

77. *Gymnosporangium kernianum* on *Juniperus* sp. (Courtesy B. Geils)

78. *Arceuthobium campylopodum* on *Pinus ponderosa* in California. (Courtesy R. Scharpf)

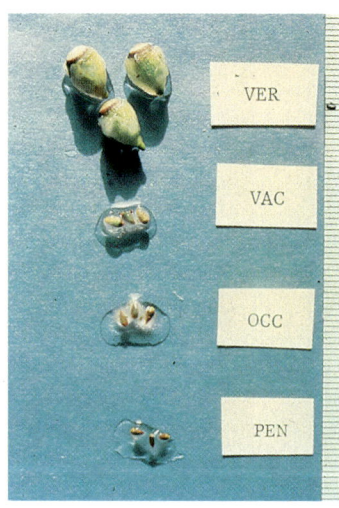

79. Small, male shoots of *Arceuthobium douglasii* on *Pseudotsuga menziesii.* (Courtesy R. Scharpf)

80. Dwarf mistletoe seeds (top to bottom): *Arceuthobiium verticilliflorum, A. vaginatum, A. occidentale,* and *A. pendens.* (Courtesy B. Geils)

81. *Arceuthobium globosum,* one of the largest dwarf mistletoes, emerging from the bole of *Pinus montezumae* in Mexico. (Courtesy B. Geils)

82. A female fruiting plant of *Arceuthobium minutissimum,* the smallest dwarf mistletoe, on *Pinus wallichiana* in Bhutan. (Courtesy R. Scharpf)

83. Witches'-brooms caused by Douglas-fir dwarf mistletoe in Oregon resulting in dieback of the upper crowns. (Courtesy E. Hansen)

84. Black, mycelial stromata of *Caliciopsis arceuthobii* (Ascomycetes) on aborted fruits of Douglas-fir dwarf mistletoe. (Courtesy B. Geils)

85. *Phoradendron juniperinum* on *Juniperus occidentalis* in California. (Courtesy R. Scharpf)

86. *Viscum abietis* on *Abies alba* in Germany. (Courtesy W. Forstreuter)

87. *Viscum laxum* on *Pinus sylvestris* in Germany. (Courtesy W. Forstreuter)

88. Red-flowered *Psittacanthus* sp. parasitizing *Pinus oocarpa* in Mexico. (Courtesy R. Scharpf)

89. Larch canker on European larch. (Courtesy C. Delatour)

90. Larch canker on European larch in France centered on a short shoot. (Courtesy E. Hansen)

91. Larch canker (*Lachnellula willkommii*) in New Brunswick. (Courtesy J. Sutherland)

92. Damage in a red pine plantation caused by *Sphaeropsis* sp. (Courtesy G. Stanosz)

93. Resinous canker caused by *Sphaeropsis sapinea*. (Courtesy G. Stanosz)

COLOR PLATES

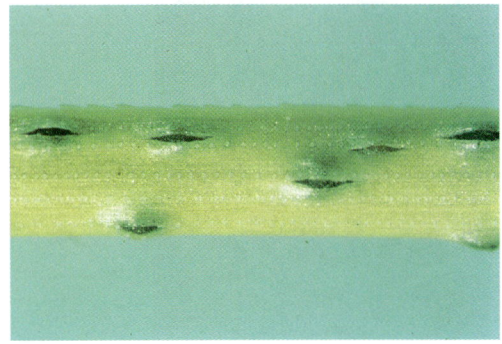

94. Pycnidia of *Sphaeropsis sapinea* on a needle. (Courtesy G. Stanosz)

95. Pycnidia of *Sphaeropsis sapinea* on cone scales. (Courtesy G. Stanosz)

96. Shoot dieback of *Pinus sylvestris* in Sweden caused by *Gremmeniella abietina*. (Courtesy J. Stenlid)

97. Dieback of *Picea abies* caused by *Gremmeniella abietina*. (Courtesy J. Stenlid)

98. Pitch canker on southern pine in Florida. (Courtesy E. Hansen)

99. Branch dieback of Monterey pine in California caused by pitch canker. (Courtesy L. D. Dwinell)

100. Pitch canker on the bole of eastern white pine. (Courtesy L. D. Dwinell)

101. Resin-soaked wood associated with pitch canker on slash pine. (Courtesy L. D. Dwinell)

102. Cross section of Atropellis canker on lodgepole pine. (Used by permission from the Canadian Forest Service, Pacific Forestry Centre)

COLOR PLATES

103. Apothecia of *Atropellis piniphila*. (Used by permission from the Canadian Forest Service, Pacific Forestry Centre)

104. Cytospora canker on Douglas-fir. (Used by permission from the Canadian Forest Service, Pacific Forestry Centre)

105. Lachnellula canker on Douglas-fir injured by sapsuckers. (Courtesy E. Hansen)

106. Seiridium canker disease of cypress in England. (Courtesy E. Hansen)

107. Sirococcus dieback of hemlock. (Used by permission from the Canadian Forest Service, Pacific Forestry Centre)

108. Pycnidia of *Sirococcus strobilinus* on hemlock. (Used by permission from the Canadian Forest Service, Pacific Forestry Centre)

109. Dieback of *Chamaecyparis nootkatensis* caused by *Kabatina thujae*. (Used by permission from the Canadian Forest Service, Pacific Forestry Centre)

110. Dermea canker on Douglas-fir. Note the contrast between the sunken brown lesion and healthy green bark. (Used by permission from the Canadian Forest Service, Pacific Forestry Centre)

111. Apothecia of *Cenangium ferruginosum* on pine. (Used by permission from the Canadian Forest Service, Pacific Forestry Centre)

112. Bud and shoot necrosis associated with Ramichloridium dieback of lodgepole pine in England. (Used by permission from the Canadian Forest Service, Pacific Forestry Centre)

113. Cankers caused by *Botryosphaeria piceae* on *Picea sitchensis*. (Used by permission from the Canadian Forest Service, Pacific Forestry Centre)

114. Cork bark on *Abies lasiocarpa*, caused by *Dermea rhytidiformans*. (Used by permission from the Canadian Forest Service, Pacific Forestry Centre)

115. Fruit bodies of *Sclerophoma pithyophila* on a Douglas-fir twig. (Used by permission from the Canadian Forest Service, Pacific Forestry Centre)

116. Grovesiella canker on noble fir Christmas tree in Oregon. (Courtesy E. Hansen)

117. Spore tendrils of *Phomopsis occulta*. (Used by permission from the Canadian Forest Service, Pacific Forestry Centre)

118. Cankers caused by *Sclerophoma* sp. (above), *Phomopsis lokoyae* (left), and *Xenomeris* sp. (right) on *Thuja plicata*. (Used by permission from the Canadian Forest Service, Pacific Forestry Centre)

119. *Sirococcus conigenus* sporulating on spruce cone scales. (Courtesy J. Sutherland)

120. Apothecium of *Calocypha fulgens* on the forest floor. (Courtesy J. Sutherland)

121. Mycelium of *Calocypha fulgens* on stratified seed. (Courtesy J. Sutherland)

122. Needle rust caused by *Chrysomyxa ledicola* in Alaska. (Courtesy E. Hansen)

123. Needle rust caused by *Pucciniastrum epilobii* in British Columbia. (Courtesy K. Lewis)

124. *Coleosporium* species on *Pinus*. (Courtesy E. Hansen)

125. Spruce broom rust caused by *Chrysomyxa arctostaphyli* on *Picea engelmannii* in Utah. (Courtesy E. Hansen)

126. Witches'-brooms caused by the fir broom rust, *Melampsorella caryophyllacearum,* on *Abies alba* in Germany. (Courtesy E. Hansen)

127. Fir broom rust on *Abies lasiocarpa* in Wyoming. (Courtesy E. Hansen)

128. Rhabdocline needle cast on Douglas-fir in Oregon. Note the susceptible tree among the healthy ones. (Courtesy E. Hansen)

129. Rhabdocline needle cast on Douglas-fir. (Courtesy E. Hansen)

COLOR PLATES

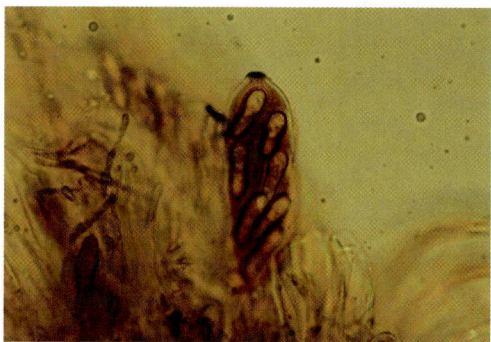

130. Ascus tip of *Rhabdocline weirii* stained blue in iodine. (Courtesy J. Stone)

131. *Rhabdocline weirii* on Douglas-fir. Ascomata form on the lower needle surface and conidiomata of the anamorph, *Rhabdogloeum pseudotsugae,* on the upper needle surface. (Courtesy J. Stone)

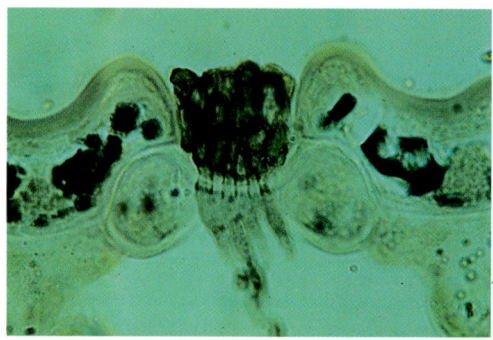

132. Developing pseudothecium of *Phaeocryptopus gaeumannii* occluding stoma of Douglas-fir. (Courtesy J. Stone)

133. Pseudothecia of *Phaeocryptopus gaeumannii* on the lower surface of a Douglas-fir needle in Oregon. (Courtesy J. Stone)

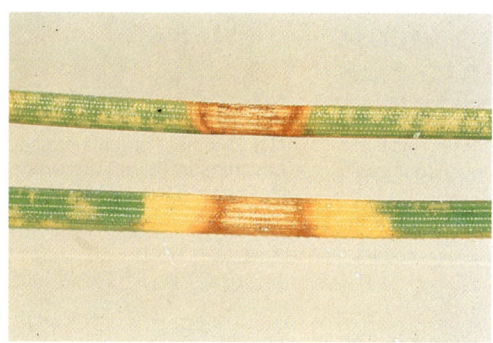

134. Brown spot needle blight. (Courtesy R. Patton)

135. Brown spot needle blight on longleaf pine in the grass stage. (Courtesy R. Patton)

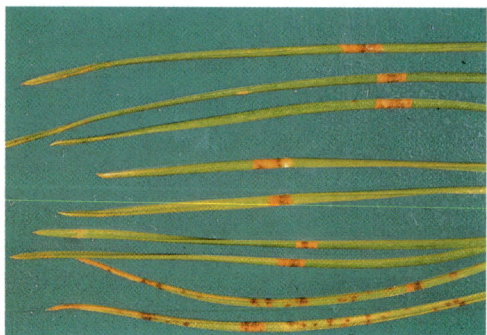

136. Dothistroma needle blight on Austrian pine in Wisconsin. (Courtesy R. Patton)

137. Dothistroma needle blight on western white pine in British Columbia. (Used by permission from the Canadian Forest Service, Pacific Forestry Centre)

138. Defoliation caused by Dothistroma needle blight on the lower crown of radiata pine planted beyond its native range in northern California. (Courtesy E. Hansen)

139. *Ploioderma lethale* on pine in Florida. (Courtesy J. Stone)

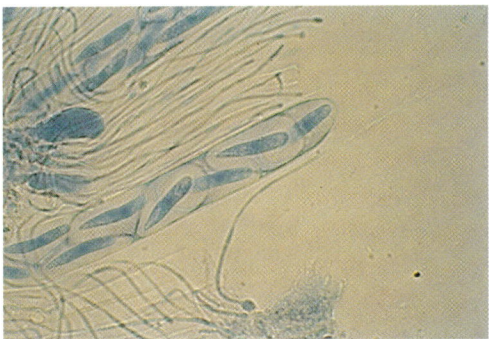

140. Asci of *Ploioderma lethale* with ascospores on pine in Florida. (Courtesy J. Stone)

141. Systemic infection of ponderosa pine by *Elytroderma deformans* in Oregon. (Courtesy L. Roth)

142. *Elytroderma deformans* hysterothecia on ponderosa pine. (Courtesy L. Roth)

143. Damage to lodgepole pine caused by *Lophodermella concolor* in British Columbia. (Used by permission from the Canadian Forest Service, Pacific Forestry Centre)

144. Damage to Corsican pine caused by *Lophodermella sulcigena* in Scotland. (Courtesy E. Hansen)

145. Resistant ponderosa pine in an off-site plantation damaged by Lophodermella needle cast in Oregon. (Courtesy E. Hansen)

146. *Lophodermella morbida* hysterothecia on ponderosa pine in Oregon. (Courtesy E. Hansen)

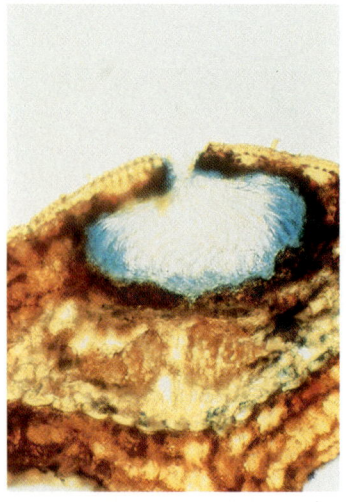

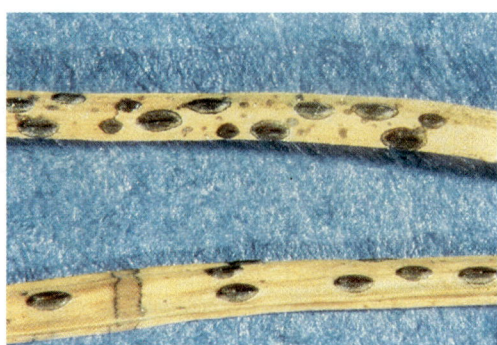

147. Sectioned ascocarp of *Lophodermella cerina* on pine in Florida. (Courtesy J. Stone)

148. Three *Lophodermium* species on pine (left to right): unknown, *L. pinastri,* and *L. seditiosum.* (Courtesy G. Chastagner)

149. *Lophodermium nitens* on lodgepole pine in Oregon. Note the small pycnidia of the conidial stage, *Leptostroma nitens,* and zone lines separating the different infections. (Courtesy J. Stone)

150. Lirula needle blight on concolor fir in Arizona. (Courtesy E. Hansen)

151. *Lirula macrospora* on Sitka spruce in British Columbia. (Used by permission from the Canadian Forest Service, Pacific Forestry Centre)

152. Bud blight of Japanese spruce caused by *Gemmamyces piceae* in Wales. (Courtesy E. Hansen)

153. Pycnidia of *Rhizosphaera kalkhoffii* on a pine needle in Florida. (Courtesy J. Stone)

154. Larch needle cast, caused by *Meria laricis*, on western larch in Oregon. (Courtesy E. Hansen)

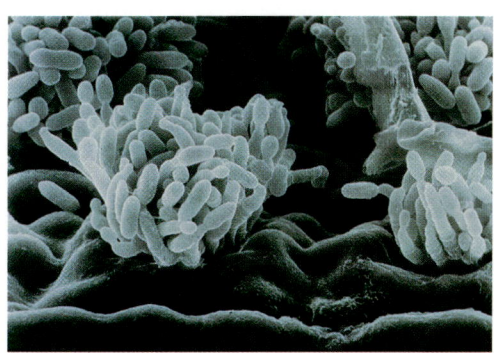

155. Conidiomata of *Meria laricis* emerging from western larch stomata. (Courtesy J. Stone)

156. Keithia blight of western red cedar, caused by *Didymascella thujina*, in British Columbia. (Used by permission from the Canadian Forest Service, Pacific Forestry Centre)

157. Black mildew (*Maurodothina farrae*) on old needles of grand fir in British Columbia. (Used by permission from the Canadian Forest Service, Pacific Forestry Centre)

158. Herpotrichia brown felt blight on Englemann's spruce in Wyoming. (Courtesy E. Hansen)

159. Extensive yellow-cedar decline at Slocum Arm, Chichagoff Island, in southeastern Alaska. (Courtesy E. Hansen)

160. Yellow-cedar decline in Alaska centered around a muskeg. (Courtesy E. Hansen)

161. Austrocedrus decline of *Austrocedrus chilensis* in Argentina. (Courtesy G. Filip)

COLOR PLATES

162. Austrocedrus decline in Argentina. Note the band of dead and dying *Austrocedrus* sp. between the *Nothofagus* forest below and the mixed forest above. (Courtesy G. Filip)

163. Ice layer from irrigation applied for frost protection on Douglas-fir seedlings in Oregon. (Courtesy E. Hansen)

164. Fumigation of a bare-root nursery in Oregon with methyl bromide and chloropicrin. (Courtesy E. Hansen)

165. Damping-off of ponderosa pine seedlings in Oregon. (Courtesy E. Hansen)

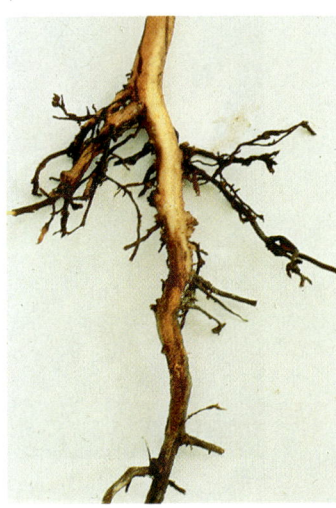

166. Phytophthora root rot of Douglas-fir. (Courtesy E. Hansen)

167. Symptoms of poorly drained nursery soil and Phytophthora root rot in Oregon. (Courtesy E. Hansen)

168. Swiss needle cast on sheared Douglas-fir Christmas tree. (Courtesy E. Hansen)

169. Lecanosticta needle blight (*Mycosphaerella dearnessii*) on Scot's pine Christmas trees in Wisconsin. (Courtesy R. Patton)

TABLE 8. Other Canker Fungi

Pathogen	Hosts
Colpoma crispum	*Tsuga, Pseudotsuga,* and *Pinus* spp.
Dermea balsamea	*Abies, Tsuga,* and *Pinus* spp.
D. tetrasperma	Conifers
Gelatinosporium spp.	Conifers
Griphosphaeria corticola	*Pseudotsuga* spp.
Grovesiella spp.	*Abies* spp. (Plate 116)
Leucostoma kunzei	Conifers
Nectria fuckeliana	*Abies* and *Pinus* spp.
Nipterella tsugae	*Tsuga* spp.
Nitschkia molnarii	*Pseudotsuga* and *Tsuga* spp.
Pestalopezia tsugae	*Tsuga* spp.
Pestalotiopsis funerea	Cupressaceae
Pezicula livida	*Pseudotsuga* and *Tsuga* spp.
Phomopsis occulta	*Larix* and *Picea* spp. (Plate 117)
Phragmopora pithya	Conifers
Phyllostictina hysterella	*Taxus* spp.
Sageria tsugae	*Tsuga* spp.
Sirodothis spp.	Conifers
Therrya pseudotsugae	*Pseudotsuga* spp.
T. tsugae	*Tsuga* spp.
Xenomeris abietis	*Pseudotsuga* and *Tsuga* spp. (Plate 118)

Other True Fir Cankers

Dermea rhytidiformans Funk & Kuijt (Ascomycetes, Leotiales) is associated with "cork-bark" disease of *Abies lasiocarpa* and causes large, grossly swollen corky ridges of bark proliferation. The small, dark brown apothecia are found in the corky ridges (Plate 114).

Nectria fuckeliana C. Booth (Ascomycetes, Hypocreales) is associated with cankers and subsequent death of *Abies concolor* in northwestern North America and Europe. The presence of small, reddish perithecia in fissures of dead bark is a diagnostic feature. The conidial state, *Cylindrocarpon cylindrioides* Wollenweb. (Hyphomycetes), is frequently isolated from cankered tissue.

Phellinus cancriformans (M. Lars. et al.) M. Lars. & Lomb. (Basidiomycetes, Hymenochaetales) (Plate 50), invades phloem and causes cankers on *Abies concolor* in Oregon and California, causing losses in merchantable product that exceed 13% of the standing volume. The more common *P. pini* var. *pini* is associated with stem and butt decay of conifers with no accompanying cankers.

Common Opportunistic Canker Fungi

Many other species of weakly pathogenic fungi are commonly associated with bark damaged by drought, frost, or sunscald or other forms of physical damage. The presence of these fungi may in fact indicate adverse growing conditions, because on healthy vigorous trees, the worst symptoms encountered are small, latent branch cankers (Table 8).

Sclerophoma species (Coelomycetes), primarily *S. semenospora* Funk and *S. pithyophila* (Corda) Höhn. (Plate 115), are very commonly associated with postdamage conifer tip blights, dieback, and diffuse cankers. The small, brown to black pycnidia have no ostioles and are produced in dead tissues and needles. The unicellular, ellipsoid conidia are released as environmental factors gradually erode the pycnial structure.

Selected References

de Hoog, G. S., Rahman, M. A., and Boekhout, T. 1983. *Ramichloridium, Veronaea* and *Stenella*: Generic delimitation, new combinations and two new species. Trans. Br. Mycol. Soc. 81:485-490.

Funk, A. 1981. Parasitic microfungi of western trees. Can. For. Serv. Pac. For. Res. Cent. Inf. Rep. BC-X-222.

Hahn, G. G. 1943. Taxonomy, distribution, and pathology of *Phomopsis occulta* and *P. juniperovora*. Mycologia 35:112-129.

Hayes, A. J. 1973. The occurrence of *Crumenula sororia* Karst. on lodgepole pine in the United Kingdom. For. 46:125-138.

Karlman, M. 1976. Contortan och svamparna. Skogen 63:628-630.

Larsen, M. J., Lombard, F. F., and Aho, P. E. 1979. A new variety of *Phellinus pini* associated with cankers and decay in white firs in southwestern Oregon and northern California. Can. J. For. Res. 9:31-38.

Nag Raj, T. R. 1993. Coelomycetous Anamorphs with Appendage-Bearing Conidia. Mycologue Publications, Waterloo, Ontario, Canada.

Oulette, G. B. 1972. *Nectria macrospora* (Wr.) Oulette sp. nov. (= *N. fuckeliana* var. *macrospora*): Strains, physiology and pathogenicity, and comparison with *N. fuckeliana* var. *fuckeliana*. Eur. J. For. Pathol. 2:172-181.

Peterson, G. W. 1973. Infection of *Juniperus virginiana* and *J. scopulorum* by *Phomopsis juniperovora*. Phytopathology 63:246-251.

Reich, R. W., and van der Kamp, B. J. 1993. Frost, canker, and dieback of Douglas fir in the central interior of British Columbia. Can. J. For. Res. 23:373-379.

Schoeneweiss, D. F. 1969. Susceptibility of evergreen hosts to the juniper blight fungus, *Phomopsis juniperovora*, under epidemic conditions. J. Am. Soc. Hortic. Sci. 94:609-611.

Schultz, M. E., and Parmeter, J. R., Jr. 1990. A canker disease of *Abies concolor* caused by *Nectria fuckeliana*. Plant Dis. 74:178-180.

Smerlis, E. 1973. Pathogenicity tests of some discomycetes occurring on conifers. Can. J. For. Res. 3:7-16.

Smith, R. S., Jr., McCain, A. H., Srago, M., Krohn, R. F., and Perry, D. 1972. Control of Sirococcus tip blight of Jeffrey pine seedlings. Plant Dis. Rep. 56:241-242.

Sutherland, J. R., Shrimpton, G. M., and Sturrock, R. N. 1989. Diseases and insects in British Columbia forest seedling nurseries. For. Can. For. Res. Dev. Agreement (FRDA) Rep. 065.

Uotila, A. 1990. Infection of pruning wounds in Scots pine by *Phacidium coniferarum* and selection of pruning season. Acta For. Fenn. 215.

Wall, R. E., and Magasi, L. P. 1976. Environmental factors affecting Sirococcus shoot blight of black spruce. Can. J. For. Res. 6:448-452.

(Prepared by B. Callan)

Cone and Seed Diseases

Spruce Cone Rust

Other name: inland spruce cone rust
Causal organism: *Chrysomyxa pirolata* G. Wint. in Rabenh.
Hosts: Spruce (*Picea* spp.) and several species of wintergreen (*Pyrola* spp. and single-delight, *Moneses uniflora*)
Distribution: north temperate zone of North America and Eurasia where both spruce and alternate hosts occur

Symptoms and Diagnosis

Diseased cones become brownish earlier in the summer than healthy cones, open prematurely, and release yellow orange aeciospores.

Disease Cycle

Spermagonia and aecia are produced on spruce cone scales, while uredinia and telia are produced on the alternate hosts. Basidiospores produced on germinating teliospores on alternate hosts infect female cones in early summer, coinciding with pollination time. Aeciospores produced on cones do not spread the disease to other cones but infect alternate hosts.

Effects on Tree and Forest, Ecological Role

Spruce cone rust does not affect the health of the tree directly, but heavy infections within high-value seed orchards are economically significant. Entire cones are usually infected, and usually no seeds form in diseased cones. Partially infected cones may produce some seeds, but they are significantly lighter and germination is reduced.

In high-value seed orchards, management strategies that include eradication of alternate hosts in and around orchards and protective fungicide sprays during infection periods should be effective.

A closely related species, *C. monesis* Ziller, occurs mainly on Sitka spruce (*Picea sitchensis*) and is found on the Queen Charlotte Islands of British Columbia (Canada) and along the western coast of North America from Oregon to Alaska.

Selected References

Hunt, R. S. 1978. Spruce cone rusts in British Columbia. Can. For. Serv. Pac. For. Res. Cent. Pest Leafl. 50.

Sutherland, J. R. 1981. Effects of inland spruce cone rust, *Chrysomyxa pirolata* Wint., on seed yield, weight and germination. Can. For. Serv. Res. Notes 1:2, 8-9.

Sutherland, J. R., Hopkinson, S. J, and Farris, S. H. 1984. Inland spruce cone rust, *Chrysomyxa pirolata*, in *Pyrola asarifolia* and cones of *Picea glauca*, and morphology of the spore stages. Can. J. Bot. 62:2441-2447.

Sutherland, J. R., Miller, T., and Salinas-Quinard, R., eds. 1987. Cone and seed diseases of North American conifers. N. Am. For. Comm. Publ. 1.

Ziller, W. G. 1974. The tree rusts of western Canada. Can. For. Serv. Publ. 1329.

(Prepared by B. Geils)

Southwestern Pine Cone Rust

Causal organism: *Cronartium conigenum* Hedgc. & N. Hunt

Hosts: Chihuahua pine (*Pinus leiophylla* var. *chihuahuana*) and at least 15 other species of hard pines in Mexico

Distribution: Mexico and Central America and southern Arizona in the United States

Symptoms and Diagnosis

Infected cones are two to four times and occasionally up to 10 times larger than normal cones. The cones are fleshy, and the surface is only slightly differentiated into scales. Aecia are produced over the entire surface area, making the infected cones bright orange. Stem infections also occur and produce lobed galls resulting from hypertrophied phloem tissues. These galls are distinctly different from those caused by western gall rust, which result from hypertrophy of xylem tissues.

Disease Cycle

Cronartium conigenum is a heteroecious rust that alternates between several species of pine and several species of oak (*Quercus* spp.). Spermagonia and aecia are produced on spruce cone scales, and uredinia and telia are produced on the alternate hosts.

Effects on Tree and Forest

Southwestern pine cone rust is the most damaging tree rust on pine in Mexico and Central America. It halts seed production, thus significantly reducing regeneration.

Selected Reference

Sutherland, J. R., Miller, T., and Salinas-Quinard, R., eds. 1987. Cone and seed diseases of North American conifers. N. Am. For. Comm. Publ. 1.

(Prepared by B. Geils)

Southern Pine Cone Rust

Causal organism: *Cronartium strobilinum* (Arth.) Hedgc. & Hahn

Hosts: slash pine (*Pinus elliottii* var. *elliottii*) and longleaf pine (*P. palustris*); alternate hosts, several species of evergreen oaks (*Quercus* spp.)

Distribution: southern United States along the Atlantic and in the Gulf Coastal Plain

Symptoms and Diagnosis

Cones infected by southern pine cone rust are three to four times larger than healthy cones. Infected cone scales are reddish orange during aeciospore production and reddish brown after spores are produced and infected cones fall from the tree.

Disease Cycle

Cronartium strobilinum is a host-alternating, heteroecious rust. Spermagonia and aecia are produced on pine cone scales, and uredinia and telia are produced on the alternate hosts.

Effects on Tree and Forest

Southern pine cone rust is important in seed orchards. In addition to the direct losses, infections increase the activities of coneworms and cause additional losses in production of viable seed.

Selected References

Sutherland, J. R., Miller, T., and Salinas-Quinard, R., eds. 1987. Cone and seed diseases of North American conifers. N. Am. For. Comm. Publ. 1.

Ziller, W. G. 1974. The tree rusts of western Canada. Can. For. Serv. Publ. 1329.

(Prepared by B. Geils)

Cone and Seed Fungi

Many saprophytic or weakly pathogenic fungi occur on conifer cones and seeds from the time of cone collection and processing, during seed extraction, and subsequently during seed testing and before nursery sowing. Besides the many species of *Alternaria, Aspergillus, Penicillium, Trichoderma,* and similar fungi, which are widespread on cones but whose role is largely undefined, Sutherland et al list *Schizophyllum commune* Fr.:Fr. on Douglas-fir cones and *Hericium abietis* (Weir ex Hubert) K. A. Harrison on mountain hemlock cones. Depending upon the fungi involved, diseased cones may have wet or dry mold and are commonly covered externally, and sometimes internally, with single-colored or multicolored mycelium and spores and have a moldy odor. Since the mold develops after cone maturation, the numbers and size of cones and seeds are not affected; however, seeds are often rotted, and excessive pitch may hinder seed extraction. Usually such fungi occur on cones that have been improperly handled during collection or subsequent storage, e.g., those that have been stored wet without proper ventilation or have been allowed to overheat.

Many of these fungi also occur on seeds. Prochazkova, who assayed more than 8,100 conifer seed lots, found species of *Penicillium* in more than 85% of the Norway spruce seed lots; *Aspergillus* in 73 and 39% of the Norway spruce and Scot's pine seed lots, respectively; *Trichothecium roseum* (Pers.:Fr.) Link in about 70% of the Scot's pine seed lots; and *Rhizopus stolonifer* (Ehrenb.:Fr.) Vuill. in 40% of the Scot's pine seed lots. Fungi that were especially prevalent on Douglas-fir seeds included species of *Alternaria, Botrytis, Penicillium, Trichoderma,* and *Trichothecium*. Such fungi frequently grow from seeds during germination tests or pathogen assays or on stratified (cold-treated) seeds before sowing. While these fungi may be pathogenic, they may also simply indicate poor-quality seed lots, such as those that contain unripe or improperly handled seeds. Perhaps they protect seeds from pathogens.

Pathogenic fungi (e.g., the pitch canker fungus, *Fusarium moniliforme* J. Sheld. var. *subglutinans* Wollenweb. & Reinking on slash pine; and *Lasiodiplodia theobromae* (Pat.) Griffon & Maubl., the perfect state of *Diplodia gossypina* Cooke, and *Diplodia pinea* (Desmaz.) J. Kickx on cones of *Pinus muricata, P. nigra, P. sabiniana, P. sylvestris,* and Douglas-fir) frequently damage cones before they are picked or later during storage. Rot from the pitch canker pathogen is often restricted to the distal portion of the cone, and *D. pinea* pycnidia are distinct on Austrian pine cones. To date, control of such fungi, many of which are seedborne and subsequently affect seedlings, depends mostly on treatment of infested seed lots with fungicides rather than on controlling the pathogens on cones.

While the list of seedborne, pathogenic fungi continues to grow, currently most is known about the seed or cold fungus, *Caloscypha fulgens* (Pers.) Boud.; *Sirococcus conigenus* (DC.) P. Cannon & Minter (syn. *S. strobilinus* G. Preuss), the cause of Sirococcus blight (Plate 119); and several species of *Fusarium*, including the aforementioned pitch canker fungus. The seed or cold fungus, so called because it occurs on seeds (especially of *Picea* and *Abies* spp.) and grows at low temperatures, is acquired when cones come in contact with forest duff where the fungus lives (Plate 120). So far, the pathogen has been reported as being seedborne only in North America, but since it occurs in Europe, seeds may also be affected there. Seed lots originating from squirrel caches most often contain the pathogen, which mummifies seed contents. Once a seed lot is infested, the fungus spreads during seed stratification (Plate 121) or after sowing under prolonged cool, wet conditions. Only dormant seeds are affected; thus, there is no threat to seedlings. Sometimes a dry crust of whitish mycelium occurs on diseased seeds, and in culture the fungus may produce either indigo or orangish pigments. Management practices include avoiding cones from squirrel caches, not stratifying infested seed lots, sowing infested seed lots under warm conditions, and treating seeds with fungicides. Seedborne inoculum of Sirococcus blight is a problem mainly of container-grown spruce in western North America. However, the fungus also occurs on seeds of Norway spruce in Europe. Seed lots become infested when old cones, where the fungus fruits, are included in cone collections. Seedborne inoculum results in death of germinating seeds and seedlings. The dead seedlings remain upright, and the needles die from the bases upward. Primary inoculum is produced by pycnidia on the killed tissues, and secondary spread occurs via irrigation water. Management practices include avoiding collection of old cones, treating seeds with fungicides, and applying fungicides in nurseries. Although many fusaria (e.g., *F. oxysporum, F. solani, F. moniliforme,* and *F. "roseum"*) have been isolated from conifer seeds, particularly Douglas-fir seeds, and can cause seed decay, their role in seedling disease etiology, (e.g., damping-off) is still unclear, perhaps because host stress, such as that caused by high temperatures, is a prerequisite for disease development. Practices used to manage such fusaria include avoiding cones from squirrel caches, applying fungicides to seeds, and using running-water rinses during and after seed stratification.

Selected References

James, R. L. 1986. Diseases of conifer seedlings caused by seedborne *Fusarium* species. Pages 267-271 in: Proc. Conifer Tree Seed Inland Mountain West Symp. R. C. Shearer, comp. U.S. Dep. Agric. For. Serv. Gen. Tech. Rep. INT-203.

Mittal, R. K., Anderson, R. L., and Mathur, S. B. 1990. Microorganisms associated with tree seeds: World checklist 1990. For. Can. Petawawa Nat. For. Inst. Inf. Rep. PI-X-96.

Motta, E., Annesi, T., and Forti, E. 1993. First report of *Sirococcus conigenus* seedborne in Norway spruce. Plant Dis. 77:1169.

Prochazkova, Z. 1991. The occurrence of seed-borne fungi on forest tree seeds in the years 1986-1991. Commun. Inst. For. Cech. 17:107-123.

Sutherland, J. R., Miller, T., and Salinas-Quinard, R., eds. 1987. Cone and seed diseases of North American conifers. N. Am. For. Comm. Publ. 1.

Timonin, M. I. 1964. Interaction of seed-coat microflora and soil micro-organisms and its effects on pre- and post-emergence (survival) of some conifer seedlings. Can. J. Microbiol. 10:17-32.

(Prepared by Z. Prochazkova
and J. R. Sutherland)

Foliage Diseases

Needle and Broom Rusts

The needle rusts of conifers are allies of the more familiar and more damaging stem rusts in the genus *Cronartium* and have similar, but more compressed, life cycles. The broom rusts are distinguished by the fact that they systemically infect the conifer host and often induce spectacular witches'-brooms. The genera are known from all of the conifer-growing areas of the world where both alternate hosts are present. In some cases, the fungus is able to overwinter on the telial host and may exist beyond the range of the aecial, or conifer, host.

Life Cycle and Identification

The needle rusts all have similar life cycles, and both heteroecious and autoecious species as well as microcyclic species are found in most genera. Uredia, telia, and basidiospores are formed on the alternate hosts. Basidiospores, formed on overwintering teliospores on the alternate hosts (various angiosperms or ferns), infect current-year needles of the conifer host, often during the spring. Pycnia and aecia usually form during the same year, and infected needles are cast before the next growing season. In other cases, including in the life cycle of microcyclic species, infection of needles occurs during the fall and aecia are formed the next spring. Some needle rusts are known only from the aecial stage on the conifer host, while other species are known only from the telial host and have presumptive conifer hosts waiting to be discovered.

Nine genera are usually recognized (Table 9), and some authors segregate three or four additional genera. As with all rust fungi, generic taxonomy is based primarily on the morphology of the telial stage. All the needle rust fungi (except

Gymnosporangium spp.) have sessile teliospores and are in the family Melampsoraceae (or order Melampsorales) in older classifications. Identification in the aecial stage on the conifer host, even to genus, is often a challenge. The aecial stages of most genera and species of needle rusts are in the genus *Peridermium*. Host-based keys and host indexes are practical necessities. Many species cannot be distinguished without inoculation studies to determine the telial host.

Species concepts in these genera, as in the rust fungi as a whole, are not uniform and continue to evolve. Autoecious species with obvious affinities to heteroecious siblings are one problem. Differences in host range are another. In some groups, morphologically similar strains with different telial (or aecial) hosts are considered distinct species, while in other groups they are distinguished as formae speciales. The problem is exacerbated by very incomplete knowledge of host range in some groups. The application of molecular systematics should help greatly in identifying species limits and relationships in these fungi.

Damage and Management

Needle rusts other than broom rusts are seldom associated with tree mortality, and any growth loss resulting from early needle loss is usually slight or transient. While infection may be spectacular in some settings, with more than 90% of the current-year needles attacked and shed prematurely, damage is limited because only current-year needles are infected and infected needles remain functional through most of the growing season. Furthermore, environmental conditions favorable for epidemic development seldom occur 2 years in a row.

Seedlings and small trees growing in the midst of large populations of the alternate host, however, may suffer repeated and extreme infection and considerable growth loss. This can be important in nurseries and young plantations, and the cosmetic effect of yellowing and defoliation reduces the value of Christmas trees. While fungicides could be used to protect the young trees from rust infection, in practice, removal of the alternate hosts through cultivation or with herbicides for some distance around the young trees is the preferred strategy.

Spruce Rusts, *Chrysomyxa* spp.

While most species of *Chrysomyxa* at least potentially alternate between spruce and ericaceous hosts, a few complete their life cycles on spruce (microcyclic species *C. weirii* H. Jacks., *C. deformans* (Dietel) Jacz., and *C. abietis* (Wallr.) Unger. and *Ceropsora* spp. in the Himalayas). Many species exist in part or even all of their ranges without spruce and are perennial in the telial host. Several species are systemic in spruce shoots or cones. *C. ledicola* Lagerh. is perhaps the most common of the spruce needle rusts, occurring circumboreally wherever *Picea* species and *Ledum* species (Labrador tea) coincide (Plate 122). In years with weather favorable for rust development (cool and moist) during the period of shoot elongation, infection can reach spectacular levels, and entire stands of spruce appear yellow from the abundant production of aecia. Growth loss results from such epidemics, but conditions are seldom favorable for the fungus 2 years in a row.

The microcyclic *C. weirii* (North America and Asia) and *C. abietis* (Europe) complete their life cycles on spruce needles. Only pycnia and telia are produced; the fungi overwinter as telia on living needles on the tree. Because there is no alternate host, these rusts have less rigorous requirements for completing the disease cycle, and epidemics can develop quickly.

Pucciniastrum spp. and Segregates

The genus *Pucciniastrum* is easier to identify and understand if several species are removed to other genera. In the limited sense, Pucciniastrum needle rusts are characterized by their rather delicate, cylindrical aecia. They can thus be distinguished from *Chrysomyxa* spp. on spruce. *Pucciniastrum* spp. on *Abies* are distinguished from the other fir needle rusts by their distinct (though delicate) peridermioid aecia that form on current or 1-year-old needles. *P. epilobii* G. Otth (fir-fireweed rust) occurs throughout the northern hemisphere wherever *Abies* and *Epilobium* spp. coexist (Plate 123). Infection levels may exceed 90% of the needles in young trees growing amid dense stands of fireweed, as after logging and burning.

Thekopsora species are distinguished primarily on the telial host. *T. aureolatum* (*Picea/Prunus*, haggrost) is noteworthy as a cone rust on *Picea abies* but may also infect the growing stem of that host and cause twisting and distortion. *Calyptospora goeppertiana* Kühn (*Abies/Vaccinium*, blueberry broom rust) is again distinctive on the telial host and for lacking uredia and for the spongy, broomed, telial stage. *Naohidemyces* spp. were recently described to accommodate two hemlock needle rusts (*Tsuga*/Ericaceae) that have uredinioid, not peridermioid, aecia.

Pine Needle Rusts, *Coleosporium* spp.

Nearly all of the needle rusts found on species of *Pinus* are in the genus *Coleosporium* (Plate 124). Pines are reported as hosts of some Melampsora needle rusts but only by artificial inoculation. There are 80 or more described species, but many are morphologically indistinguishable, differing only in telial host. Thus, *C. tussilaginis* (Pers.) Lév in C. d'Orb. is either a host-specific rust on *Tussilago farfara* or a wide-host-range rust that attacks both composites and campanulas, depending on one's species concept. There are many such examples in the genus. Mycologists from China, home to more than half the described species, are suggesting that currently described spe-

TABLE 9. Needle Rust Fungi

Genus Segregates	Aecial Form	Number of Species	Hosts (Aecial/Telial)	Identification
Chrysomyxa	Peridermioid	20	*Picea*/Ericaceae	Aecia robust, often confluent
Coleosporium	Peridermioid	80	*Pinus*/dicots, especially composites	Aecia mature on previous year's needles
Ceropsora	Microcyclic	1	*Picea*	Telia orange red on *Picea* spp.
Hyalopsora	Peridermioid	7	*Abies*/Polypodiaceae	Aeciospores like those of *Milesina*, but yellow
Melampsora	Caeomoid	80	Various/Salicaceae for conifer rusts	Aecia without peridia
Melampsorella	Peridermioid	3	*Abies*/*Cerastium* and *Stellaria*	Systemic and brooming
Melampsoridium	Peridermioid	6	*Larix*/Betulaceae	Delicate aecia
Mikronegeria	Caeomatoid	2	*Araucaria, Austrocedrus*/*Nothofagus*	...
Milesina	Peridermioid (Milesia)	34	*Abies*/"ferns"	White aeciospores
Pucciniastrum sensu latu	Peridermioid, uredinoid	...	*Abies, Picea, Tsuga*/various	...
Calyptospora	Peridermioid	1	*Abies*/*Vaccinium*	No uredia
Naohidemyces	Uredinoid	2	*Tsuga*/Ericaceae	Aecia urediumlike
Pucciniastrum sensu strictu	Peridermioid	30	*Abies, Picea*/various	Aecia delicate, single
Thekopsora	Peridermioid	15	*Tsuga, Picea*/*Prunus*	...
Uredinopsis	Peridermioid	26	*Abies*/"fern"	White aeciospores

cies overlap significantly and often have rather plastic host ranges. They seem to be suggesting that many species could be consolidated into two series, those on diploxylon and those on haploxylon pines.

Pine needle rusts, like other conifer needle rusts, do not generally cause serious damage to the pine. Young trees growing among dense stands of the alternate hosts, often weedy composite species, may suffer growth loss. In some situations, control, usually aimed at the alternate host, may be appropriate.

Melampsora and *Melampsoridium* spp.

Melampsora and *Melampsoridium* spp. are morphologically unrelated but have similar hosts. Both genera are much more significant on their telial hosts (*Salix* or *Populus* and *Betula* spp., respectively) than on conifers. Needle rusts encountered on *Larix* or *Pseudotsuga* spp. are likely to be *Melampsora* species (or *Melampsoridium*). *M. occidentalis* H. Jacks. causes significant needle loss and stunting on Douglas-fir when cottonwoods and the conifer are planted in close proximity, e.g., in nurseries and occasionally in plantations. A few species, e.g., *M. pinitorqua* Rostr. in Europe (*P. sylvestris/Populus*) and the autoecious *M. farlowii* (Arth.) J. J. Davis in eastern North America, infect expanding shoots during the spring, causing twisting and dieback.

Fir/Fern Rusts

The three genera of fern rusts, *Hyalopsora*, *Uredinopsis*, and *Milesina*, are considered by many uredinologists to be the most primitive of the rusts for two reasons, their simple morphology and their "primitive" fern and conifer hosts. They are distinguished from other needle rusts on *Abies* spp. by their white aeciospores (*Uredinopsis* and *Milesina* spp.) and their unusually conspicuous pycnia (*Milesina* spp.), but species within the genera are very difficult to distinguish without artificial inoculation of ferns. Some species are very widespread (*U. pteridis* Dietel & Holw. in Dietel, *Abies*/bracken fern), but damage is seldom reported.

Broom Rusts and Systemic Needle Rusts

The broom rusts differ from other needle rusts in that they colonize the conifer host systemically. Mycelium grows from needles into twigs and apical meristems. The fungi follow the growing tips, colonizing the newly formed needles from the inside and often stimulating the growth of dense witches'-brooms. Systemically infected twigs are stunted and have conspicuously yellow needles that bear masses of aecia. Needles are shed after one season, and brooms may appear bare until budburst the next spring. A few species in three genera may grow systemically in the conifer (Table 10), not including the Gymnosporangium rusts covered separately.

Broom rusts are more damaging than other needle rusts, although it is not clear that growth loss results without infection of the main stem. Brooms starting in the leader or on lateral branches close to the main stem may become bole infections, thus affecting the tree's height and causing distortion of the wood grain. Large brooms are vulnerable to breakage, and the resultant wounds are open to infection by decay fungi.

Spruce broom rust (*Chrysomyxa arctostaphyli* Dietel) is known only from North America, where it is associated with extensive decay and mortality in spruce stands in the Rocky Mountains (Plate 125). Disease incidence is about 15%, and infected trees have a three-fold increase in mortality and are prone to breakage and decay.

Fir broom rust (*Melampsorella caryophyllacearum* J. Schröt) occurs in most forested areas of the world where *Abies* species grow (Plate 126) and causes economically important losses in both Europe and North America. Disease incidence may reach 70% of the trees, and in Germany, 10–15% of the trees may have main-stem infections. Brooms lead to loss in height growth, breakage, and death (Plate 127). In Europe, damage from fir broom rust is reduced by removal of trees with bole infections during stand thinnings.

Damage from *Peridermium cedri*, probably a *Melampsora* sp., on *Cedrus deodara* in the Himalayas, has also been reduced through stand thinning and pruning. The systemic shoot rusts, such as that caused by *Chrysomyxa woroninii*, on spruce throughout the world are less damaging than the broom rusts because the infected twigs die without forming brooms. The fungus is thus annual on spruce.

Selected References

Cummins, G. B., and Hiratsuka, Y. 1983. Illustrated Genera of Rust Fungi. American Phytopathological Society, St. Paul, MN.

Hiratsuka, N. 1958. Revision of the Taxonomy of the Pucciniastraceae. Kasai, Tokyo.

Kaneko, S. 1981. The species of *Coleosporium*, the causes of pine needle rusts, in the Japanese Archipelago. Rep. Tottori Mycol. Inst. 19.

Peterson, R. S. 1963. Effects of broom rusts on spruce and fir. U.S. Dep. Agric. For. Serv. Res. Pap. INT-7.

Peterson, R. S. 1974. Rust fungi with *Caeoma*-like sori on conifers. Mycologia 66:242-255.

Sato, S., Katsuya, K., and Hiratsuka, Y. 1993. Morphology, taxonomy, and nomenclature of *Tsuga*-Ericaceae rusts. Trans. Mycol. Soc. Jpn. 34:47-62.

Xue, Y., Shao, L. P., Cheng, D. S., and He, B. Z. 1995. Some findings in infection series and host range of *Coleosporium*. Pages 81-87 in: Rusts of Pines. Proc. Rusts Pines IUFRO Working Party S2.06.10 Conf., 4th. S. Kaneko, K. Katsuya, M. Kakishima, and Y. Ono, eds. Institute of Agriculture and Forestry, University of Tsukuba, Tsukuba, Japan.

Ziller, W. G. 1974. The tree rusts of western Canada. Can. For. Serv. Publ. 1329.

(Prepared by E. Hansen)

Needle Blights and Needle Casts

Diseases discussed in this section are caused by a diverse group of ascomycetous fungi, or their conidial anamorphs, that occur on conifer foliage, buds, and shoots. These fungi represent a range of severity as pathogens from severe defoliators to

TABLE 10. Broom and Systemic Rust Fungi of Conifers

Fungal Species	Hosts (Aecial/Telial)	Distribution	Disease name
Melampsorella caryophyllacearum	*Abies/Stellaria, Cerastium*	Worldwide	Fir broom rust, "Radertanne"
Melampsora (*Peridermium*) *cedri*	*Cedrus deodara/Salix*?	Himalayas	Peridermium witches'-broom
Mikronegeria alba	*Austrocedrus/Nothofagus*	South America	White rust
Chrysomyxa			
arctostaphyli	*Picea/Arctostaphylos*	North America	Spruce broom rust
deformans	*Picea* (microcyclic)	Himalayas	Shoot rust
woroninii	*Picea/Ledum*	Circumboreal	Spruce shoot rust

inconspicuous parasites. Many have extended incubation periods during which infected host tissue remains symptomless. Some are more virulent on introduced hosts or those outside their native ranges, which has forced greater attention on seed provenance in reforestation. The types of diseases caused by these fungi are usually categorized according to symptoms and fungal life cycles.

Several of the foliar diseases of conifers have been termed "needle casts" or "needle blights," depending on whether the disease symptoms or the causal organisms were emphasized. Needle cast usually refers to a disease that causes premature needle abscission but has also been used more generally to refer to foliage diseases caused by the so-called needle cast fungi, the genera of ascomycetes that comprise an ecologically similar, closely related group of conifer needle parasites originally included in Hypodermataceae sensu Darker (Fig. 10). Other serious needle cast diseases are caused by nonhypodermataceous fungi, such as *Phaeocryptopus* and *Rhabdocline* spp.

Symptoms of several diseases caused by hypodermataceous fungi, and consequently sometimes referred to as needle cast diseases, paradoxically include prolonged retention of necrotic needles, e.g., Lirula and Isthmiella needle casts of spruce and fir. Retention of dead needles appears to be a specialized adaptation in diseases caused by pathogens with protracted infection cycles. These diseases have also been termed needle blights.

In this section, needle cast will be used to describe diseases whose prominent or diagnostic symptoms include premature needle abscission and needle blight to describe diseases in which killed needles remain attached, regardless of the taxonomic or ecological similarities of the pathogens.

Rhytismataceae

Contemporary systematic concepts have led to the redisposition of most of the genera formerly placed in Hypodermataceae to Rhytismataceae (Fig. 10). Ascocarps of hypodermataceous pathogens usually mature and release ascospores during the spring, coinciding with bud break and the flush of new needles. A universal feature of this group is that ascospores, although variously shaped, have a conspicuous gelatinous sheath or appendage that facilitates dispersal by rain and adhesion to the needle cuticle. Many of these fungi form a conidial state prior to or concurrent with maturation of the ascogenous state. In most instances, e.g., *Lirula macrospora*, *Lophodermium* spp., and *Ploioderma* spp., the conidial states are thought to function as spermatia and not as infective propagules.

Selected References

Darker, G. D. 1932. The Hypodermataceae of conifers. Contrib. Arnold Arbor. Harv. Univ. 1:1-131.

Darker, G. D. 1967. A revision of the genera of the Hypodermataceae. Can. J. Bot. 45:1399-1444.

Hunt, R. S., and Ziller, W. G. 1978. Host-genus keys to the Hypodermataceae of conifer leaves. Mycotaxon 6:481-496.

Minter, D. W. 1995. The Rhytismatales on conifers from Europe. Pages 65-84 in: Shoot and Foliage Diseases in Forest Trees. P. Capretti, U. Heiniger, and R. Stephan, eds. Stampa Tipografia Bertelli, Firenze, Italy.

Peterson, G. W., tech. coord. 1986. Recent research on conifer needle diseases. U.S. Dep. Agric. For. Serv. Gen. Tech. Rep. WO-50.

(Prepared by J. Stone)

Rhabdocline Needle Casts

Other name: Douglasienschütte (German)
Causal organisms: *Rhabdocline pseudotsugae* Syd. subsp. *epiphylla* A. K. Parker & J. Reid; *R. weirii* A. K. Parker and J. Reid subsp. *oblonga* A. K. Parker and J. Reid; and *R. weirii* subsp. *obovata* A. K. Parker & J. Reid (anamorph *Rhabdogloeum pseudotsugae* Syd.)
Host: *Pseudotsuga menziesii* (Douglas-fir)
Distribution: North America and Europe

Some of the most common and injurious needle pathogens of Douglas-fir are species of *Rhabdocline*, causal agents of Rhabdocline needle cast disease (Plate 128). The genus, which originally contained a single species, *R. pseudotsugae*, was revised in the 1969 work by Parker and Reid to a complex of two main species, *R. pseudotsugae* and *R. weirii*, each with several subspecies. The pathogens are apparently endemic throughout the natural range of Douglas-fir but have been introduced into other parts of North America and Europe where Douglas-fir is planted as an exotic species. Attention to planting stock from appropriate seed sources in recent years has virtually eliminated severe losses to the disease in forest plantations, although the disease can be a chronic problem in Christmas tree plantations and seed orchards.

Symptoms

Symptoms of infection first appear as chlorotic flecks in late summer to autumn. These flecks gradually enlarge, forming mottled, chlorotic patches on current-year needles. The contrast between the infected patches and the uninfected green portions of the needles gradually intensifies during the winter until early spring, when the needles have a distinctly blotchy appearance (Plate 129). By early summer, the needles become straw colored and abscise. Ascomata initially are bright orange on attached green needles and gradually darken over several

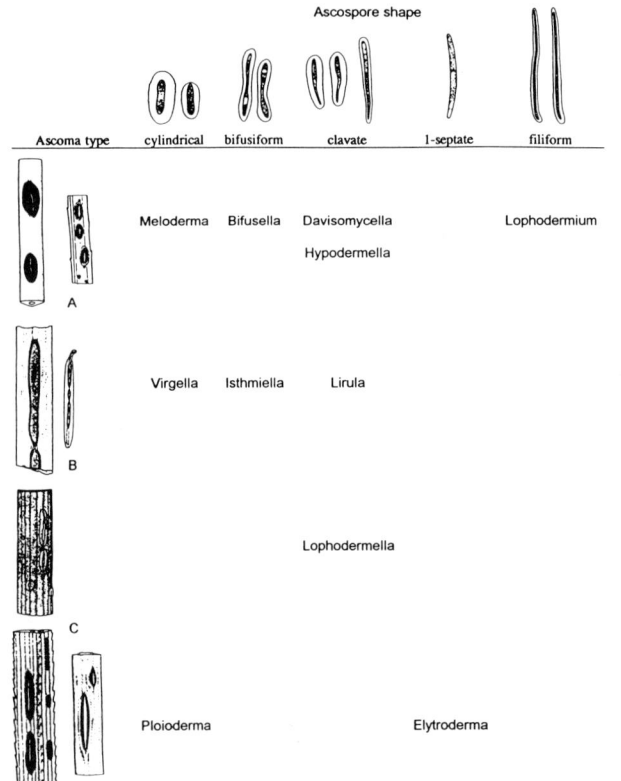

Fig. 10. Generic ascoma and ascospore characteristics of hypodermataceous fungi on conifer needles. **A,** Black and oval to elliptical; zone lines and perimeter lines may be present. **B,** Black; occur on leaf veins. **C,** Concolorous and oval to elongate. **D,** Black and narrow. (Adapted from Darker, 1932 and 1967)

weeks. The ascomata erupt through the epidermis in mid- to late spring while the needles are still attached. Tiny flaps of epidermis cover the ascomata when they are dry and are forced open as the hydrated ascomata expand. The ascomata can be small and circular to long, narrow strips on either side of the needle midrib, except those of *R. weirii* subsp. *oblonga*, which span the entire needle width. Conidial cirri are conspicuous on the upper surfaces of needles affected by *R. weirii* subsp. *weirii*.

Diagnosis

Parker and Reid separated *R. pseudotsugae* and *R. weirii* on the basis of the reactions of ascus tips to Melzer's iodine: *R. weirii* has I+ asci and *R. pseudotsugae* I− asci (Plate 130). *R. weirii* subsp. *weirii* is most easily identified on the basis of its regular association with a conidial state, *Rhabdogloeum pseudotsugae*, which produces acervular conidiomata on the upper surfaces of needles while the ascomata of the *Rhabdocline* state form on the lower surfaces (Plate 131). Ascomata of *R. weirii* subsp. *oblonga* are rectangular and span the entire width of the needle across the midvein. The two subspecies of *R. pseudotsugae* are differentiated by the occurrence of the ascomata on the upper (subsp. *epiphylla*) or lower (subsp. *pseudotsugae*) needle surfaces.

Asci of all species are eight spored. Ascospores are initially hyaline, aseptate, and constricted in the middle and have a hyaline gelatinous sheath. Ascospores become two celled with one brown and one hyaline cell before germination; in the subspecies *obovata*, this occurs while the ascospores are still in the asci. Conidia of the *Rhabdogloeum pseudotsugae* anamorph of *R. weirii* resemble the ascospores but have apical gloeoid appendages. A third species, *R. parkeri*, is widespread in Douglas-fir in the Pacific Northwest but is generally asymptomatic and produces ascomata only on abscised needles during midwinter. However, the conidial state of *R. parkeri*, which is virtually identical to the larch needle cast pathogen, *Meria laricis* (discussed in the section Foliage Diseases of Other Conifers), can frequently be found sporulating on naturally senesced needles and on needle galls caused by the insect *Contarinia pseudotsugae*, giving the galled needles the appearance of a fungal disease infection.

Biology and Disease Development

Sporulation generally coincides with bud break; ascospores (and conidia of *R. weirii* subsp. *weirii*) are rain dispersed. Severity of disease is associated with cool, rainy conditions during early shoot emergence (mid-April to early June) the previous spring. Ascocarps are devoid of ascospores by late summer. There is apparently a strong genetic component to host susceptibility to *R. pseudotsugae*, but the existence of two species and several subspecies complicates comparative susceptibility studies.

Disease Management

Injury from these pathogens has been reduced recently by attention to seedling provenance. Both the interior and coastal forms of Douglas-fir are susceptible, but the interior variety (*Pseudotsuga menziesii* var. *glauca*) is much more susceptible than the coastal form. Serious damage is usually confined to Christmas tree farms, particularly those in which interior varieties of Douglas-fir are planted. Occurrence of the disease in forest plantations can be very disjunct, and trees adjacent to or near affected trees can remain completely without symptoms. Incidence of the disease in natural stands of coastal variety Douglas-fir is scattered and usually insignificant. Severely diseased trees should be removed. In nurseries and Christmas tree plantations, the disease can be controlled with fungicides.

Selected References

Chastagner, G. A., Byther, R. S., and Riley, K. L. 1990. Maturation of apothecia and control of Rhabdocline needle cast on Douglas-fir in western Washington. Pages 87-92 in: Recent Research on Foliage Diseases. W. Merrill and M. E. Ostry, eds. U.S. Dep. Agric. For. Serv. Gen. Tech. Rep. WO-56.

Hoff, R. J. 1987. Susceptibility of inland Douglas-fir to Rhabdocline needle cast. U.S. Dep. Agric. For. Serv. Res. Note INT-375.

Parker, A. K., and Reid, J. 1969. The genus *Rhabdocline* Syd. Can. J. Bot. 47:1533-1545.

(Prepared by J. Stone)

Swiss Needle Cast

Other names: rouille suisse du Douglas (French); Schweizer Douglasiennadelschütte and Adelopus Douglasiennadelschütte (German)

Causal organism: *Phaeocryptopus gaeumannii* (T. Rohde) Petr. (syn. *Adelopus gaeumannii* T. Rohde)

Hosts: *Pseudotsuga menziesii* (Douglas-fir) and *P. macrocarpa*

Distribution: western United States and Canada; introduced into central and eastern North America, Europe, and Australasia

Phaeocryptopus gaeumannii parasitizes its host by utilizing the products of photosynthesis. In diseased trees, it significantly reduces assimilation rates of the foliage and impairs needle function to the point that chlorosis develops and premature needle casting ensues. Where it occurs naturally, *P. gaeumannii* is in balance with its host, which retains healthy, green foliage for up to 7 years. Defoliation does occur occasionally during wet summers and in areas at the geographic and altitudinal extremes of the natural distribution. However, where Douglas-fir has been planted outside its natural range,

Fig. 11. Pseudothecia of *Phaeocryptopus gaeumannii* on a Douglas-fir needle. (Courtesy J. Stone)

infection levels are often very high and premature needle casting is common although not always obvious, since disease severity can vary within infected plantations. Many stands appear comparatively healthy but nevertheless retain a reduced complement of older foliage, and crowns may be slightly discolored and thinned. Where the disease is more severe, crowns become intensely chlorotic and retain little but the current foliage. In wood-production stands, the main economic effect is reduced growth increment. A significant decrease in wood volume production was found to have occurred in infected plantations in both Germany and New Zealand several years after introduction of the pathogen. In Christmas tree stands in the United States and Canada, chlorosis and needle casting seriously reduce the market value of the product.

Symptoms and Diagnosis

Fungal fruit bodies (pseudothecia) are formed in two diffuse, sooty bands along each side of the lower living needle surface (Fig. 11). Each fruit body is visible with a hand lens as a tiny black sphere (up to 0.1 mm in diameter) seated above a needle stoma (sometimes capped by the dislodged white stomatal wax plug) or, if immature, as a black speck that fills the stomatal chamber (Plate 132). Mature fruit bodies are more densely distributed, and hence more easily seen, on older needles (Plate 133). Infected foliage may be green or discolored from pale green to uniform or mottled yellow green to yellow and eventually browns from the tips. Symptoms are more severe on older foliage. Premature needle casting begins with the oldest foliage; diseased trees retain needles for less than 5 years. When the disease is not severe, crowns of infected trees are green and dense or pale green and slightly thinned, and the disease may be barely noticeable. When the disease is severe, crowns may be bright yellow and retain only the current year's foliage. Severely defoliated crowns occasionally produce epicormic shoots.

Disease Cycle

Fruit bodies begin to produce ascospores during the spring (in March in the northern hemisphere), and production continues for up to 6 months. The highest incidence coincides approximately with the 2-month period following bud burst. Spores are released intermittently during periods of rainfall, mostly from 1-year-old foliage. Infection of newly flushed foliage occurs during a period of 6–8 weeks during May and June while shoots are still expanding. Susceptibility drops markedly once foliage is fully extended, although some needles may be infected during the following year. Ascospores germinate on and penetrate the needle surface. The fungal hyphae grow extensively among and within needle cells, but no haustoria are produced and no host tissue response is elicited. Fruit body initials become visible in stomatal chambers when needles are 4–6 months old, and diffuse strands of superficial hyphae extend out over the needle surface. A proportion of the initials develop further, forcing out the stomatal plugs as they grow and reaching maturity in the spring. Additional fruit bodies are produced in subsequent seasons on older needles. No asexual repeating phase has yet been confirmed, and it appears that there is only one infection cycle per season. Infected needles become chlorotic sometime after the first year and are eventually shed prematurely. Crown discoloration is most intense during the autumn and winter.

Contributing Factors

P. gaeumannii requires moisture for spore release and infection of host foliage. Rainfall is a key factor in the development of the disease, which is significant mainly in regions where the spring rainfall is substantially greater than that in much of the natural Douglas-fir distribution range. Temperature is less of a factor than needle wetness. Spore release occurs at temperatures between 5 and 25°C; 20°C is optimal. The disease is also affected by the genetic origin of the host. Douglas-fir from the Rocky Mountain region in the North American interior is reported to be more susceptible than trees from near the coast. At the provenance level, Douglas-fir seed lots collected from dry zones where rainfall is insufficient to support infection are particularly susceptible when planted on sites with significant levels of spring rainfall. Where spring rainfall is substantial (a 3-month mean that exceeds 250 mm), even resistant seed lots succumb to the high inoculum pressures generated. Under these circumstances, more than 80% of needles can become infected, although even at this level of infection, there is variation in disease expression (chlorosis and needle cast). Another factor may be stress. Infected trees growing on adverse sites where local edaphic or other conditions are unfavorable are more likely to succumb to the disease. This may explain the extreme severity of symptoms shown by infected trees in certain localities. Age may also predispose stands to disease, but this factor is not well understood. Infection has been found on trees of all ages from 1 to at least 70 years old but is less well studied in older plantations.

Disease Management

Swiss needle cast disease is controlled on young trees by spraying new foliage with a fungicide during the infection period. At least two sprays are recommended in Christmas tree plantations, the first shortly after bud burst and the second 3–4 weeks later. Regular spraying in production plantations with longer rotations is not economically justifiable.

Improvement in Douglas-fir plantation health is achievable through tree breeding, since between- and within-provenance variation in disease susceptibility has been demonstrated. Both resistance and tolerance to infection should be included among selection factors in programs designed to produce pedigree seed for specific sites. The most promising source of breeding material appears to be the coastal strip from British Columbia south to northern California. Where plantations are established within the natural Douglas-fir range, efforts should be made in the interim to restock sites with seed that originates from the same location.

Thinning has been advocated for management of infected wood-production stands. Precommercial thinning is now practiced in New Zealand in order to achieve a sufficiently large log size at the subsequent production thinning. However, thinning does not reduce the high incidence of infection or recoup the volume increment lost per hectare through disease. In Christmas tree stands, precautions are needed to prevent the spread of infection via shearing equipment.

Selected References

Hood, I. A. 1982. *Phaeocryptopus gaeumannii* on *Pseudotsuga menziesii* in southern British Columbia. N.Z. J. For. Sci. 12:415-424.

Hood, I. A., Sandberg, C. J., Barr, C. W., Holloway, W. A., and Bradbury, P. M. 1990. Changes in needle retention associated with the spread and establishment of *Phaeocryptopus gaeumannii* in planted Douglas-fir. Eur. J. For. Pathol. 20:418-429.

McDermott, J. M., and Robinson, R. A. 1989. Provenance variation for disease resistance in *Pseudotsuga menziesii* to the Swiss needle cast pathogen, *Phaeocryptopus gaeumannii*. Can. J. For. Res. 19: 244-246.

Michaels, E., and Chastagner, G. A. 1984. Seasonal availability of *Phaeocryptopus gaeumannii* ascospores and conditions that influence their release. Plant Dis. 68:942-944.

Skilling, D. D. 1981. Control of Swiss needle cast in Douglas-fir Christmas trees. Am. Christmas Tree J. 25(3):34-37.

Strittmatter, W. 1974. Ökologische und biologische Studien an der Baumart Douglasie im Zusammenhang mit dem Auftreten von Phaeocryptopus gäumanni (Rohde) Petr. Schriftenr. Landesforstverwaltung Baden-Württemberg 43.

(Prepared by I. Hood)

Brown Spot Needle Blight

Other name: Lecanosticta needle blight
Causal organism: *Mycosphaerella dearnessii* Barr. (syns. *Scirrhia acicola* (Dearn.) Siggers, *Dothidea acicola* (Dearn.) Morelet, *Oligostroma acicola* Dearn., and *Systremma acicola* (Dearn.) F. A. Wolf & Barbour; anamorph *Lecanosticta acicola* (Thuem.) Syd. in Syd. & Petr.)
Hosts: pines, especially longleaf pine and Scot's pine
Distribution: North America, southern Europe, South America, South Africa, and China

Brown spot needle blight is most widely known as one of the major obstacles to increased production of longleaf pine in the southeastern United States and also as a major cause in the central states of making Scot's pine Christmas trees unmarketable.

Brown spot needle blight has been reported on at least 28 species of pines. It occurs mainly in North America but is known also in scattered localities in Europe (e.g., Austria, Greece, and Yugoslavia), South America, South Africa, and southeastern China. In North America, it occurs in all coastal states from North Carolina to Texas and in Arkansas, Missouri, Pennsylvania, Ohio, Tennessee, and Oregon; on lodgepole and jack pine in Manitoba; and on ponderosa and Scot's pines in the central plains and Great Lakes regions.

The Fungus

The causal fungus has long been known by the names of the ascigerous stage, *Scirrhia acicola,* and the conidial stage, *Lecanosticta acicola*. In 1984, however, Evans gave new teleomorph designations to both the brown spot and the Dothistroma blight fungi. The perfect stage of the brown spot fungus is now known as *Mycosphaerella dearnessii*.

Conidia from acervuli in black stromata are exuded in a sticky mass under moist conditions and disseminated by splashing rain. They are cylindrical, curved, and olive green to brown and have one to four septa. Ascospores are hyaline, oblong-cuneate, and unequally bicellular and have two prominent, brown, oil drops in each cell. They are discharged during moist periods and disseminated by wind. There appears to be a clear physiological difference between the isolates from the northern and southern United States but little difference between the isolates from the southern United States and China.

Infection and Symptom Development

The fungus produces two types of necrotic lesions on the needles. 1) The common circular lesions coalesce to form oblong areas, which are straw yellow to light brown and have a darker border. 2) Less common is a "bar spot," an amber yellow band 2–3 mm long and infiltrated with resin in which a small brownish spot becomes localized on one side (Plate 134). A diseased needle usually has a dead tip, a central portion with several spots in green tissue, and a green basal portion. Secondary infections are produced throughout the growing season. As the season progresses, internal advance of the fungus and coalescence of the lesions kill most of the needle tissue, eventually causing defoliation.

In the southern United States, black stromata, which produce acervuli and conidia, appear about 2 weeks or longer after development of the lesions. Pseudothecia of the sexual stage may form 6–8 weeks after conidial formation. Primary infections are initiated near needle tips by ascospores or conidia during the spring, usually about April. Secondary cycles of infection by conidia or ascospores may occur throughout the year, and conidia intensify the disease and spread the fungus in local areas.

In the north-central and midcentral United States, conidia are the only spores found. Infection occurs during the spring or early summer, and needle spots appear in August or September. Soon thereafter, needles turn brown as they die and are cast, and stromata develop on the killed tissue. Some needles can become infected in the fall by conidia released from infected needles retained on the tree. The following spring, these needles die and are cast during June and July.

Cells collapse in affected mesophyll, resulting in large intercellular spaces. Hyphae are sparse, and a sharp definition exists between affected mesophyll and unaffected cells at the edges of symptomatic tissue. This localized collapse and very limited presence of hyphae suggest the possibility that a toxin is produced during the host-pathogen interaction, as is known to occur in Dothistroma blight.

Damage

Brown spot causes defoliation. In longleaf pine, no active growth in height occurs until the seedling diameter reaches about 2.5 cm at the ground line, regardless of age. This period is called the "grass stage." Any factor, such as defoliation, that keeps seedlings from reaching the maximum diameter for height growth is critically important. Brown spot infection reduces height growth and subsequent vigor of the seedling by extending the grass stage to 10 to 20 years or longer (Plate 135). In Scot's pine Christmas tree plantations, needle browning and thinning of crowns by defoliation make the trees unmerchantable.

Disease Management

Repression of brown spot is important in the management of longleaf pine in the southern United States. In nurseries, effective control of the disease may be obtained by the application of recommended foliar fungicidal sprays or root dips. In young natural stands or plantations of longleaf pine that have not yet begun active height growth, prescribed burning of the site kills the fungus and its spores without harming the trees because of their remarkably fire-resistant terminal buds. Burning of established stands during the dormant season may be necessary whenever the mean infection level reaches 20% and until active height growth begins and the majority of seedlings reach a height of about 45 cm. Genetic control of the disease through development of resistant strains has been progressing, and limited supplies of resistant stock are gradually becoming available.

In Christmas tree plantations of Scot's pine in the north-central states, effective protection is provided by the application of recommended fungicidal sprays. Shearing should be avoided during wet weather. Only the more resistant long-needle cultivars of Scot's pine, such as Austrian Hills or German, should be grown, and the highly susceptible short-needle cultivars, such as Spanish or French Green, should be avoided.

Selected References

Evans, H. C. 1984. The genus *Mycosphaerella* and its anamorphs *Cercoseptoria*, *Dothistroma* and *Lecanosticta* on pines. Mycol. Pap. 153. Commonwealth Mycological Institute, Kew, England.
Kais, A. G. 1975. Environmental factors affecting brown-spot infection on longleaf pine. Phytopathology 65:1389-1392.
Maple, W. R. 1976. How to estimate longleaf seedling mortality before control burns. J. For. 74:517-518.
Skilling, D. D., and Nicholls, T. H. 1974. Brown spot needle disease: Biology and control in Scotch pine plantations. U.S. Dep. Agric. For. Serv. Res. Pap. NC-109.

(Prepared by R. F. Patton)

Dothistroma Needle Blight

Other names: red band needle disease; maladie des bandes rouges (French)

Causal organism: *Mycosphaerella pini* Rostr. in Munk (syn. *Scirrhia pini* Funk & A. K. Parker; anamorph *Dothistroma pini* Hulbary, syn. *Dothistroma septospora* (Doroguine) Morelet)

Hosts: many species of *Pinus*, especially Monterey, ponderosa, and Austrian pines in plantations

Distribution: worldwide

Since its description on needles of Austrian pine in 1941 in Illinois, Dothistroma needle blight has become known as an important disease of pines worldwide, especially on pine plantations in East Africa, Chile, and New Zealand. More than 30 pine species are known to be hosts, and in North America alone it has been found on 29 pine species and hybrids, including two-, three-, and five-needle pines. Less susceptible species also reported as hosts include Douglas-fir, Sitka spruce, and European larch. In the United States, Austrian (Plate 136), lodgepole, and ponderosa pines are particularly susceptible, and plantations of these and other species have sustained severe damage (Plate 137). Also, Monterey pine in plantations on the west coast of North America have been severely damaged (Plate 138), but the disease is essentially unknown in the natural stands of this species.

This disease has been most damaging in exotic pine plantations, chiefly of Monterey pine, in the southern hemisphere, particularly East Africa, South Africa, Chile, and New Zealand. In Europe, where the fungus has been relatively unimportant so far, the disease has been recorded from England, France, Spain, Romania, and Yugoslavia. It has occurred also in Russian Georgia, India, Japan, Australia, and in South America in Argentina, Brazil, Uruguay, and Chile. This disease has been termed a classic example of the introduction and spread of an exotic pathogen in an area with favorable environment and susceptible hosts.

The Fungus

Mycosphaerella pini is known mainly from the most commonly seen imperfect stage, *Dothistroma pini*. The sexual stage, originally placed in the genus *Scirrhia*, is generally rare and was found first in Alaska, British Columbia, California, and Oregon but has now been reported from Europe, Africa, and New Zealand. Evans noted the similarity of this species to the brown spot fungus, renamed the perfect stage *Mycosphaerella pini*, and noted that differences between the anamorphs are most useful in separating the two species. Although three varieties have been proposed on the basis of conidial length, Evans considered such separation to be invalid. However, he did follow Morelet's name for the anamorph, *Dothistroma septospora*.

Infection and Symptom Development

In the United States and Canada, yellow to tan spots that become translucent, water-soaked chlorotic bands develop typically from September to November. These bands may turn brown to reddish brown, most distinctively on needles of pines in the western United States, giving the name "red band" to the disease. Infected needles become necrotic, first at the tip and then over an extensive length, and are cast prematurely, usually during the year following infection. In other areas of the world, symptoms develop several months after maturation of spores in the spring, whatever months of the year that occurs.

Conidia appear in black stromata that are erumpent through the epidermis in needle spots or necrotic portions of the needle. The spores typically begin to mature during the spring (sometimes even late in the winter) and are disseminated by rain splash anytime during the growing season. Conidia are hyaline (an important characteristic that distinguishes *M. pini* from the brown spot fungus) and cylindrical to slightly curved and have one to five but usually three septa. The ascostromata and ascospores are similar to those of the brown spot fungus.

Conidia germinate under moist conditions, and germ tubes are positively directed toward stomata. A branched, often four-footed appressorium develops in the stomatal antechamber. After penetration between the guard cells by a fine hyphal strand, a substomatal vesicle is typically formed from which one or more infection hyphae develop for a localized distance into the mesophyll. A toxin, dothistromin, a difuranthraquinone, diffuses into the tissue in advance of the hyphae, killing the cells and resulting in their collapse and the production of typical symptoms.

Damage

The most immediate effect on trees is the reduction in growth in both height and diameter. Growth reduction is measurable when 25% of the foliage has been killed, and with 50% loss, diameter increment is halved and height growth reduced. Trees less than 10 years old are most susceptible, and although severe defoliation may result in death, growth reduction is the more significant effect. In Africa, temperature and moisture conditions so favored disease development that productivity of the conifer plantations was markedly reduced, especially of Monterey pine, one of the most important but also most susceptible species. In New Zealand, the disease threatened the forest industry based on exotic pine species, mostly the highly susceptible Monterey pine, planted on 400,000 ha. An effective aerial spray program, however, lessened the anticipated eventual impact. In Chile, the disease has not proved to be chronically serious, possibly because wet springs, which favor infection, occur only about once every 10 years. In North America, the disease has gained some notoriety and caused damage to ornamental, shelterbelt, and Christmas tree plantings but has not been deemed important enough for large-scale surveys of damage and impact.

Important factors governing infection and disease development are temperature (optimal temperatures are 12–24°C) and duration of free moisture influenced by rainfall. Trees in regions with an annual rainfall of more than 1,270 mm are likely to suffer from severe defoliation.

Disease Management

Control of Dothistroma needle blight by protective chemical sprays has proved feasible. In the central United States, two applications, one in May and the other in June, of copper fungicides effectively prevent infection of both first- and second-season needles on plantings of Austrian and ponderosa pine. In East Africa, results of field trials were promising, but technical difficulties were encountered in aerial spray programs. In New Zealand, however, a successful aerial spray program was developed with copper-based fungicides in which sprays were applied when the overall infection level of unsuppressed green crowns reached 25% (15% in high-hazard regions). As many as seven or eight applications may be required until Monterey pines become resistant at about 20 years of age. By the late 1980s, cost-benefit analyses had not determined how much net production was saved by spraying, but the conclusion was that spraying should be continued in order to avoid the unknown consequences of leaving large, contiguous areas of forest unsprayed.

Ultimate control will be achieved through the use of disease-resistant stock. Austrian pine and ponderosa pine seed from several geographic areas with a useful degree of resistance have been identified for use in plantings in the Great Plains. In New Zealand, the aerial spray program was so efficient that large-scale research for resistant material was hindered, but with progeny trials underway, seed from resistant parents may become available in commercial quantities.

Selected References

Evans, H. C. 1984. The genus *Mycosphaerella* and its anamorphs *Cercoseptoria*, *Dothistroma*, and *Lecanosticta* on pines. Mycol. Pap. 153. Commonwealth Mycological Institute, Kew, England.

Gibson, I. A. S. 1972. Dothistroma blight of *Pinus radiata*. Annu. Rev. Phytopathol. 10:51-72.

Peterson, G. W., and Walla, J. A. 1978. Development of *Dothistroma pini* upon and within needles of Austrian and ponderosa pines in eastern Nebraska. Phytopathology 68:1422-1430.

Van der Pas, J. B., Bulman, L., and Horgan, P. G. 1984. Disease control by aerial spraying of *Dothistroma pini* in tended stands of *Pinus radiata* in New Zealand. N.Z. J. For. Sci. 14:23-40.

(Prepared by R. F. Patton)

Cyclaneusma Needle Cast

Causal agent: *Cyclaneusma minus* (Butin) DiCosmo, Peredo, & Minter (syn. *Naemacyclus minor* Butin)

Hosts: diploxylon pines, including Scot's (*Pinus sylvestris*), radiata or Monterey (*P. radiata*), Austrian (*P. nigra*), lodgepole (*P. contorta*), mugho (*P. mugo*), ponderosa (*P. ponderosa*), and Jeffrey's (*P. jeffreyi*)

Distribution: Europe, North America, and Australasia

Cyclaneusma needle cast is thought to be responsible for the premature casting of needles on numerous species of pine, although some research indicates that the causal fungus is probably an endophyte and that environmental stress may be the most important factor in symptom development. Generally, since only the 2- and 3-year-old needles are prematurely cast, the disease is considered to be a significant problem only in Christmas tree plantations and nurseries and not in natural forest stands. Trees that suffer repeated defoliation may retain only a 1-year complement of needles, which reduces their value as Christmas trees and adversely affects growth.

The Pathogen

Apothecial fruiting bodies (200–600 μm long) are elliptical, subhypodermal, and the same color as the dead needles. Spores are released when apothecia are exposed by the opening of the two flaps of needle tissue. Asci are cylindric with an acutely pointed tip and contain eight spores. Ascospores are hyaline, filiform, and bent like a boomerang and have two septa near the middle about 8 μm apart. Conidia are similar to those of *Phomopsis* spp. and are produced in immersed, globose pycnidia.

Symptoms

Normally, symptoms first appear late in the summer and fall on needles that are 2 years old and older. Light green to yellow spots develop on infected needles and expand to form brown bands on yellow needles. About 1 month after symptoms first appear, off-white fruiting bodies the same color as the needle begin to appear as elliptical swollen areas below the epidermis. During wet weather, the epidermis of the needle covering the swelling fruiting body swings open like double doors pushed fully open. Spores are then released by the exposed apothecia.

This disease could be confused with other fungus needle cast diseases caused by *Lophodermium* spp., *Lophodermella* spp., *Mycosphaerella pini*, and *Elytroderma deformans* and with damage caused by environmental stresses, air pollution, and the pine needle sheathminer (*Zelleria haimbachi*).

Biology and Disease Development

The life cycle of the pathogen of this disease and the role of environmental stress are not precisely defined because of the complexities involved, including an extended period for infection to occur, a long incubation period before symptoms become evident, the susceptibility of needles of various ages, the existence of numerous susceptible hosts, and differences in environments where this disease is present.

Spores are forcibly discharged from fruiting bodies anytime during the growing season. Periods of high humidity and moisture trigger these releases, which peak 4–6 hr after the onset of rain. Spores are produced on both attached and fallen needles. When free moisture is present on needles, the spores germinate and the fungus is able to enter through stomata. Current-year needles are infected in June and 1-year-old and older needles during April through November.

Infections are known to occur even at temperatures as low as 2°C. Infected needles will not show symptoms for 10–15 months, at which time they will yellow during the fall. Fruiting bodies with spores begin to form in the fall. However, several weeks of warm weather during the spring is required for ripening, after which the mature spores can be liberated.

Disease Management

Cultural practices that minimize prolonged periods of leaf wetness, such as weed control and adequate tree spacing in plantations, help reduce disease pressure. Plantings should be established with disease-free seedlings. Nutrient deficiencies and moisture stress should also be avoided, if possible.

Fungicides have shown efficacy in protecting needles from infection. However, because of the extremely long and apparently unpredictable infection periods throughout the growing season, efficient timing of fungicide applications is not easily accomplished. Multiple applications have achieved control but often do not improve tree grade or density of foliage. Also, poorly understood environmental conditions may often cause heavy needle casting late in the fall, regardless of levels of infection.

Selected References

Gadgil, P. D. 1985. Cyclaneusma needle cast. For. Res. Inst. For. Pathol. N.Z. Publ. 11.

Minter, D. W. 1995. The Rhytismatales on conifers from Europe. Pages 65-84 in: Shoot and Foliage Diseases in Forest Trees. P. Capretti, U. Heiniger, and R. Stephan, eds. Stampa Tipografia Bertelli, Firenze, Italy.

Wenner, N. G., and Merrill, W. 1990. Control of Cyclaneusma needle cast on Scots pine in Pennsylvania. Pages 27-33 in: Recent Research on Foliage Diseases. W. Merrill and M. E. Ostry, eds. U.S. Dep. Agric. For. Serv. Gen. Tech. Rep. WO-56.

(Prepared by G. Chastagner)

Other Foliage Diseases of Pines

Ploioderma Needle Casts

The most injurious needle pathogen of hard pines in the southern United States, *Ploioderma lethale* (Dearn.) Darker, affects several species including *Pinus clausa*, *P. echinata*, *P. elliottii*, *P. glabra*, *P. palustris*, *P. rigida*, and *P. serotina*. The fungus infects current-year foliage, and the first symptoms appear late in the winter on 1-year-old foliage. Infected needles become brown at the tips and in irregular, gray brown bands separated from the green, healthy portions of the needle by a darker brown margin. The brittle, necrotic portions break off before the entire needle is killed, leaving green needle bases attached. Ascocarps (hysterothecia) develop on the necrotic needles during late spring and are narrow, black, and embedded in the needle and open by a longitudinal split (Plate 139).

P. hedgcockii (Dearn.) Darker infects *P. elliottii*, *P. taeda*, *P. clausa*, *P. virginiana*, *P. echinata*, *P. palustris*, and *P. rigida* and differs from *P. lethale* (Plate 140) in that the asci have four spores (there are initially eight spores, but four fail to develop). The ascomata develop on green needles, which turn yellow and are prematurely shed.

P. lowei Czabator infects *P. elliottii* var. *elliottii*. The ascomata are circular to oval, 1–2 mm long, conspicuously raised above the needle surface, and produced on naturally senesced and prematurely killed green needles. Asci contain eight spores arranged in two rows and are initially aseptate but become cylindrical with one septum. The related *P. pedatum* (Darker) Darker occurs on *P. radiata* in California.

Fig. 12. Witches'-brooms caused by *Elytroderma deformans* on ponderosa pine in Oregon.

Elytroderma Needle Disease

Elytroderma deformans (Weir) Darker, causes witches'-brooms (Plate 141) and severe, damaging defoliation of *P. ponderosa* and *P. jeffreyi* in western North America and has also been recorded from *P. banksiana, P. contorta, P. edulis, P. cembroides,* and *P. attenuata.* Severe infections can kill trees. The disease cycle differs from that of other needle casts in that the fungus invades via needles and then invades and persists in shoots and twigs, causing the characteristic growth abnormality. The early disease symptom is a red brown discoloration of 1-year-old needles on individual branches in the spring. The bases of needles on "flagging" branches remain green and attached, but the brittle, necrotic needles may break off during the winter. Twigs and branches are gradually killed as each year's new foliage dies through successive cycles. The fungus invades the branches, which may form brooms (Fig. 12). "Resin cysts," elongated, resinous lesions in the shoots, are diagnostic for *E. deformans.*

Unlike those of *Davisomycella* and *Lophodermium* spp., the black clypeus of *E. deformans* is very narrow and does not completely cover the hymenium, only the central fissure (Plate 142). The ascospores are large and cylindrical with a median septum. Transverse zone lines usually are not present on infected needles. *E. torres-juanii* Diamanidis & Minter affects *P. halepensis* and *P. brutia* in Spain and Greece. Two-year-old needles may be partially discolored. Ascocarps appear on 3- to 4-year-old needles. Defoliation of *P. brutia* caused by *E. torres-juanii* has been especially severe in Greece.

Lophodermella Needle Casts

A number of serious needle blights of pines are caused by species of *Lophodermella. L. arcuata* (Darker) Darker affects five-needle pines (*P. albicaulis, P. flexilis, P. lambertiana,* and *P. monticola*) in the western United States. One-year-old needles turn brown along their entire length late in the spring. Ascomata mature and release ascospores that infect the current-year needles during late summer. The ascomata are inconspicuous, concolorous, and slightly raised above the surface of the needle. The host preference and geographic range are unique and suffice to diagnose this pathogen.

L. montivaga Petr. and *L. concolor* (Dearn.) Darker both affect *P. attenuata, P. banksiana,* and *P. contorta* in the Rocky Mountains and Cascades; *L. concolor* also affects *P. contorta* along the Pacific coast (Plate 143). Both species can cause severe defoliation and sometimes result in tree death. The ascomata of *L. concolor* are small and inconspicuous, ovoid, and concolorous and open by a longitudinal split along a row of stomata. Needles are infected in midsummer and appear dwarfed and stunted, become red brown, and are shed during May and June of the following spring. Severity of the disease has been linked to the contributory effects of drought stress, root diseases, and bark beetle infestation. Ascomata are commonly invaded by *Hemiphacidium longisporum* Ziller & Funk along the Pacific coast.

Symptoms caused by *L. montivaga* help to distinguish this pathogen from *L. concolor.* The distal end of the needle is killed, and the browned portion is separated from the healthy basal portion of the needle by an indistinct, broad, zone line. Ascomata frequently coalesce to form elongated groups. The ascospores are narrower (3–4 μm) than those of *L. concolor* (6–8 μm). *L. montivaga* resembles the European species *L. sulcigena* (Rostrup) Höhnel, which is pathogenic on *P. mugo, P. nigra,* and *P. sylvestris* (Plate 144).

L. morbida Staley & Bynum has been reported as an aggressive pathogen of *P. ponderosa* and *P. attenuata* planted in off-site areas in western Oregon, Washington, and northern California (Plate 145). The ascomata are conspicuously dark brown and usually 1–6 mm long, although they are commonly up to 20 mm long. The subhypodermal position of the ascocarp in cross section differentiates *L. morbida* from species of *Davisomycella* (Plate 146). Infection occurs during mid-June, and needles are usually cast 14–15 months later.

L. cerina (Darker) Darker (Plate 147) affects *P. elliottii* and *P. taeda* in the gulf states as well as *P. contorta* and *P. ponderosa* in Arizona, New Mexico, Colorado, and California. The symptom appearing first late in the fall on the southern pines is a reddish brown discoloration of the foliage; ascomata are produced early in the spring and mature by late March. Affected needles have pale gray brown bands bearing the inconspicuous, concolorous apothecia. Ascospores (68–78 × 3–3.5 μm) are long, clavate, and aseptate and can be used to distinguish this pathogen from *L. montivaga* in the southern Rocky Mountain states. *L. cerina* commonly occurs in association with *Ploioderma lethale* and *P. hedgcockii* in the gulf states. Defoliation caused by the pathogen can be severe but is rarely fatal.

Ascocarps of *L. montivega* and *L. morbida* are commonly parasitized by *Hendersonia pinicola* Wehmeyer, a fungus that may also be a primary pathogen of *P. contorta,* and *H. acicola* Tub. invades ascocarps of *L. cerina.*

Bifusella and Meloderma Needle Blights

Bifusella linearis (Peck) Höhn. causes a needle blight (Bifusella blight or tar spot needle cast) of five-needle pines (*P. strobus, P. flexilis, P. monticola,* and *P. albicaulis*) and is widely distributed in the northern United States and Canada. Two- and 3-year-old needles die back from the tip, leaving attached green bases. The fungus is readily identified by its conspicuous, raised, shiny, black ascomata (0.5–15 mm), which are subcuticular and formed on the abaxial surfaces of necrotic needles. The ascospores are bifusiform, strongly constricted in the middle, and arranged spirally in the ascus. Two other species of *Bifusella* affect pines in the western United States. *B. pini* (Dearn.) Darker occurs on *P. contorta, P. edulis, P. flexilis,* and *P. monophylla* in the southwest and Rocky Mountains and *B. saccata* (Darker) Darker on *P. albicaulis* in the Sierra Nevada.

Meloderma desmaziersii (Duby) Darker causes a needle blight of a wide variety of *Pinus* species, including *P. banksiana, P. monticola, P. radiata, P. strobus, P. resinosa, P. rigida, P. flexilis,* and *P. wallichiana.* Infection occurs during the summer, and scattered yellow spots appear the following spring. The spots coalesce, and needles turn red brown late in the spring. Ascocarps begin to appear in early spring and mature by early summer (May–June). The ascocarps (0.5–1.0 mm) are distinctly black, oval, and raised and have a central slit outlined by the colorless lip cells. Pycnidial conidiomata (presumed spermatia) are usually produced with the ascocarps.

Lophodermium Needle Casts

Many species of *Lophodermium* occur on pines, but only *L. seditiosum* Minter, Staley, & Millar is considered a serious pathogen (Plate 148). Many other species infect healthy needles as early-colonizing parasites that cause no symptoms before they sporulate on senescent or naturally abscised needles. *L. seditiosum* attacks two- and three-needle pines in Europe and also has been found in the United States. Needles are killed late in their first growing season, and the fungus sporulates during late summer or early fall of the second year. Infected needles exhibit a characteristic discoloration and "drooping syndrome." Fruiting can also occur on foliar and cone debris, and control measures include removal of debris and late-summer foliar sprays.

Lophodermium species are characterized by conspicuous, subcuticular or subepidermal, shiny, black, ovoid to ellipsoid ascomata raised above the needle surface. The ascomata have a median central split with lighter colored "lip cells." They are outlined by a darker perimeter line, and sometimes black or brown transverse zone lines are present on needles. Ascospores are long and filiform with a gelatinous sheath. Conidial states of *Lophodermium* (*Leptostroma* spp.) are produced with the ascogenous states and are also commonly produced in cultures isolated from surface-sterilized, healthy needles. In North America, *L. nitens* Darker is a common species on five-needle pines (Plate 149), and *L. molitoris* Minter and *L. conigenum* (Brunaud) Hilitzer are common on hard pines in the north. *L. australe* Dearn. occurs on several southern pines in the southeastern United States, where it might be confused with *Ploioderma lethale*. *L. australe*, however, has long, filiform ascospores and fruits on naturally abscised needles and those killed by other causes.

Davisomycella Needle Casts

Davisomycella spp. produce black ascocarps that superficially resemble those of *Lophodermium* and *Bifusella* spp. The well-developed black clypeus and subepidermal placement of the ascocarps differentiate it from *Lophodermella* spp., which have similar clavate ascospores but lack the clypeus and have subhypodermal ascocarps. *D. ampla* (J. J. Davis) Darker is considered the most injurious of the species, causing periodic defoliation of *P. contorta* and *P. banksiana*. Symptoms of Davisomycella needle cast are browned tips or necrotic bands at the distal ends of green, 2- and 3-year-old needles. Ascocarps (0.75–1.5 mm) are subepidermal and broadly ellipsoid and open by a median longitudinal split. The ascocarps lack the characteristic perimeter line of *Lophodermium* spp., have no lip cells, and are not associated with zone lines. Ascospores are clavate and aseptate. Ascocarps of *D. ampla* on *P. contorta* along the Pacific coast of North America are commonly invaded by *Sarcotrochila macrospora*. Another important species found primarily on ponderosa pine is *D. medusa* (Dearn) Darker, which resembles *D. ampla* but has longer ascocarps. The long, narrow ascocarps of *D. medusa* might be confused with those of *Elytroderma deformans*, but the cylindrical, one-septate ascospores of *E. deformans* distinguish the two species.

Selected References

Childs, T. W. 1968. Elytroderma disease of ponderosa pine in the Pacific Northwest. U.S. Dep. Agric. For. Serv. Res. Pap. PNW-69.

Czabator, F. J., Staley, J. M., and Snow, G. A. 1971. Extensive southern pine needle blight during 1970-1971, and associated fungi. Plant Dis. Rep. 55:764-766.

Funk, A. 1985. Foliar fungi of western trees. Can. For. Serv. Pac. For. Res. Cent. Inf. Rep. BC-X-265.

Harvey, G. M. 1976. Epiphytology of a needle cast fungus, *Lophodermella morbida*, in ponderosa pine plantations in western Oregon. For. Sci. 22:223-230.

Hoff, R. J. 1985. Susceptibility of lodgepole pine to the needle cast fungus *Lophodermella concolor*. U.S. Dep. Agric. For. Serv. Res. Note INT-349.

Jalkanen, R. 1986. *Lophodermella sulcigena* on Scots pine in Finland. Commun. Inst. For. Fenn. 136.

Minter, D. W. 1981. *Lophodermium* on pines. Mycol. Pap. 147. Commonwealth Mycological Institute, Kew, England.

(Prepared by J. Stone)

Foliage Diseases of Other Conifers

Abies spp.

Needle blights of *Abies* spp. are caused mainly by hypodermataceous ascomycetes in the genera *Virgella*, *Isthmiella*, and *Lirula*. All are characterized by long, shiny, brown to black hysterothecia that form on needles that are 2 years old and older and often extend the length of the needle along the midrib (Plate 150). *V. robusta* (Tub.) Darker has cylindrical ascospores; species of *Isthmiella* have dumbbell-shaped (bifusiform) ascospores; and *Lirula* spp. have clavate ascospores. *V. robusta* resembles *L. abietis-concoloris* (Mayor ex Dearn.) Darker but can be easily distinguished by the arrangement of pycnidia in two parallel rows on either side of the midvein. Pycnidia of *L. abietis-concoloris* are produced in a single row along the midvein. The related species *L. punctata* (Darker) Darker has irregularly scattered pycnidia on the upper needle surface and four-spored asci. *V. robusta* is found mainly in western North America and affects *A. amabilis*, *A. concolor*, *A. magnifica*, and *A. procera*; *L. abietis-concoloris* affects *A. amabilis*, *A. fraseri*, *A. grandis*, *A. lasiocarpa*, and *A. procera*.

I. abietis (Dearn.) Darker, with eight-spored asci, and *I. quadrispora* Ziller, with four-spored asci, both affect *A. lasiocarpa*. *I. faullii* (Darker) Darker affects balsam fir and may be the most destructive of the hypodermataceous needle pathogens on *Abies* spp. The conidiomata are produced on the upper needle surface, either along the midvein or more commonly in sinuous parallel rows on either side of the midvein. Conidiomata appear on browned needles in midsummer 1 year after infection. Ascocarps begin to develop at the same time but do not mature until the following year. The fungi *Leptosphaeria faulii* Darker (anamorph *Coniothyrium faulii* Darker) and *Lophomerum autumnale* (Darker) Magasi in Ouellette & Magasi sometimes invade the ascocarps of all three species of *Isthmiella*, preventing maturation of ascospores. *L. faulii* has been suggested as a possible means of biological control of *I. faulii*.

Picea spp.

Lirula macrospora (R. Hartig) Darker (cause of Lirula or Lophodermium needle blight) affects many species of spruce and is readily recognized because of its distinctive narrow, elongate, dark brown to black ascomata that form centrally along the midrib on 2-year-old and older needles (Plate 151). The fungus infects newly emerged needles, which are killed late in the growing season and turn pale yellow to reddish brown and finally pale brown and have a distinct black band at the petiole end. Killed needles remain attached and are shed 2–3 years after infection. Injury by *L. macrospora* is seldom serious, although affected trees may be unsightly and growth may be reduced. Early fruiting of *L. macrospora* initially appears as small, oval blisters, which are colorless at first and later become black, scattered irregularly over the undersides of the killed needles. These blisters are replaced over the course of the next year with the developing ascomata, which have a shiny, black covering layer of fungal tissue and may extend the entire length of the needles. Ascospores are clavate, hyaline, and single celled. Ascomata of *L. macrospora* may be invaded by *Sarcotrochila piniperda* (Rehm) Korf, which has ellipsoid, initially hyaline ascospores that form one septum and turn brown with age.

Isthmiella crepidiformis (Darker) Darker resembles *L. macrospora* and causes a similar foliar browning on black, Engelmann's, and white spruce. Infected needles remain attached, and ascocarps (3 × 0.5 mm), which develop on the lower surfaces of 2-year-old and older needles, are conspicuous, shiny, black, and smaller than those of *L. macrospora*. Ascospores of *I. crepidiformis* are constricted in the middle, hyaline, and aseptate.

Other fungi associated with foliage disease and discoloration in spruce are *Phaeocryptopus nudus* (Peck) Petr., which causes a browning of attached needles (also on *Abies* spp.); *Stigmina verrucosa* (Morgan-Jones) Sutton, a hyphomycete that causes yellow spots on the foliage of *P. glauca, P. engelmannii,* and *P. mariana;* and the coelomycete *Diedickea piceae* Bonar, known only from northern California, which causes a yellow discoloration of spruce needles.

Bud blight of spruce species in northern Europe is caused by *Gemmamyces piceae* (Borthwick) Casagrande (Plate 152), but this fungus so far has not been recorded in North America. In Canada, bud necrosis and dieback is caused by *Camarosporium strobilinum* Bomm., Rouss., & Sacc. on subalpine fir and white spruce and by *Dichomera gemmicola* Funk & Sutton on Sitka, white, and Engelmann's spruce and Douglas-fir. Both fungi can cause severe dieback.

Rhizosphaera species are found on foliage of *Picea, Abies, Pinus, Pseudotsuga,* and *Tsuga* species (Plate 153). They commonly occur as saprobes on needles killed by other causes, but under some conditions, *Rhizosphaera* spp. can apparently cause needle browning, premature needle abscission, and a characteristic girdling and breakage of spruce needles. *R. kalkhoffii* Bubák causes serious needle blights of *Picea pungens* and several other spruce species and *Pinus ponderosa* and *P. thunbergii* in North America and Europe, but injury is greatest in nurseries, in Christmas tree plantations, and on hosts planted outside their native ranges. Injury in natural and reforested stands is generally negligible. The initial symptom is browning of the needles, often more distinct on the distal portion, with a distinct band that separates the diseased and healthy portions. On *Picea pungens*, infected needles turn purple brown and later form shrunken, brown, necrotic bands, usually at the distal ends of needles, on which the pycnidia form. The spherical pycnidia are pale to dark brown and associated with the stomata in regular rows and typically bear the waxy stomatal plug on top.

Larix spp.

A serious needle blight of *Larix* spp. is caused by *Hypodermella laricis* Tub. throughout the northern hemisphere (Fig. 13). It kills both needles and spur shoots. Killed needles are retained after normal needle abscission, and infection of newly emerging needles occurs during the spring. Symptoms of Hypodermella needle blight are first visible as a red brown discoloration of all needles on a spur. *H. laricis* is identifiable from the conspicuous, raised, shiny, black, elliptical hysterothecia that form medially along undersides of attached dead needles during midsummer. A distinct black line also forms at the petiole end of the needle. Ascocarps mature during the winter and open by a median longitudinal split at bud break the following spring. The asci are usually four spored (rarely eight); ascospores are hyaline and club shaped and have a thick, gelatinous sheath. The ascocarps are often accompanied by an acervular conidial (presumed spermatial) state, *Leptothyrella laricis* (Dearn.), which has hyaline, pyriform conidia.

Larch needle cast caused by the conidial fungus *Meria laricis* Vuill. is identifiable by early wilting, browning, and

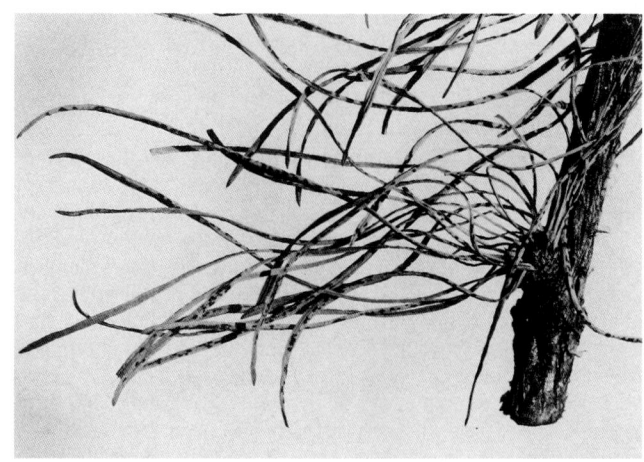

Fig. 13. Hysterothecia of *Hypodermella laricis* on western larch.

premature abscission of needles, which begin soon after needles flush in the spring and continue throughout the growing season as a result of repeated reinfection (Plate 154). Conidia (8–14 × 3–5 μm) are produced in small, pinkish sporodochia that emerge from the stomata (Plate 155). The conidia may be very inconspicuous on withered needles but can be detected more readily in a transparent adhesive tape impression taken from needles hydrated in 70% ethanol, mounted on a microscope slide in lactoglycerol with aniline or trypan blue, and viewed under a compound microscope. The conidia are hyaline, aseptate, and constricted in the middle and become septate with one brown and one hyaline cell upon germination.

Larch needle cast caused by *Mycosphaerella laricina* (R. Hartig) Mig. in Thomé is more common in Europe, but the pathogen was introduced into North America relatively recently. So far, it has not been found west of the Rocky Mountains in the United States or Canada. The disease appears first as small, brown spots that develop on needles that remain attached to the twigs but abscise prematurely during late summer. The anamorph, *Cerceseptoria* sp., develops in the brown spots during midsummer, forming pustules that erupt through the needle epidermis. Black ascocarps (pseudothecia) are immersed in the needles, and elliptical, two-celled ascospores are produced on fallen needles the following spring, coinciding with needle flush.

Cupressaceae

Keithia blight of *Thuja plicata* and *T. occidentalis* is caused by *Didymascella thujina* (E. J. Durand) Maire (Plate 156). Although the disease can be severe in young trees and nurseries and entire shoots can be killed, mature trees apparently tolerate moderate amounts of the pathogen, and it is common and widespread. Symptoms appear first as scattered browning of individual scale leaves on 1-year-old shoots during late winter and early spring. Apothecia begin to erupt through the epidermis, which often remains attached as a small flap, in late spring. Apothecia (0.5–2 mm long) are oval to elliptical, cushionlike, and initially light yellow brown but become darker olive brown. Asci are two spored and have hyaline, ovoid, one-septate ascospores with unequal cells surrounded by a gelatinous sheath. The larger cell becomes dark brown, and the smaller cell remains hyaline while still in the ascus. Characteristic pits are left in the scale leaves when the spent apothecia drop out during the fall and winter. A related species, *D. chamaecyparidis* (J. F. Adams) Maire occurs on *Chamaecyparis thyoides* in eastern North America, and *D. tetramicrospora* Pantidou & Darker occurs on naturally abscised foliage of junipers in western North America.

Cercospora needle blight, caused by *Asperisporium sequoiae* (Ellis & Everh.) Sutton & C. S. Hodges (syn. *Cercospora sequoiae* Ellis & Everh.), is a destructive foliar disease of several coniferous hosts including *Juniperus, Thuja, Chamaecyparis, Taxodium, Cryptomeria* and *Cupressus,* and *Sequoia* and *Sequoiadendron* spp. The fungus is found throughout the southeastern United States, where plantations of *Cupressus arizonica* may be severely affected. The fungus also causes severe damage to *Cryptomeria japonica* in Japan and Brazil resulting from twig and stem cankers. Injury to other hosts is confined to foliage. Older trees are rarely killed, but seedlings, particularly *Cryptomeria* spp., may be killed by the disease in 1–3 years. Symptoms first appear during the summer as a discoloration and necrosis of foliage on the lower branches near the stem that gradually progress upward, leaving only the outermost foliage green. Diseased shoots drop during October through November. Cushionlike conidiomata (100–300 µm long) of *A. sequoiae* appear early in the summer on dead foliage, mainly on the upper surfaces of the needles and scales. Conidia (36–60 × 5 µm) are medium brown, zero to eight phragmoseptate, and verrucose, have blunt bases with prominent thickening, and are tapered toward the apex. The conidia are produced throughout the growing season, mainly during the late spring and summer; dispersal of conidia and establishment of infections occurs during rainy periods. The related *Asperisporium juniperinum* (Georgescu & Badea) Sutton & Hodges is associated with a blight of *Juniperus communis* in the north-central United States and southern Ontario.

A serious needle blight of *Juniperus virginiana* in the southeastern United States and of *J. communis* in the north-central United States and southern Canada is caused by *Pseudocercospora juniperi* (Ellis & Everh.) Sutton & Hodges (syn. *Cercospora sequoiae* var. *juniperi* Ellis & Everh.). The pathogen is similar to *A. sequoiae* and was formerly considered a variant within this species. The conidia (42–57 × 2–3 µm) of *P. juniperi*, which are pale brown and narrower and more finely verrucose than those of *A. sequoiae* and lack a thickened basal scar, and the more prominent conidiomatal stroma serve to differentiate this fungus from *A. sequoiae*. *P. juniperi* also occurs infrequently on *C. arizonica* in the southeastern United States and on *C. japonica* in Japan. Disease symptoms are similar to those caused by *A. sequoiae*.

Because many members of Cupressaceae have needles reduced to scales appressed along the stem, distinction between foliar pathogens and twig blights or cankers is sometimes arbitrary. *Kabatina juniperi*, for example, is discussed in the section on canker diseases.

Tsuga spp.

There are relatively few needle diseases of significance on *Tsuga* spp. *Fabrella tsugae* (Farl.) Kirscht. on *T. heterophylla* and *T. canadensis* in the Pacific Northwest and northeastern United States and *Korfia tsugae* (Cash & R. W. Davidson) J. Reid & Cain on *T. canadensis* in the southern Appalachians are closely related to *Rhabdocline* spp. *F. tsugae* is occasionally reported to cause outbreaks of needle cast disease, but it is also frequently found fruiting on naturally senescent older needles.

Black Mildews and Sooty Molds

The black mildews and sooty molds form superficial mats of melanized hyphae on needle surfaces. The black mildews are obligately parasitic ascomycetes in the order Meliolales, which has a mainly tropical and subtropical distribution. Several species occur on conifers in the temperate zones and usually affect foliage of young trees in dense shade or very humid sites (Plate 157). *Rasutoria abietis* (Dearn.) Barr (syn. *Epipolaeum abietis* (Dearn.) Shoemaker) occurs on *Abies* spp.; *R. tsugae* (Dearn.) Barr affects *Tsuga canadensis, T. heterophylla,* and *T. mertensiana;* and *R. pseudotsugae* (V. Miller & Bonar) Barr affects Douglas-fir. All form dense mats of superficial hyphae on needles, penetrate the needles through the stomata, and form clusters of perithecia on the undersides of needles. Affected needles may be yellowed and have a conspicuous splotchy appearance caused by the dark brown to black colonies. Although unsightly, the affected foliage is apparently able to tolerate the pathogens, which rarely cause severe injury. However, they may contribute to premature abscission of older needles. *Eupelte farrae* (Pirozynski & Shoemaker) Arx & E. Müller (syn. *Maurodothina farrae* Pirozynski & Shoemaker; Asterinaceae), which penetrates epidermal cells by fine connections through hyphopodia that also bear conidia, and *Asteridiella pitya* (Sacc.) Hansf. occur on *Abies* and *Taxus* spp., respectively. *Stomiopeltis pinastri* (Fuckel) Arx forms minute, circular, superficial patches ("flyspecks") on needles of Douglas-fir and several other conifers. The fungus is entirely superficial; the disklike colonies detach easily from the cuticle and cause no internal needle damage.

Sooty molds, such as *Limacinia alaskensis* Sacc. & Scalia, typically occur in association with insect honeydew or foliar exudates on a variety of hosts. The fungi form thick, black, superficial mycelia on foliage and stems. *Atichia glomerulosa* (Achar. ex H. Mann) Stein in Cohn (teleomorph *Seuratia millardetii* (Racib.) Meeker), another epiphytic fungus, forms gelatinous stromatic mats on foliage and bark of many conifers, particularly in very humid locations.

Felt Blights and Snow Blights

Snow blights (white snow molds) and felt blights (brown snow molds) are associated with snow cover of foliage and typically affect lower branches and young trees (Plate 158). These psychrophilic parasites, with optimal growth temperatures near 0°C, rapidly colonize needle surfaces early in the spring under the snow. The brown felt blights are caused by *Herpotrichia coulteri* (Peck) Bose, which affects primarily pines, and *H. juniperi* (Duby) Petr., which affects many genera of Pinaceae and Cupressaceae. *H. parasitica* (R. Hartig) Rostr. causes a similar blight of *Abies alba* in Europe, but most reports of its occurrence in North America are unconfirmed. The mycelium begins as a superficial, epiphytic growth on snow-covered foliage and later becomes endoparasitic and causes the needles to brown and fall. The killed needles are typically bound together by the mass of brown mycelium (hence the name brown felt), which often remains attached to branches. The felty mycelium becomes dormant during the summer but can survive summer desiccation and resumes growth during the winter and spring as a perennial blight. Needles and twigs are killed the first spring after felt formation, and ascocarps mature during the second summer. Significant losses caused by felt and snow blights can occur in northern nurseries when late snow cover persists in the spring.

Snow blights similarly begin as superficial coverings of white mycelia growing on snow-covered foliage. Unlike the mycelium of the felt blights, however, the superficial mycelia do not persist after snow melt. A number of fungal species can be involved, depending on the host. The most common are *Phacidium abietis* (Dearn.) J. Reid & Cain; *P. sherwoodiae* DiCosmo, Nag Raj, & Kendrick; and *Sarcotrochila balsameae* (J. J. Davis) Korf (syn. *Stegopezizella balsameae*) (J. J. Davis) Syd. in Syd. & Petr., which affect primarily Douglas-fir and true firs, although *P. abietis* can also attack white pine and several spruces. *P. dearnessii* DiCosmo, Nag Raj, & Kendrick and *P. taxicola* Dearn. & House in Dearn. affect *Taxus* spp.; *Sarcotrochila piniperda* (Rehm) Korf affects spruce; *Lophophacidium hyperboreum* Lagerberg affects primarily spruce but also balsam fir and white pine; and *Hemiphacidium planum* (J. J. Davis) Korf affects pines. Reports of *P. infestans* P. Karst., the primary snow blight pathogen in Europe, from North America are considered unconfirmed. Needles affected by snow blights are initially discolored yellow, gradually turn red to red brown and finally gray, and remain attached. The apothecia of *Phacidium*, *Hemiphacidium*, and *Sarcotrochila* spp. are circular and erupt through irregular ruptures or flaps of the host epidermis to expose the orange to brown hymenium. *L. hyperboreum* has elongate apothecia (hysterothecia) that open by longitudinal splits.

Selected References

Anonymous. 1974. Keithia disease of western red cedar, *Thuja plicata*. Br. For. Comm. Leafl. 43.

Bjorkman, E. 1963. Resistance to snow blight (*Phacidium infestans* Karst.) in different provenances of *Pinus silvestris* L. Stud. For. Suec. Publ. 5.

Funk, A. 1985. Foliar fungi of western trees. Can. For. Serv. Pac. For. Res. Cent. Inf. Rep. BC-X-265.

Peterson, G. W., and Wysong, D. S. 1968. Cercospora blight of junipers: Damage and control. Plant Dis. Rep. 52:361-362.

Sutton, B. C., and Hodges, C. S. 1990. Revision of *Cercospora*-like fungi on *Juniperus* and related conifers. Mycologia 82:313-325.

(Prepared by J. Stone)

Abiotic Diseases

Abiotic factors, such as various climatic agents, can cause tree injuries directly and induce growth abnormalities, decline, and dieback. Abiotic agents also predispose conifers to biotic diseases. The relative importance of biotic and abiotic diseases in the northern hemisphere changes by latitude. Biotic diseases are of great importance in southern locations, but the role of abiotic diseases increases rapidly toward northern and alpine timber lines. In these environments, trees are subject not only to climate-related injuries, but also to low thermal sum during the growing season. Abiotic diseases are important for natural selection, and damage from climate extremes is a normal component of some ecosystems.

In recognizing causal agents of damage, one can have in mind that biotic agents are generally specific to a certain conifer species or a few related species, whereas damage that occurs at the same time on several tree species, conifers as well as broad-leaved species and possibly shrubs and herbs, most probably is of abiotic origin.

Edaphic and Aquatic Factors

Root condition is very important for conifer trees and their mycorrhizal symbionts. Root function and tree health are affected by factors such as soil texture, nutrition, and aeration. Rapid changes in any of these factors may cause changes in tree physiology, if not in the appearance of the tree. In addition to causing direct visible symptoms, edaphic factors can predispose trees to biotic agents.

Low levels of nutrients are not necessarily pathological phenomena, especially if nutrients are balanced. Nutrient deficiency or imbalance can be triggered by several factors, such as excess nitrogen fertilization, root dieback, frost, and pruning (including natural pruning by organisms such as *Gremmeniella abietina*). Industrial fertilizers (mainly phosphorus and potassium and sometimes nitrogen) can be used to treat deficiencies but do not necessarily prevent nutritional imbalances, which can lead to malformation and death of terminal buds, rounded tops with multiple leaders, and in severe cases to death of an entire tree. Furthermore, malformed buds and shoots are often killed by frost.

Long-term decreases in pH affect the availability of nutrients, which can lead to various pathological problems. For example, forest damage described during the early 1980s in central Europe was associated with magnesium and potassium deficiency caused by lowered pH resulting from acid rain. Several types of dieback have been described from the Hawaiian islands, where lava causes low soil pH and results in dieback and death of the broad-leaved ohia tree (*Metrosideros polymorpha*).

Lack of water, or drought, does not normally cause pathological phenomena in naturally forested areas. However, during very long dry periods, conifers may die in groups, particularly on sites where the soil is shallow and coarse. Newly planted seedlings and germlings tend to wither if dry periods last very long. Furthermore, droughts predispose forests to fires. Winter buildup of surface ice during cold, snowless spells may also cause drought in the soil and asphyxiation of roots that try to grow under an ice lens when the growing season begins. This typically occurs in stream-side patches of less than 1 ha.

Excess water is a well-recognized problem, in both undrained peatlands and mineral soils, particularly if the climate results in formation of thick humus. After clear-cutting, the ground water level tends to rise. For this reason, plowing has been recommended for areas with old forests with thick humus layers if they are to be clear-cut and reforested.

In coastal areas, forests nearest the coast may be exposed to high levels of salt from ocean storm spray or flooding, resulting in tree injury. Storm sprays can also affect needles in the interior of the country.

Climatic Factors

Climatic factors cause considerable and widely reported injuries of various kinds throughout coniferous forests. They cause direct visible injuries and can also affect the resistance of trees to various biotic agents. Climatic factors, particularly temperature, largely determine the location of tree lines.

Low temperatures can injure conifers throughout the year, depending on tree hardiness, the actual temperatures, and the rate of cooling. It is not only late and early frosts during the growing season, but various types of frost-related phenomena during the dormant season that cause injuries to conifers.

Late frosts (spring frosts) are common everywhere. If they occur after bud burst, they damage the very susceptible newly flushed shoots. Species susceptibility to late frosts varies geographically. In Scandinavia, spruce species are most susceptible, whereas in North America, true firs and Douglas-fir tend to be most susceptible to late frosts. Symptoms range from frost rings to pale green needles on bent shoots to death of the current year's shoots.

Early frosts on unhardened conifers during August and September cause bark damage, needle injuries, and discoloration. Early frost damage is most common on introduced conifers and provenances that originate from areas with milder climates and shorter day lengths. These provenances tend to grow longer and start their hardening processes later than indigenous species and local provenances. For example, some Sitka spruce provenances have suffered severe early frost discoloration when introduced into Scotland. Similarly, during a short, cold summer, trees cannot assimilate enough for their hardiness or defense mechanisms to develop sufficiently. This means that they may suffer from frost or other abiotic injuries directly or they may become predisposed to various pathogens. For example, infections of *G. abietina* are most often associated with growing seasons that are unfavorable to Scot's pine.

In northern boreal forests, a summer with a low temperature

sum (and therefore low light intensity) normally results in severe yellowing of the oldest needles, especially on pines. At the end of warm summers, this needle loss may not occur. Drought is also known to cause loss of older needles. However, this often happens earlier in the season.

Although conifers are known to be very hardy during the winter, most frost injuries occur outside the growing season. Chronic injuries occur during the winter and acute injuries during the spring, when the trees are losing hardiness. During midwinter when trees are at their hardiest, heavy frosts can cause plant tissues to lose water. The water loss is promoted by continuous wind, and it can happen anywhere, but especially under alpine conditions. This process, called winter drying (or winter desiccation), occurs slowly throughout the winter months.

Freeze-drying (frost drought) occurs mainly during the spring and affects already clearly dehardened shoots and needles. Typical freeze-drying injury requires that specific weather conditions occur. Solar intensity has to be high enough to warm the shoots and needles in the canopy, particularly in the upper canopy. The warmed foliage begins to transpire and lose water, which cannot be replaced because the soil is frozen. However, the resultant drought is not severe enough or long enough to cause visible damage to water-deficient shoots and needles, unless this situation is followed by a windless, frosty night during which the air temperature drops to -10 to $-20°C$. This causes total dehydration of the water-deficient cells and leads to cell and tissue collapse, death, and necrosis. Only 1 day and night is needed for severe freeze-drying injuries. If injuries occur late in April, the symptoms can be seen a few days thereafter. The topmost part of the canopy and the youngest shoots within the branches are most susceptible to freeze-drying. All tree species grown in the boreal zone are susceptible to freeze-drying, but introduced conifers, such as lodgepole pine (*Pinus contorta*) in Scandinavia, are particularly so. Susceptibility to freeze-drying damage seems to be genetic, because symptoms appear in the same trees repeatedly. Damage on Scot's pine, the most important conifer species in Scandinavia, is common and prominent and is made worse by fertilization with nitrogen. Pines with brown leaders can be found nearly every year and are more numerous two to three times per decade. In fact, about 1% of natural pines in southern and western Lapland have dry tops, most caused by freeze-drying.

The symptoms of both winter drying and freeze-drying appear during the spring before bud burst and are restricted to parts of the plants that were above the snow cover at the time of the injury. Dead shoots become blue stained by early summer; the primary colonizer of frost-damaged shoots in Fennoscandia is *Sclerophoma pithyophila*. Shoot death caused by *G. abietina* does not occur until 2–3 weeks after Scot's pine has undergone bud burst in the middle of June, and the shoots infected do not become blue stained. Instead, their wood tissue is yellow to yellowish green.

The "red belt" phenomenon in its classical form consists of damage located in a band at certain elevations on a mountain slope. The damage is most dramatic if the injured trees are conifers, because dark reddish brown foliage results. Conifers above and below the belt are uninjured. Red belt damage is common in both Europe and North America. However, in North America, the belt is less distinctive. Red belt is caused by rapid changes in winter air temperature, resulting in the freezing and thawing of tissues of needles, shoots, and often buds within a few minutes. There is always a preceding temperature inversion between the valley bottom and the slope above. An air stream or uplifting wind lifts the cold air at the valley bottom to the warmer upper slopes so quickly that normal hardening procedures cannot occur. Warm Chinook winds often contribute to the red belt phenomenon in the Rocky Mountains. Red belt injuries normally occur during the winter months, and symptoms can be seen during late February to early March. Although defoliated, conifers generally survive red belt damage, because buds in the leader shoot and upper branches tolerate the rapid temperature changes.

Sometimes during late autumn and rarely during the winter, low temperatures occur when there is no snow cover on the ground. Lack of insulating snow results in a decrease in soil temperature. With air temperatures of -35 to $-40°C$, the subsurface root layer can be exposed to temperatures as low as -20 to $-25°C$. Pine and spruce root cells begin to die at $-10°C$, and at temperatures below $-10°C$, roots near the surface can be killed by frost. If the root damage is moderate, trees will survive, but there may be exceptional needle yellowing and height growth may cease 2–3 months earlier than normal. Furthermore, these trees may show reduced increment and have shortened needles and top dieback caused by water- and nutrient-supply problems. Premature needle yellowing is caused by translocation of nutrients from the oldest needles to younger, more functional needles. Conifers most likely to suffer root damage are those in places where root systems are least protected against decreases in soil temperature, including forest tree nurseries and sites such as dry, sandy pine heaths (especially if the natural protective vegetation layer is missing). Seedlings in containers are also susceptible.

High temperatures can also cause damage to trees. Although temperatures seldom reach levels that cause proteins to coagulate, this can happen in germlings established on burned sites where the soil-surface temperature may rise dangerously high. High temperatures also result in the droughty conditions that were discussed earlier. Increased solar radiation and increased stem temperatures of trees in newly cut forest edges that face south or southwest can result in sunscald. The symptoms of sunscald are bark splitting and cambial death, particularly when warm daytime conditions are combined with cold night conditions.

High winds are another source of damage that can range from the felling of a few scattered trees to extensive areas of blowdown. Windblown trees can result in an increase in bark beetle populations and other forest pest problems. In addition, winds can cause root injuries to trees that remain standing, leading to invasion by decay fungi and bacteria. In areas where winds are strong and continuous, such as Scotland, grounds have been classified according to wind-throw risk. Strong winds combined with hail injure the bark and cambium of trees, especially in the southern hemisphere. Hail not only decorticates stems, but also breaks shoots.

Snow is a natural and important feature of boreal coniferous ecosystems. Despite its insulating properties and the water it supplies in the spring, snow can also be damaging. Throughout the alpine coniferous forests and areas at relatively high altitudes, the canopies of trees are subjected to snow packing nearly every year. Breakage can occur, depending on the snow load and the weather. In northern Finland, spruce and pine forests suffer from snow breakage at elevations as low as 200 m. Maximum snow loads measured there have been more than 3,000 kg of dry snow per single Norway spruce (19 m tall).

Breakage of tree tops, especially of Norway spruce, favors invasion by decay fungi. Because of the risk of snow damage at high-elevation forest sites, the introduction of exotic conifer species is not recommended. For example, it is thought that snow prevents Scot's pine from growing to merchantable sizes in areas with high snow levels.

Anthropogenic Factors

Atmospheric deposition affects all coniferous forests worldwide, although the amount of deposition varies greatly. Atmospheric pollutants can have a positive effect on conifer health, but most often the effect is negative and occurs either directly through deposition on the foliage or indirectly through accumulation in the rhizosphere. The type, level, and duration of pollution episodes strongly influence the reaction of conifer trees. For example, long-term exposure to a pollutant at a low level may not cause visible injuries, whereas short-term exposure to high levels of sulfur dioxide can cause obvious discoloration of the foliage. Damage caused by atmospheric pollutants is best recognized in the vicinity of emission sources; farther away, their impacts are diminished (see Decline Diseases).

When properly used, herbicides kill weeds without damaging the desired conifer species. However, when used in excess or at inappropriate times, injury can occur. For example, glyphosate oversprayed in Scot's pine plantations may cause yellowing and dwarfing of current shoots and needles if applied before the young tissues lignify. Some herbicides can even be toxic to conifers.

Needle and shoot death, similar to that caused by frost and drought, can be seen on road-side trees when deicing salts have been heavily used during the winter. The damage is most dramatic on lower branches near the road before the conifer growing season begins.

Effect of Global Climate Change on Abiotic Diseases

It has been predicted that the mean global temperature will increase 2–4°C during the next 100 years. Various scenarios and models produce different "futures," but most models predict that boreal coniferous zone temperatures will also increase. This increase in temperature could mean better growing conditions for conifers and result in raising the elevation and latitude limits for conifer growth. However, it also suggests that more damage could result from greater rates of fungal growth and expansion of ranges.

Models are limited in their predictive abilities. It is unknown how the temperature increase might affect seasonal temperatures and fluctuations. Increasing climatic extremes may have negative effects on tree health. For example, an increase of 2–4°C in mean temperature can easily be reached by milder-than-normal winters and cooler-than-normal summers. This means that there will be less snow to protect roots during the winter and less water in the spring when new growth begins. With a parallel increase in winter respiration, this may lead to decline of trees, particularly shallow-rooted conifers such as Norway spruce. If milder winters, which accelerate dehardening, are followed by cold spells, conifers as well as broad-leaved plants will most certainly suffer temperature-related injuries. Recent climatic injuries of Scot's pine on the northern timberline in Fennoscandia indicate that temperature extremes, especially during the spring but also during autumn, have to be seriously considered in the future.

Selected References

Christersson, L., and von Fircks, H. 1990. Frost and winter desiccation as stress factors. Aquilo Ser. Bot. 29:13-19.
Hall, R., Hoefstra, G., and Lumis, G. P. 1972. Effects of deicing salt on eastern white pine: Foliar injury, growth suppression and seasonal changes in foliar concentrations of sodium and chloride. Can. J. For. Res. 2:244-249.
Hüttl, R. F., and Fink, S. 1988. Diagnostic fertilizer trials for revitalizing damaged Norway spruce (*Picea abies* Karst.) stands in southwest Germany. Forstwiss. Centralbl. 107:167-183.
Jalkanen, R. 1990. Nitrogen fertilization as a cause of dieback of Scots pine in Paltamo, northern Finland. Aquilo Ser. Bot. 29:25-31.
Jalkanen, R. 1993. Abiotic and biotic diseases of the northern boreal forests in Finland. Pages 7-21 in: Forest Pathological Research in Northern Forests with a Special Reference to Abiotic Stress Factors. R. Jalkanen, T. Aalto, and M. L. Lahti, eds. Finn. For. Res. Inst. Res. Pap. 451.
Jalkanen, R., and Kurkela, T. 1990. Needle retention, age, shedding and budget, and growth of Scots pine between 1865 and 1988. Pages 691-697 in: Acidification in Finland. P. Kauppi, P. Anttila, and K. Kenttämies, eds. Springer-Verlag, Berlin.
Kolari, K. K. 1979. Micro-nutrient deficiency in forest trees and dieback of Scots pine in Finland: A review. Folia For. Res. Pap. 389.
Nicoll, B. C., Redfern, D. B., and McKay, H. M. 1996. Autumn frost damage: Clonal variation in Sitka spruce. For. Ecol. Manage. 80:107-112.
Roll-Hansen, F., and Roll-Hansen, H. 1987. Skogskader i Farger. Landbruksforlaget, Oslo.
Tranquillini, W. 1982. Frost-drought and its ecological significance. Pages 379-400 in: Physiological Plant Ecology. II. Water Relations and Carbon Assimilation. Encyclopedia of Plant Physiology, vol. 12B. O. Lange, P. Nobel, C. Osmond, and H. Ziegler, eds. Springer-Verlag, Berlin.
Venn, K., and Aamlid, D. 1990. Drought of spruce trees in frozen soils in Norway. Aquilo Ser. Bot. 29:87-90.

(Prepared by R. Jalkanen)

Decline Diseases

Since the 1960s, tree declines have advanced from an obscure topic pursued by a limited number of forest pathologists to one of the primary environmental issues of today, and 10–20 new technical publications appear per month. However, the topic is plagued by a lack of agreement on what defines a decline and a general rush to attach the term to any disorder of trees in order to obtain the maximum emotional response from an environmentally sensitive public. This short review builds on the hypothesis that while the term "decline" describes a condition of deteriorating tree health without reference to causality, there is a need for a more precise phrase, "decline disease," to describe problems that have the decline symptomology (i.e., increasing dieback over time) but that also have unique etiologies that differ from simple biotic and abiotic diseases. Recent reviews of tree decline concepts (e.g., the various articles included in *Forest Decline Concepts*) provide a base of information for synthesizing an integrated concept of decline diseases. Decline diseases are contrasted with diseases with a single causal agent and are characterized by gradual deterioration leading to the death of canopy-dominant, mature trees caused by a specifically ordered combination of preconditioning, triggering, and secondary factors.

The germ theory, which originated about a century ago, has been the foundation of our profession. It directs us to search for a constant association between a specific biotic agent and a particular disease and to verify the pathogenicity of the specific agent through controlled inoculations. Similarly, demonstration of constant association and replication of damage through controlled exposures has evolved as the recognized procedure for demonstrating simple causal relationships with abiotic disease and injury-inducing agents.

It is inappropriate to group a number of simple known and/or unknown diseases under a single term, e.g., the German "Waldsterben." This is a complex of unrelated diseases and injuries with single causal agents rather than a complex disease or a decline disease.

A fundamental characteristic of a decline disease is the absence of a primary causal agent or pathogen. Unfortunately, even a diligent researcher may miss a simple causal agent, and therefore some disorders that are described as declines are later better understood as simple biotic diseases. For example, live oak decline and part of the ash dieback problem (ash yellows) are presently recognized as caused by *Ceratocystis fagacearum* and a phytoplasma (mycoplasmalike organism), respectively. These two diseases do not meet the first criterion for a decline disease, i.e., the absence of a specific pathogen, and therefore have little in common with most decline diseases other than the symptom of a gradual deterioration of the crown.

A second characteristic of a decline disease is the preconditioning (sometimes called predisposition) associated with relative "age" and the canopy position of the affected trees, which is a phenotypic expression of the interaction of genetic and edaphic factors. Manion suggests that viruses should also be considered predisposing factors, because they function as genetic determinants in trees. Decline diseases occur in canopy-dominant, mature trees. Mueller-Dombois emphasizes this point with his cohort senescence model. Populations of trees are simplified and synchronized to senesce together by various perturbations and edaphic factors of the site. Manion uses the term predisposition in his decline spiral and suggests further that decline disease is an ecosystem-stabilizing process whereby competitive-dominant trees die back and decline to allow stress-dominant individuals to survive and contribute to the gene pool of future generations.

The third characteristic of a decline disease is related to the inciting, perturbation, or stress event. A more appropriate term might be "triggering event," a short-term, widespread event such as drought, tropical storm, winter injury, or insect defoliation. Triggering events occur randomly and infrequently during the lifetime of a population of trees. Auclair et al suggest that a universal mechanism to account for historic episodes of forest dieback is xylem cavitation initiated by extreme freezing or moisture fluctuations. These events trigger dieback as a general response in the tree population. Some trees recover, while others, during the subsequent decades, die back farther (decline) and die. The point to recognize is that dieback is a general adjustment response to these stress-inducing events. It is not a diagnostic symptom of decline.

Dieback may lead to recovery or to decline. The trees that decline are those that are preconditioned by age and crown position. These trees are highly dependent on annual growth of vascular xylem tissues to transport massive amounts of water and minerals from the roots to large, demanding crowns. Defoliation and dieback cause trees to direct energy to refoliation rather than to growth of xylem tissues. A stress event may also redirect energy to reproduction in mature trees and thereby further limit xylem growth. Understory trees do not have such demanding crowns, and younger trees do not have the option of reproduction. Therefore, understory trees and younger trees may recover after a dieback-triggering event, while mature, overstory trees, preconditioned by highly demanding crowns, die back farther because of the decreased transport capacity. Crown dieback places further demands on energy to establish defensive zones to limit invasion by microorganisms into living tissues from the many interfaces with the undefended dead tissues. Dieback in preconditioned trees, therefore, feeds back in a nonreversible loop to further dieback and decline. It also predisposes trees to infection and infestation by organisms of secondary action, further limiting recovery.

The fourth feature of a decline disease includes the contributing factors that are variously referred to as biotic factors, saprogens, or secondary action organisms. These microbes and insects are not sufficiently aggressive to overcome the defenses of healthy trees but are adapted to infect or infest stressed trees that have limited energy reserves for defense. Although these biotic agents may be constantly associated with declining trees, it is generally not possible to induce decline by inoculation or infestation experiments. By invading or infesting vital root and stem tissue, these agents further hasten the dieback and may provide the coup de grace for preconditioned trees that have been subjected to the stresses of a triggering event.

In summary, decline diseases gradually lead from senescence to death of populations of mature, dominant trees. There is no single biotic or abiotic cause for death, since none of the associated factors alone can be demonstrated to produce the symptoms. Competitive dominant trees are preconditioned over time to depend on the annual production of xylem transport tissues to maintain a large crown. Triggering events initiate crown dieback, refoliation, seed production, defense, and possibly other energy-demanding processes that limit growth of xylem transport tissues. The reduction in xylem transport tissues limits recovery of mature, dominant trees, resulting in further dieback and predisposition to invasion by organisms of secondary action. Contributing (secondary action) organisms invade or infest vital root and stem tissues, further compounding the dieback leading to death. Therefore, decline diseases are characterized by a gradual dieback over time (decline) of mature, canopy-dominant populations of trees 1) preconditioned by the long-term interaction of age and edaphic factors, 2) triggered by a short-term, infrequent weather or biotic event that affects all trees but that initiates compounding effects on preconditioned trees, and 3) further compounded by the action of secondary, contributing organisms that provide the coup de grace.

Selected References

Auclair, A. N. D., Worrest, R. C., Lachance, D., and Martin, H. C. 1992. Climatic perturbation as a general mechanism of forest dieback. Pages 38-58 in: Forest Decline Concepts. P. D. Manion and D. Lachance, eds. American Phytopathological Society, St. Paul, MN.

Houston, D. R. 1992. A host–stress–saprogen model for forest dieback–decline diseases. Pages 3-25 in: Forest Decline Concepts. P. D. Manion and D. Lachance, eds. American Phytopathological Society, St. Paul, MN.

Kandler, O. 1992. The German forest decline situation: A complex disease or a complex of diseases. Pages 59-84 in: Forest Decline Concepts. P. D. Manion and D. Lachance, eds. American Phytopathological Society, St. Paul, MN.

Manion, P. D. 1991. Tree Disease Concepts. Prentice-Hall, Englewood Cliffs, NJ.

Mueller-Dombois, D. 1992. A natural dieback theory, cohort senescence as an alternative to the decline disease theory. Pages 26-37 in: Forest Decline Concepts. P. D. Manion and D. Lachance, eds. American Phytopathological Society, St. Paul, MN.

Skelly, J. M., and Innes, J. L. 1994. Waldsterben in the forests of central Europe and eastern North America: Fantasy or reality? Plant Dis. 78:1021-1032.

(Prepared by P. D. Manion)

Recent Forest Declines in Europe

Other names: Waldsterben, Tannensterben
Causal agents: controversial and varied, including drought and point source pollution

During the 1980s, the literature on forest health was dominated by accounts of a new type of decline that was believed to be affecting trees throughout Europe. Initially, reports were confined to the conifers Norway spruce (*Picea abies*) and silver fir (*Abies alba*), but later reports (from the mid-1980s) also included broad-leaved trees. The first records of the problem came from Bavaria and Baden-Württemberg in southern Germany, and problems were later reported from other areas in Germany (e.g., the Harz Mountains). Subsequently, reports of forest decline appeared from a variety of European countries, especially after the establishment of an international inventory of crown condition, which supposedly revealed both the massive extent of the decline and its rapid spread through Europe. "Acid rain" was usually considered to be the culprit. The decline was believed to be different from the air pollution injury that had been observed in industrial areas since the nineteenth century. The pollution injury was generally restricted to specific areas in the neighborhood of major sources of pollution and was characterized by direct foliar injury.

Symptoms

Early descriptions of the declines were generally very inadequate. The two most important symptoms (and the only ones assessed in the international surveys of "forest health") were defoliation and discoloration. Neither of these symptoms was new, although it was often argued that their widespread occurrence and rapid spread were novel. The discoloration was based on the yellowing of needles reported from certain areas and found to be caused by a magnesium deficiency. The yellowing typically occurred in older needles and often developed during the spring prior to bud burst. This symptom can be attributed to the withdrawal of magnesium from the older needles and its allocation to new growth. Defoliation is based on visual assessment of the transparency of the crown and is therefore a subjective measure of what the observer believes should be present. It is not based on any quantitative assessment of the needle longevity or the proportion of foliage with premature abscission.

It is worth examining the claims of a general, rapid, and unprecedented forest decline in detail, because they form the foundation of the belief in the existence of a major problem in European forests today and have also been the justification of a huge amount of research on tree ecology and physiology in both Europe and North America.

Extent of the Decline

Although the extent of the defoliation and discoloration was taken as evidence of the decline in Europe, in 1992 discoloration was reported on only 7.5% of the 56,882 conifers sampled. Of this, 5.6% involved trees scored as having 10–25% of the remaining foliage discolored. The nature of the discoloration was not given; the field surveys reported only whether discoloration of any form was present. There was no information on what proportion of trees in Europe would be expected to be discolored in mid- to late summer when the assessments were conducted.

In Europe, 24.3% of the trees were scored as having more than 25% defoliation. Early reports used a threshold of 10% defoliation as an indication of damage of some form, but this threshold was increased to 25% in the late 1980s as it became apparent that the 10–25% category was an interim phase. A second threshold of 60% defoliation was set as the point at which trees would not recover. In fact, after 10 years of measurements, it is apparent that trees with this level of defoliation can recover, and the entire concept of a general defoliation threshold, at which damage can be defined, is viewed with increasing suspicion.

The national and international figures for defoliation and discoloration are misleading because there are major differences in the standards used within and between countries. It is now clear that defoliation is assessed in different ways in different countries and that the results obtained for one country cannot be reliably compared with the results obtained in another. Although this has been suspected for some time and was demonstrated as early as 1986, international reports have coninued to produce tables comparing the extent of defoliation in different countries, albeit with a footnote pointing out that direct comparisons between countries are not possible.

Spread of the Decline

The increase in the number of reports of defoliation from different countries during the 1980s has been considered evidence of a spread in forest decline. However, this ignores the simple fact that until 1983, no country had undertaken a formal, systematic survey of defoliation and that during the next 7 years, more and more countries joined the international program. Consequently, the increase in the number of reports can be directly related to the number of countries participating in the program. To date, no geostatistical assessment has been made in the changes in defoliation through time on a European scale, although maps have been plotted showing the changes site by site from one year to the next.

Since the early 1990s, new methods have been developed for looking at the long-term needle retention of conifers. In Scot's pine, a method was developed to enable reconstruction of the numbers of needles held by the lead shoot throughout the life of the tree. A similar method exists for needle retention in Norway spruce. Both methods provide the possibility of determining the "normal" needle retention for conifers, but to date they have been little used. With this method, a decline in needle retention of Scot's pine at two sites in England since the 1980s has been documented. However, given the relatively small sample, it is not yet clear how widespread this trend is.

Actual Declines

The confusion that has arisen as a result of the misinterpretation of the international survey results should be separated from the clear evidence that exists for the sporadic declines that have occurred in some species at some sites. Two seem to be particularly important: the declines of silver fir and Norway spruce in eastern France and southern Germany. These declines should be clearly distinguished from the important declines that have occurred in other parts of Germany, southern Poland, the Czech Republic, northern Russia, and other areas as a result of very high levels of sulfur dioxide. These declines are well documented in the older literature.

The declines in France and Germany were characterized by marked yellowing of older needles and by defoliation. In Germany, analyses of yellowed needles revealed magnesium deficiency as the cause of the yellowing, and fertilization with magnesium resulted in recovery of the trees. In France, the situation appeared a little more complex. The yellowing was also caused by magnesium deficiency, but drought stress and overstocking within the stands also appeared to play roles.

Causes

There has been a huge amount of literature published on the causes of the recent forest decline. This work can be broadly divided into chamber studies, field studies, and epidemiological studies. The majority of experiments have been concerned with the role of air pollution in defoliation and discoloration.

Chamber studies. For physical reasons, chamber studies are restricted to relatively short-term experiments on the effects of specific pollutants or combinations of pollutants on seedlings or young trees. This work has shown that the doses required to induce the symptoms seen in the field are generally higher than ambient levels. However, major differences in the responses to pollution exist between and within species, and it is difficult to generalize. A number of problems have arisen in chamber studies, and recent work has suggested that the responses of young trees grown in chambers may be very different from the responses of larger trees exposed to the same pollutant doses. However, the results from such experiments strongly suggest that air pollution (particularly ozone) at levels experienced in Europe today could be having an effect on a number of physiological processes.

Field studies. Because of the cost and time involved, there have been many fewer field studies than chamber studies. However, field studies are much more useful, since they can be directly related to what is happening in the forest. To date, field studies in Switzerland and Germany have suggested that air pollutants are not the primary causes of the defoliation or discoloration seen in trees. However, deposition of pollutants, particularly nitrogen and sulfur, may be having an important effect on soil chemistry and biology, which subsequently affects trees.

Epidemiological studies. Epidemiological studies provide a method of generating and screening hypotheses, but they need to be supported by experiments. A number of studies have been conducted and have tended to reveal relationships between defoliation and a variety of different environmental factors. For example, three groups of Sitka spruce stands in England appeared to share a number of crown characteristics that were associated with higher than average levels of pollution. However, the stands were all located in the northern part of the country, and some other environmental agent, such as winter injury, may account for the clustering of symptomatic stands.

In Germany, a study of the spatial correlations between defoliation and air pollutants involved a variety of environmental parameters, including soil and weather conditions and sulfate and nitrate deposition. A statistically significant positive correlation was found between defoliation and parameters such as tree age, elevation, and soil hydrology, but no correlation was found between defoliation and air pollution. Contrary to expectations, defoliation was lowest on acidic podzolic soils and highest on well-buffered, neutral to basic, calcium-rich soils, such as rendzinas and cambisols.

Current Knowledge

The present state of knowledge about declines of conifers in Europe can be summarized as follows. A variety of different declines have occurred in the past, and it is likely that they will continue to occur. These declines invariably involve a single species and are restricted to specific areas. Careful analysis of each situation often reveals a likely cause. Large-scale forest declines, involving a variety of species and the general breakdown of the forest ecosystem, are restricted to the areas surrounding point sources of air pollution. The presence of trees with low crown density does not necessarily mean that a state of decline exists because the "normal" crown condition of conifers has not yet been documented and the assertion that all trees should have no loss of density contradicts the present state of knowledge concerning forest dynamics.

The possibility of a decline in the health of forests is a subject that arouses intense emotional feelings in much of Europe. Consequently, the subject has received a huge amount of media coverage that has not always been accurate. At the same time, environmental pressure groups have used forest health as a general indicator of the state of the environment and have tended to interpret the results of any surveys as evidence that their policies are vindicated. As a result, the issue has become firmly entrenched within the popular press, and although scientific interest in the phenomenon has been replaced by research on climate change and elevated levels of carbon dioxide, it is still widely thought that forests are continuing to decline.

Selected References

Innes, J. L. 1993. Forest Health: Its Assessment and Status. CAB International, Wallingford, England.
Jalkanen, R. E., Aalto, T. O., Innes, J. L., Kurkela, T. T., and Townsend, I. K. 1994. Needle retention and needle loss of Scots pine in recent decades at Thetford and Alice Holt, England. Can. J. For. Res. 24:863-867.
Kandler, O. 1990. Epidemiological evaluation of the development of Waldsterben in Germany. Plant Dis. 74:4-12.
Schulze, E.-D., Lange, O. L., and Oren, R., eds. 1989. Forest Decline and Air Pollution. A Study of Spruce (*Picea abies*) on Acid Soils. Springer-Verlag, Berlin.
Skelly, J. M., and Innes, J. L. 1994. Waldsterben in the forests of central Europe and eastern North America: Fantasy or reality? Plant Dis. 78:1021-1032.

(Prepared by J. L. Innes)

Northeastern Subalpine Red Spruce Decline

Other names: red spruce decline
Causal agents: site-specific biotic agents, injury-inducing events, and naturally induced decline disease
Species affected: red spruce (*Picea rubens*)
Distribution: the mountains of the northeastern United States

During the 1980s, the remnant of a logging- and fire-decimated red spruce population experienced a reawakening of interest and importance that is almost unparalleled in modern forest history. The attention was generated not because of present or future economic value, but because of a perception that increased mortality and decreased growth were caused in some way by acidic deposition. This section will reexamine the evidence for atmospheric deposition as a factor in a regional decline of red spruce and introduce an alternative hypothesis that the observed symptoms are the consequence of a complex of site-specific biotic agents, injury-inducing events, and naturally induced decline disease.

Red spruce is a component of the eastern spruce-fir forest type, which is most prominent on the subalpine upper slopes of the northern Appalachian and Adirondack Mountains where the climate is too harsh for most hardwoods, except paper birch. The spruce-fir forest type occupies the transition zone between the northern hardwoods and the boreal forest.

Symptoms and Causes

As with most decline diseases, a diverse array of nonspecific symptoms is associated with red spruce decline. Many specific symptoms are associated with particular damaging agents on a site-specific basis.

Dendrochronological evidence supports a widespread, recent pattern of reduced growth in red spruce. Although it is often stated that the growth loss began in the 1960s, the onset varies widely among individual trees (from 1944 to 1980 in one study), suggesting that local site-specific events, not a regional change, are responsible.

Needle reddening and bud death are the symptoms most often associated with red spruce decline and are most easily seen during the spring and summer, respectively. They are usually ascribed to low temperatures late in the fall and early in the winter that exceed the minimum cold tolerance of the trees. Observations during 1992 and 1993, however, suggested that damage occurred during late winter. Needle reddening and bud death are observed on red spruce at all elevations and are not specific to the high-elevation decline syndrome.

Crown condition is rated in various ways and includes needle retention and branch dieback. There is no consistent trend toward crown deterioration in the region, and higher damage ratings are not associated with increased mortality. Branch and crown dieback are most apparent in overstory trees in open, poorly stocked, high-elevation stands, suggesting that wind- and ice-induced damage may be important.

Root deterioration has been studied at only a few sites and seems to be associated with root abrasion resulting from bole movements during strong winds. Root deterioration was correlated with crown dieback on one site but not another.

Witches'-brooms (clusters of branches) have been noted on both living and dead spruce in decline areas. In many cases, they result from the emergence of recovery shoots from branches where dormant buds have been released by the death of the terminal bud. In other cases, they are caused by dwarf mistletoe (*Arceuthobium pusillum*) infection. Careful examination is necessary to distinguish the causes.

Death of red spruce is the presumed end result of decline. Mortality levels in many high-elevation stands are 50% and greater, in contrast to a regional average for red spruce and other species that is close to 15%. Specific agents can usually be identified as causes of canopy gaps in the spruce-fir forest. Root-decay fungi are especially important at low elevations, and wind stress causes most gaps at high elevations. In mature red spruce stands, the spruce beetle often causes the deaths.

A major hurricane on November 25, 1950, caused unprecedented damage to mountain forests in New York State. Some tree deaths observed in earlier surveys were undoubtedly the direct consequences of that storm. Today, trees that survived the 1950 storm continue to die in some situations as a result of dynamic interactions between changes in intertree competition and exposure to wind set up by the 1950 storm.

Acidic Deposition

The speculation that acidic deposition is responsible for the decline of high-elevation spruce is based on two recent studies involving exclusion chambers. DeHayes et al suggested that fall and winter cold tolerance (critical temperature) of red spruce seedlings at a mountain site in Virginia was reduced by 3–5°C by exposure to ambient pollutants. A significant reduction was observed only in the October and November samples. The most cold-tolerant sample, with a critical temperature of −46.1°C, was an unchambered sample taken in January. Vann et al suggested that the cold tolerance of trees protected from high-elevation ambient air and cloud water in New York was approximately 10°C greater than that of unprotected trees. Exclusion chambers were placed around branches of four naturally occurring mature trees, but only two trees were responsive to the treatments. The January data showed that the critical temperature for exposed needles of these two trees was −26 to −29°C, while the needles of the various protected treatments on these same trees had a critical temperature of −34 to −46°C. It is hard to understand how these trees could have survived without protection, since temperatures below this threshold occur almost every winter in this region.

The limited sample size and inconsistencies within and between these two papers do not warrant the very definitive interpretation that acidic deposition affects winter cold tolerance and therefore contributes to the decline of red spruce. The evidence for the involvement of air pollutants in the decline of red spruce, after more than a decade of intensive research, is still very inconclusive.

Red Spruce: A Natural Disturbance-Induced, Gap-Phase Ecosystem

It is important to recognize the dynamic, climatically driven history of the northeastern subalpine forest and the relatively recent and tenuous position of red spruce. Since the last ice age (about 12,000 years ago), pollen records indicate that the Adirondack Mountains have been successively dominated by white spruce, alder, hemlock, and white pine. About 2,500 years ago, the subalpine red spruce-balsam fir-paper birch forest of today became dominant. Red spruce has a long life expectancy, extreme shade tolerance, and slow growth rate. Balsam fir has a shorter life expectancy but faster growth rate. Paper birch rapidly colonizes the site after a major disturbance. As the birch mature, wind and ice damage trigger decline disease and death, allowing balsam fir an opportunity to dominate for a period until its senescence and decline allows red spruce to dominate and eventually, in turn, to senesce and decline.

Wind, ice, and winter freezing injury are important preconditioning factors in the premature aging of trees in this subalpine region. Major events synchronize populations of trees by triggering decline of a senescing portion of the population and subsequent replacement with another species or a younger age group of the present species. A major triggering event such as the 1950 hurricane had an impact on all the surviving trees. The full impact of a triggering event on preconditioned trees may not be recognized for decades, but the senescence, decline, and death of individual trees allow wind and ice to more directly impact the survivors. The dynamic decline in red spruce is therefore a continual process integrated with balsam fir and paper birch in a natural disturbance-induced, gap-phase ecosystem that attains stability only when viewed in the long term.

Selected References

Anonymous. 1988. Proceedings of the US/FRG research symposium: Effects of atmospheric pollutants on the spruce-fir forests of the eastern United States and the Federal Republic of Germany. U.S. Dep. Agric. For. Serv. Gen. Tech. Rep. NE-120.

Cox, S., and Miller-Weeks, M. 1991. Damage agents associated with visual symptoms on red spruce and balsam fir in the northeastern United States. U.S. Dep. Agric. For. Serv. Rep. NA-TP-01-91.

DeHayes, D. H., Thornton, F. C., Waite, C. E., and Ingle, M. A. 1991. Ambient cloud deposition reduces cold tolerance of red spruce seedlings. Can. J. For. Res. 21:1292-1295.

Eagar, C., and Adams, M. B., eds. 1992. Ecology and Decline of Red Spruce in Eastern United States. Ecol. Stud. 96. Springer-Verlag, New York.

Manion, P. D., and Castello, J. D. 1993. Snow depth identifies late winter as the "window" for freezing injury of red spruce. (Abstr.) Phytopathology 83:1351-1352.

Reams, G. A., and Van Deusen, P. C. 1993. Synchronic large-scale disturbances and red spruce growth decline. Can. J. For. Res. 23: 1361-1374.

Vann, D. R., Strimbeck, G. R., and Johnson, A. H. 1992. Effects of ambient levels of airborne chemicals on freezing resistance of red spruce foliage. For. Ecol. Manage. 51:69-79.

Worrall, J. J., and Harrington, T. C. 1988. Etiology of canopy gaps in spruce-fir forests at Crawford Notch, New Hampshire. Can. J. For. Res. 18:1463-1469.

(Prepared by P. D. Manion)

Yellow-Cedar Decline

Other names: Alaska-cedar decline, Alaska yellow-cedar decline

Causal agents: primary cause unknown; biotic factors play a minimal role; site factors are important; damage to fine roots by either freezing or soil toxicity are two leading hypotheses

Species affected: yellow-cedar (*Chamaecyparis nootkatensis*); other tree species less directly affected

Distribution: southeastern Alaska

Yellow-cedar is an ecologically important and exceptionally valuable tree species in the coastal areas of British Columbia and Alaska. Smaller, scattered populations are also found in Washington and Oregon. However, it is suffering from the most spectacular forest decline in western North America; more than 200,000 ha of unmanaged forest throughout southeastern Alaska are affected (Plate 159). Within the natural range of this tree species, decline is either absent or not severe farther to the northwest along the Gulf of Alaska or to the south through British Columbia to the Oregon-California border.

Symptoms and Diagnosis

Symptoms of yellow-cedar decline suggest a below-ground or root problem. Crowns die as a unit rather than as scattered, individual branches. Crown death in some trees is relatively sudden and results in trees with a full, but reddish brown, complement of foliage. Slow death with gradual thinning of proximal foliage throughout the crown for 10 years or more is more typical of declining cedars. Declining trees rarely recover. Root symptoms include dead fine roots (apparently the initial symptom), dead small-diameter roots, and necrotic lesions on large roots. Narrow, necrotic, cambial lesions often spread from dead roots up the bole of the tree during the final stages of death. In declining yellow-cedar forests, nearly all large cedars are now dead. Dendrochronological evidence indicates that mortality has been progressive for about 100 years. Most living Alaska-cedar trees show symptoms of decline, while other species and sometimes understory Alaska-cedar appear to be healthy.

Development of Decline

Widespread death of yellow-cedar began about 1880; many of the trees that died more than 100 years ago are still standing today. The slow deterioration of snags (dead trees) is the result of yellow-cedar's extreme resistance to wood decay, which has provided the opportunity to reconstruct the epidemiology of decline over the past century. Spread of mortality from one site to another apparently has not occurred since onset, but local advance (less than 1 m per year) from bog and semibog plant communities to adjacent stands along a gradient to better drainage has occurred (Plate 160).

Predisposing Factors and Possible Causes

Biotic factors are apparently not the primary causes of decline. The only common potentially destructive insects on yellow-cedars are *Phloeosinus* spp. (bark beetles), which frequently attack the cambium and phloem of cedars but only during the late stages of decline. More than 50 taxa of fungi have been collected or isolated from Alaska yellow-cedar, including *Armillaria* sp., but none has the capability to kill unstressed trees. *Phytophthora gonapodyides* (Petersen) Buisman was infrequently isolated from baits placed beneath cedar trees but never directly from cedar tissues. It is not the primary cause of decline, because it lacks strong pathogenicity on yellow-cedar. Four genera of parasitic nematodes were found in declining forests, but their low populations and association with healthy forests indicate that they do not cause decline. Viruses and phytoplasmas, which are unlikely threats to conifers, are currently under study. Basal scars, once thought to be symptoms of decline, are caused by the Alaska brown bear (*Ursus arctos*). More than half of yellow-cedars in many stands have old or fresh wounds, but scar incidence is not greater on dying trees than on healthy trees, nor is it greater on cedars in declining stands than in healthy stands.

Although local spread of decline, the single species directly affected, and some symptoms suggest a possible pathogenic cause, research conducted to date suggests that biotic factors play a minor causal role in decline. Site factors, especially poorly drained soils, appear to be more important. Yellow-cedar trees growing on productive, well-drained sites have been unaffected. Abiotic factors that may have triggered decline are soil toxins and freezing damage. Organic toxins that kill cedar roots could result from natural, anaerobic decomposition in the wet, highly organic soils. Another hypothesis involves a warming climate that has caused more rain than snow to fall during the winter, thereby reducing the insulating snowpack at low elevations where decline is most severe. Fine roots of yellow-cedar, which are shallow on wet sites, might then be more exposed to freezing damage when frigid, continental air moves across the region, an event that occurs periodically every winter.

Ecological Significance and Disease Management

The ecological effects of yellow-cedar decline are substantial. Most canopy-level trees die, reducing transpiration and opening the lower canopy strata and forest floor to numerous physical changes (e.g., in light, temperature, and precipitation levels). Forest structure is greatly altered by the abundance of snags in various stages of deterioration. Species composition and succession are affected as yellow-cedar forests give way to western hemlock (*Tsuga heterophylla*) and mountain hemlock (*T. mertensiana*), which raises concerns about biodiversity in a region of few tree species. However, the survival of yellow-cedar is not presently threatened. Currently, there is little effort to manage for this naturally occurring forest decline other than to salvage dead trees and evaluate yellow-cedar's regeneration. New information on the lack of site-to-site spread and the fact that no contagious agents appear to cause decline have allowed forest managers to resume the management of the valuable yellow-cedar tree on unaffected sites without fear that it will spread. Because of its early onset and occurrence in relatively pristine forests, yellow-cedar decline may be one of the best examples worldwide that a forest decline can result from a natural process.

Selected References

Hennon, P. E., Hansen, E. M., and Shaw, C. G., III. 1990. Causes of basal scars on *Chamaecyparis nootkatensis* in southeast Alaska. Northwest Sci. 64:45-54.

Hennon, P. E., Hansen, E. M., and Shaw, C. G., III. 1990. Dynamics of decline and mortality of *Chamaecyparis nootkatensis* in southeast Alaska. Can. J. Bot. 68:651-662.

Hennon, P. E., Shaw, C. G., III, and Hansen, E. M. 1990. Dating decline and mortality of *Chamaecyparis nootkatensis* in southeast Alaska. For. Sci. 36:502-515.

Hennon, P. E., Shaw, C. G., III, and Hansen, E. M. 1990. Symptoms and fungal associations of declining *Chamaecyparis nootkatensis* in southeast Alaska. Plant Dis. 74:267-273.

(Prepared by P. E. Hennon)

Pole Blight

Causal agent: apparently, the interaction of drought, poor water-holding capacity of soils, and trees in great competition and need for water

Species affected: western white pine (*Pinus monticola*)

Distribution: Idaho, western Montana, and eastern Washington in the United States and southern British Columbia in Canada

Pole blight directly affected a single tree species, western white pine, to such an extent that it threatened its future commercial use in a large portion of its range. Western white pine is a fast-growing, commercially important tree with a natural range along the Pacific coast of California to British Columbia and inland to Idaho, Montana, eastern Washington, and southern British Columbia. Pole blight killed western white pine throughout these inland forests but was not a problem along the Pacific coast, except for several locations in British Columbia. Mortality was confined to even-aged, "pole-sized" trees in stands that were about 40–100 years old. Older western white pine growing within the distribution of pole blight escaped damage.

Pole blight was first noticed in 1927 in northern Idaho. By the mid-1930s, mortality was severe enough to warrant surveys and investigations of the cause. The pole blight epidemic appeared to be intensifying during the 1930s and 1940s. By 1954, approximately 40,000 ha were known to be affected. The problem apparently subsided during the late 1950s. A complicating factor at that time was the presence of the fungus *Cronartium ribicola* J. C. Fisch., cause of the white pine blister rust disease. This fungus was accidentally introduced into Vancouver, British Columbia, in 1910 and eventually spread throughout the range of western white pine. It killed so many western white pines that the effects of pole blight were somewhat obscured. Regardless, death caused by pole blight was diminishing by the late 1950s.

Symptoms and Diagnosis

Pines affected by pole blight were characterized by a yellowing of needles, a reduction in terminal and radial growth, bole lesions with resinosis, and dead fine roots. Death in the crown usually began at the tops of trees and progressed down. Rarely, if ever, did affected trees recover. A fungus, *Leptographium* sp., was isolated from necrotic bole lesions and found to have the ability to kill seedlings and cause lesions but did not induce crown symptoms. Lesions were a secondary symptom of pole blight, however, and appeared only after trees showed a loss of vigor and reduced radial growth. The fungus *Armillaria* sp. was present but not consistently found on dying pines and was not the primary cause.

The initial symptom of pole blight may have been dying fine roots. Death of fine roots and the general incidence of pole blight were significantly correlated with low water-storage capacity of soils and shallow root depth. Pole-sized pines growing on sites with deeper soils and increased water-storage capacity had less rootlet mortality and less tree mortality. Leaphart and Stage determined that the period from 1916 to 1940 represented the most adverse (i.e., hot and dry) growth conditions for western white pine in the last 280 years. A reasonable hypothesis is that moisture stress was the primary cause of pole blight, initially causing fine roots to die and indirectly triggering other, secondary symptoms. Moisture stress resulted from the interaction of drought, soils with poor water-holding capacity, and pole-sized pines in the maximum growth phase having a great demand for water.

Ecological Significance and Disease Management

Pole blight is no longer a concern to forest managers. During the 1950s, stands exhibiting pole blight were dropped from blister rust-control programs. Combined with white pine blister rust, pole blight had a significant economic and ecological impact by reducing populations of western white pine over large areas. Pole blight has affected the species composition of today's forests by greatly reducing the abundance of western white pine several decades ago and may serve as one of the best-documented examples of how climate can interact with other factors to influence a forest ecosystem.

Selected References

Graham, D. P. 1958. Results of pole blight damage surveys in the western white pine type. J. For. 56:652-655.

Leaphart, C. D. 1958. Pole blight—How it may influence western white pine management in light of current knowledge. J. For. 56:746-751.

Leaphart, C. D., and Copeland, O. L., Jr. 1957. Root and soil relationships associated with the pole blight disease of western white pine. Soil Sci. Soc. Am. Proc. 21:551-554.

Leaphart, C. D., and Stage, A. R. 1971. Climate: A factor in the origin of the pole blight disease of *Pinus monticola* Dougl. Ecology 52:229-239.

Parker, A. K. 1951. Pole blight recorded on the British Columbia coast. Dominion Dep. Agric. For. Biol. Div. For. Pathol. Note 4.

(Prepared by P. E. Hennon)

Austrocedrus Decline

Common names: mal del ciprés de la cordillera or secamiento del ciprés (Spanish)
Causal agent: undetermined
Species affected: *Austrocedrus chilensis*, Cupressaceae
Distribution: Argentina

Austrocedrus chilensis is a conifer species endemic to the southern Andes. In Argentina, it grows in the foothills along a strip only 60–70 km wide that extends between 39°30′ and 43°35′ south latitudes. This apparently limited east-west range actually accounts for a broad moisture gradient, to which *A. chilensis* is very well adapted. It can be found as a component of the mesic *Nothofagus* forests in areas that receive more than 1,500 mm of precipitation per year in the west or forming open, pure, xeric forests that receive less than 600 mm per year in the east.

A. chilensis suffers from an extensive decline that was first noticed in 1945 (Plates 161 and 162). Since then, there has been an evident expansion and intensification of the decline throughout the entire distribution of this species in Argentina. Although *A. chilensis* also occurs in Chile, no extensive mortality has been observed there. Aboveground symptoms of the decline seem to reflect water deficiency and include chlorosis and death of needles that progress from the proximal to the distal portions of the branches, resulting in crown thinning and often in death of all the foliage. Roots appear to be the most affected parts of the tree. By the time aboveground symptoms are evident, roots show an extensive decay that sometimes results in tree fall while the foliage is still green. Patterns and types of root decay vary. Some species of basidiomycetes have been found consistently associated with wood decay of roots, but their role in the decline remains unclear. Death of cambium and phloem sometimes extends to the collar and basal portions of the trunk. Butt rot may also be present.

Mortality can reach 70–80% in areas 800 m^2 or larger, and trees of all sizes are affected. Disease incidence tends to increase on sites with high levels of precipitation and on low-elevation sites with gentle slopes that result in wet soil conditions. In the Nahuel Huapi Lake area, for example, there is no reported mortality in forests with less than 900 mm of precipitation per year. The effect that soil moisture seems to have on the incidence of mortality and the distribution of contagious, symptomatic trees suggest the involvement of soil-borne microfungi in the decline. However, despite numerous isolation attempts, no known pathogen has been identified. Moreover, recent preliminary inoculation tests with fungal isolates that could potentially be involved in the decline failed to cause disease in *A. chilensis* seedlings.

Despite the ecological and economic importance of *A. chilensis*, there have been only limited attempts to develop a management plan to predict and reduce mortality, which unfortunately is more prevalent on sites conducive to tree growth or where recreational activities are concentrated. The adaptability of this species is a factor that may play an important role in future management strategies.

Selected References

Ciesla, W. M., and Donaubauer, E. 1994. Decline and dieback of trees and forests: A global overview. FAO For. Dev. Pap. 120.

Havrylenko, M., Rosso, P. H., and Fontenla, S. 1989. *Austrocedrus chilensis:* Contribution to the study of its mortality in Argentina. Bosque 10(1):29-36.

Rosso, P. H., Baccalá, N., Havrylenko, M., and Fontenla, S. B. 1994. Spatial pattern of *Austrocedrus chilensis* wilting and the scope of autocorrelation method in natural forests. For. Ecol. Manage. 67:273-279.

(Prepared by P. Rosso)

Part II. Diseases in the Forest

Europe

The European continent has a great variety of climatic conditions and consequently a variety of forest types. Excluding Scandinavia, the forested area of Europe is 144 million hectares, about 55% of which is covered by conifers. The importance of conifers varies widely among countries, ranging from 15–20% in Hungary, Italy, and the former Yugoslavia up to 80% in Poland, Russia, and the Czech Republic.

The European conifer forest is composed basically of species native to the continent, although extensive plantings of exotic species are found. Because they are tolerant and fast growing, particularly when young, many exotic conifer species as well as native species planted beyond their historic ranges are used in afforestation for the wood-products industry. Afforestation is applied in areas where native stands were damaged (e.g., by war, storms, or fire) or not productive enough and on former agricultural soils. In many European countries, the creation of artificial stands was supported and encouraged by governments. The forest surface in Europe has increased significantly during the last century; for example, 2 million hectares have been afforested in France since 1947, and an outstanding effort was also made in the United Kingdom. The main conifer species used for afforestation include Norway spruce, Scot's pine, maritime pine, and black pines native to Europe and Douglas-fir and Sitka spruce from North America, Atlas cedar (*Cedrus atlantica*) from Morocco, and Japanese and hybrid larches.

In the European Union (excluding Fennoscandia) approximately 700,000 ha are planted to Norway spruce, nearly half of that area in Germany. Pines are also frequently used. Scot's pine, because it is a very tolerant species, was planted most often, especially in France, Germany, England, and Poland. It was used as early as the eighteenth century and covers 1–2 million hectares. Its growth and productivity are nevertheless rather poor in many cases, so foresters often replace it with more productive species. Black pines were also used throughout Europe. Laricio pine was more often used in France and the United Kingdom. Austrian black pine was propagated widely in France. The battlefields of the First World War were reafforested with Austrian black pine. Maritime pine was widely used in artificial stands from the nineteenth century in areas adjacent to the Atlantic Ocean in southern France, Spain, and Portugal. In France, maritime pine now occupies a total area of 1.3 million hectares, including the largest pure artificial forest in Europe, Landes de Gascogne (900,000 ha).

Exotic conifers are used mainly in the Atlantic region where oceanic climate and low altitude provide conditions favorable to many species. Douglas-fir (*Pseudotsuga menziesii*) is the most important exotic tree used in middle Europe. Sitka spruce (*Picea sitchensis*) is one of the most important afforestation species in the Atlantic countries of Ireland, France, Denmark, and especially the United Kingdom (500,000 ha, particularly in Scotland). Lodgepole pine (*Pinus contorta*), because it is very tolerant, has taken the place of Scot's pine for afforestation in the United Kingdom and Ireland, especially on peaty soils. Japanese larch (*Larix kaempferi*) is used in oceanic climates (a total of 190,000 ha), mainly in the United Kingdom but also in Belgium, the Netherlands, and France. Hybrid larch (*L.* × *eurolepis*) is also often considered because of its very fast growth rate.

European conifer forests generally are carefully managed for a long time, especially in the temperate zone of the continent. Regeneration of native conifer forests is usually accomplished through natural sowing, but it is often supplemented by planting. Plantations are the rule in artificial forests. The estimated annual production of conifers in Europe (outside Scandinavia) is 130 million cubic meters of logs, 82 million cubic meters of sawwood, and 75 million cubic meters of pulpwood. The most productive species (up to 10 m^3/ha/year) are favored for log production and include Norway spruce, white fir, Corsican pine, Douglas-fir, Sitka spruce, and maritime pine (under intensive conditions). In many cases, rotation ages are between 50 and 70 years, but very valuable Norway spruce and white fir trees in mountainous areas are grown on longer rotations, usually 80–100 years. Sitka spruce in Denmark and Scotland and maritime pine in France are cut earlier, often before 50 years of age.

In 1888, Robert Hartig, a professor in München, Germany, began the second edition of his well-known book on forest tree diseases with the following statement: "Changes have occurred in German forests, homogenous stands composed of even aged trees and of one tree species have replaced heterogeneous stands and coppice-with-standards. . . . Replacement of broadleaved stands by pure conifer stands especially, has induced during this century and even more during the recent years, dangers which exceed all we have known before." This historic book highlighted the silvicultural changes and the importance of conifer diseases that led to the development of forest pathology as a science. One century later, conifers are even more important in Europe, and knowledge of diseases (Table 11), including those of new exotic conifer species, has increased significantly. Needs of the wood industry have also increased, and forest health is a rising concern.

TABLE 11. Important Pathogens of Conifers in Europe

Pathogen	Hosts
Heterobasidion annosum	*Picea abies, Pinus sylvestris*
Armillaria ostoyae	*Picea abies, P. sitchensis, Pinus pinaster*, and other conifers
Phellinus pini	*Pinus* spp., *Cedrus atlantica*
Viscum album	*Abies alba, A. cephalonica*
Phaeolus schweinitzii	*Picea sitchensis*
Seiridium cardinale	*Cupressus sempervirens*
Lachnellula willkommii	*Larix decidua*
Gremmeniella abietina	*Pinus sylvestris, P. nigra*
Lophodermium seditiosum	*Pinus sylvestris*
Lophodermella sulcigena	*Pinus* spp.
Melampsora pinitorqua	*Pinus sylvestris, P. pinaster*
Endocronartium pini	*Pinus sylvestris*
Cronartium flaccidum	*Pinus sylvestris, Pinus* spp.
Melampsorella caryophyllacearum	*Abies alba*

Heterobasidion annosum is considered the most important pathogen detrimental to conifers in Europe. It affects nearly all conifer species and results in tree death or root and stem rot. It is widespread in Europe, including the Mediterranean areas. The pathogen affects mainly pine (especially Scot's pine) and spruce, and it is considered more severe in artificial stands. Deaths in pine crops may cause a loss of up to 20% of the volume. Norway spruce is nearly as susceptible as Sitka spruce. The volume rejected because of rot can be as high as 30%. The annual loss of Norway spruce in western Europe probably exceeds 3 million cubic meters. Stump treatments developed in the United Kingdom, such as the use of *Phlebia gigantea* as a biological control agent for pines and urea for other conifers, have been generally applied there for many years. Recently, borates have become preferred. In other European countries, stump treatments are not regularly used but are often recommended.

Most of the damage caused by *Armillaria* spp. to conifers throughout Europe is recorded in young plantations up to about 8–10 years old. When trees are older or are growing in natural stands, damage levels decrease. Gaps of dead trees do sometimes develop in these areas, but although damage may be impressive, it is not economically significant. The infection rate can be as high as 10–30% in Norway spruce or Sitka spruce; and in experimental plots of maritime pine, mortality can be 90%. Douglas-fir is less susceptible to *Armillaria* spp. than these other tree species. *A. ostoyae* is the main species damaging conifers in Europe. It is evident that the impact of *A. ostoyae* did not increase significantly in conifer stands suffering from forest decline as it did in white fir in France. No control of *Armillaria* spp. is currently practiced in Europe, but in eastern countries, mixing conifers with broad-leaved trees is often recommended.

The root and butt rot fungi mentioned above are the principle pathogens on spruces and firs in Europe. Other diseases, such as needle diseases, are usually present in natural forests of Norway spruce, but none have spread with significant impact into artificial forests as *H. annosum* did. Some, under subalpine conditions, have had an impact similar to that in boreal countries.

A very common disease of white fir is caused by the rust fungus *Melampsorella caryophyllacearum*. The disease can be locally important, leading to wood decay and stem breakage. Removing affected trees during thinning is the only control method used. Another common pathogen of fir is leafy mistletoe, *Viscum album*. It can be especially damaging in stands suffering from dry conditions. White and Greek firs in the mountains, especially overmature trees, may be severely infected. Wood production can be decreased by 30%. Control measures, used mainly in Switzerland and France, consist of growing white fir as near as possible to its optimal site and cutting overmature trees.

Douglas-fir suffers from no severely damaging diseases in Europe. Root and stem rots caused by *Phaeolus schweinitzii*, *Sparassis crispa,* and *Calocera viscosa* have been observed in some cases, but the incidences of such diseases are currently low, probably because the Douglas-fir forest in Europe is relatively young. Some needle diseases were introduced into Europe from North America. *Phaeocryptopus gaeumannii* (called Swiss rust in France) was a concern during the 1930s, but it is currently damaging only in wet soil and under poor growth conditions. Damage caused by *Rhabdocline pseudotsugae* is efficiently avoided by using resistant provenances of Douglas-fir (i.e., provenances from northwestern Washington).

Pines harbor the largest number of diseases among European conifers. The most damaging decay fungus is probably *Phellinus pini*. Incidence of disease increases with tree age beyond the usual length of the forest rotation, so damage is seldom seen in artificial stands. In the past, it was important on pines wounded for resin production (e.g., maritime pine and Scot's pine), but currently in natural forests were pines often grow for more than 100 years, *P. pini* is still an important cause of rot.

The main twig disease is caused by *Gremmeniella abietina*, important especially on Laricio pine and black pine. It is damaging chiefly to young trees up to about 20 years old and is sometimes a problem in nurseries. It is widespread in central and northern Europe, where severe attacks have been recorded from the beginning of this century. Under forest conditions, the main means of control is silvicultural: avoiding north-facing slopes and damp conditions and removing infected older trees.

Other diseases important on pines in Europe are the rusts. The most damaging are twisting rust and blister rust. *Melampsora pinitorqua,* which is also considered a subspecies of *M. populnea*, is widespread in Europe and results in trees with defective stems. In maritime pine, large differences in susceptibility have been observed between provenances, but no provenance has shown more resistance than the local one. No direct control of the disease is practiced. *Cronartium flaccidum* is a host-alternating blister rust. A very similar rust is caused by *Endocronartium pini,* but the fungus is autoecious and endocyclic. Blister rust is widespread throughout Europe, and the distinction between the causal agents is not always made. The most severe epidemics have been registered in plantations in Italy. A special feature of the disease is resin top disease, considered typical of *E. pini* infection. It can be seen in many countries in Europe, mainly on mature Scot's pine and seems to be increasing in the United Kingdom. No control of blister rust is practiced, other than avoiding pine plantations in high-hazard areas. Cutting of infected trees is recommended. A special mention must be made of *Cronartium ribicola*, well known in North America as the cause of white pine blister rust since its introduction at the end of the last century. In Europe, presence of the pathogen nearly precluded the use of the very valuable North American white pines in afforestation.

(Prepared by C. Delatour)

Fennoscandia

In Fennoscandia, forestry has on the whole been focused on harvesting seminatural forests and reestablishing forests of similar kind. However, production forests differ from natural ones in several respects: 1) normally, Norway spruce have been favored on the more productive sites and Scot's pine on drier sites at the expense of early succession, broad-leaved trees; 2) forest fires have been excluded; 3) very little dead wood is left in the forests; and 4) at the time of final felling, the trees are usually younger than full-grown trees in natural forests. Regeneration is typically accomplished through large-scale clear-cuttings followed by replanting. In addition, Scot's pine is often regenerated naturally from open stands of seed trees. Stands are typically thinned several times during their life spans. The degree of mechanization is high today. Almost all final cuttings are done by machine, and 75% of all thinning operations are mechanical. The main native commercial species are Norway spruce and Scot's pine. Exotic tree species grown on a larger scale include *Pinus contorta* and *Larix* spp.

The most widespread and economically damaging fungal disease in northern Europe is probably root rot caused by *Heterobasidion annosum* (Table 12). Losses are estimated to be $100–300 million annually. Locally, the economic consequences of root rot have increased substantially during recent years, since rotted logs are now rejected by most pulp mills because of the bleaching problems that developed after chlorine-based methods were abandoned.

Norway spruce suffers the most severe root rot problems in northern Europe. Over the natural range of Norway spruce, the S form of *H. annosum* is predominant. Outside this area and in the south of Sweden, east of Finland, and most of the Baltic states and western Russia, the P form is more common.

Old-time forestry operations in northern European countries involved extracting and transporting logs on top of the snow cover during the winter months. Although loggers were unaware of it at the time, this practice reduced stump infection by *H. annosum*, since very few infections are established on stumps created at temperatures below +5°C. Although winter cuttings are recommended in northern Fennoscandia, the commercial forestry industry cannot rely solely on this practice. Instead, stump treatment has been introduced into many parts of northern Europe to prevent spore infection by *H. annosum*. Urea has been the choice for chemical treatment, mainly because it is not toxic to humans. Initially, a 20% aqueous solution was used, but experiments in Sweden showed that a 30% solution provides better protection. Boron oxides (i.e., Polybor) are also used on a limited scale. Biological stump treatment with suspensions of oidia of *Phlebiopsis gigantea* has become the most widely used control measure in Fennoscandia. Strains of *P. gigantea* that will effectively colonize both spruce and pine stumps are available commercially. Stump treatments, urea, Polybor, or *P. gigantea* spores are applied to stumps mechanically at the time of felling by using specially designed spraying equipment.

The short-term biological effect of stump treatments is a reduction in *H. annosum* spore infections ranging from about 70 to 90% with 20 and 30% urea, respectively, to more than 95% with 1 g of *P. gigantea* spores per liter and 5% Polybor. Long-term effects have not been widely evaluated. A possible bonus with the biological treatment is the replacement of *H. annosum* in parts of the root systems of infected stumps.

After severe attacks, replanting a stand with tree species less susceptible to *H. annosum* has been recommended in northern Europe. The main alternatives have been Scot's pine and *Betula pendula*. However, there are two caveats to this practice. First, there are no alternative species that can match Norway spruce in terms of production on most of the sites subject to root rot. Therefore, the choice has been instead to replant with spruce and take the loss caused by decay. Second, in areas were the P form of *H. annosum* is predominant, Scot's pine is not an alternative. Shortening the rotation time is another approach taken by the forestry industry in Fennoscandia in order to minimize the losses caused by root rot. The increase in the commercial value of the stand culminates 10–20 years earlier in a severely diseased stand than in a healthy one.

European larch (*Larix decidua*) is subject to attack by a canker fungus, *Lachnellula willkommii*, in areas outside the tree's natural range in Europe. This disease is not a serious problem to Japanese or Siberian larches or their hybrids. Consequently, the forestry industry in northern Europe has handled the larch canker problem by not growing European larch. Instead, the less susceptible larch species have been used.

The introduction of *Pinus contorta* into Fennoscandia has been accompanied by disease problems. Provenances from southern areas of North America typically have problems adapting to the local climate in northern Fennoscandia, and such trees show an increased susceptibility to attack by *Gremmeniella abietina*. Devastating epidemics have been reported that resulted in the death of whole plantations. Thus, foresters are increasingly aware of the importance of choosing the appropriate provenance for the culture of *P. contorta*.

Scot's pine and to a lesser extent Norway spruce are also affected by *G. abietina*. *Pinus sylvestris* suffers from a shoot dieback in the crown. During most years, the disease is inconspicuous, but epidemics have occurred and caused extended dieback of whole stands in Fennoscandia and eastern Russia. This typically happens in overstocked, pole-size stands after cold summers. *Picea abies* is affected by a top dieback when growing as an understory of *P. sylvestris*. This disease is relatively common and is a good indicator of the presence of *G. abietina* in a stand but has limited importance for practical forestry.

It is believed that dense stands favor the spread of spores of *G. abietina* within the stand, and recently killed branches are the most active sources of spores. The recommendation has therefore been made to thin the diseased stands and selectively take out severely affected trees. A proper choice of provenance is also important for *P. sylvestris*.

During the early 1980s, pruning of pine trees became a frequent practice in the forestry industry in Fennoscandia. Because of work programs for unemployed people, the pruning was often carried out during the autumn. Many of these trees became infected with *Phacidium coniferarum*, which entered through the pruning scar and developed in the inner bark while the tree was dormant. The fungus also caused a blue stain of the wood and sometimes killed the tree. After these observations were made, autumn pruning of conifers was abandoned. This story illustrates the point that forest pathology is not a major concern of foresters in northern Europe. The risk of *P. coniferarum* infection after autumn pruning was well documented in the literature before the practice was introduced. The problems could have been avoided if this knowledge had been spread among foresters.

In parts of northern Europe where snow regularly covers the ground for more than a month every year, *Phacidium infestans* is a severe problem for *P. sylvestris* in the seedling and sapling stages. It has been recommended that seed sources from a more northern provenance, which are generally more resistant to the fungus, be used. Also, slashing is avoided during late summer and autumn, since the fungus can infect the green needles of the felled trees and from these foci infect the remaining seedlings.

Wounds caused by felling and extracting trees during thinning operations are frequent entry points for decay-causing pathogens. Wounds on roots close to the stem are the most vulnerable, and the spread of the decay is proportionate to the size of the wound. Awareness of these problems has led foresters, in collaboration with researchers at the universities, to develop less damaging machinery. The result has been machines that exert much less pressure on the soil, and consequently the amount of damage to the root systems and the number of wounds on the remaining trees have decreased. Another factor contributing to reduced wounding is the education of the staff taking part in thinning operations. Guidelines for the amount of acceptable damage are written into law. In Sweden, for example, not more than 5% of remaining trees are allowed to be damaged and the area of the scar should not exceed 15 cm^2.

There are two rust species that cause dieback of tops (Peridermium tops) of Scot's pine in northern Europe, *Cronartium flaccidum* and *Endocronartium pini*. Disease incidence in Fennoscandia is approximately 4% of mature trees, but locally more than 25% of stems are attacked. The recommendation for managing the disease has been to take out diseased trees during thinning operations.

(Prepared by J. Stenlid)

TABLE 12. Important Pathogens of Conifers in Fennoscandia

Pathogen	Hosts
Heterobasidion annosum	*Picea abies, Pinus sylvestris*
Lachnellula wilkommii	*Larix decidua*
Gremmeniella abietina	*Pinus sylvestris, P. contorta, Picea abies*
Phacidium	
coniferarum	*Pinus sylvestris, P. contorta, Picea abies*
infestans	*Pinus sylvestris*
Armillaria ostoyae	*Picea abies, Pinus sylvestris*
Phellinus pini	*Pinus sylvestris*
Lophodermium seditiosum	*Pinus sylvestris*
Lophodermella sulcigena	*Pinus sylvestris*
Melampsora pinitorqua	*Pinus sylvestris*
Cronartium flaccidum	*Pinus sylvestris*

Boreal Forest of North America

The boreal forest encompasses circumpolar, conifer-dominated forests at northern latitudes. It is the most extensive biome in North America, bounded by tundra on the north, the mountainous subalpine forests on the west, and temperate broad-leaved forests or prairie grasslands on the south. In the western hemisphere, boreal forests occupy a band extending from the interior of Alaska and the Canadian territories, across northern British Columbia to the area between Hudson Bay and the Great Lakes, and then east to Newfoundland. The western portion of the North American boreal forest covers much of Manitoba and extends northwest to Alaska.

Vegetation in the boreal forest is reflective of the harsh climatic conditions: wind, heavy snowfalls, low temperatures during the growing season, and fire. Forest cover is frequently sparse, patchy, and dwarfed because of these harsh conditions and the wet, cold soils. Compared with temperate and tropical forests, boreal forests have very little species diversity. Dominant tree species include white and black spruce (*Picea glauca* and *P. mariana*), lodgepole and jack pine (*Pinus contorta* and *P. banksiana*), balsam and subalpine fir (*Abies balsamea* and *A. lasiocarpa*), aspen and balsam poplar (*Populus tremuloides* and *P. balsamifera*), and birch (*Betula papyrifera*).

Pressure is growing to exploit the boreal forest. In western Canada, in particular, the boreal forests were once considered marginal for timber extraction. However, increased demand for timber, advances in harvesting and wood utilization technology, and reduced availability of timber in southern and coastal regions have made boreal forests important for timber supply.

Harvesting is usually done mechanically with feller bunchers, rubber-tired skidders, and limbers. Much of the harvesting, particularly in wet areas, occurs during the winter when the ground is frozen and there is a protective layer of snow. Harvesting and road transport of logs ceases during the spring thaw and is resumed in drier areas during the summer. Clearcutting is essentially the only silvicultural system used in the boreal forest; very little harvesting is done as partial cuts. Within 3 years of harvest, the clear-cuts are regenerated, in most cases by planting 1- to 2-year-old spruce or pine seedlings. Most of the wood harvested in the boreal forest is softwood and is used for construction-grade lumber and pulp.

Boreal forest ecosystems are characterized by seral stages maintained by disturbance rather than by climax forests. Fire is often the primary agent of disturbance, and frequent fires serve to increase productivity in early and midseral stages by increasing nutrient availability. Many of the boreal species are morphologically and reproductively well adapted to relatively frequent fires. Other common disturbance agents are changes in the permafrost layer, insect outbreaks, disease, windthrow, and flooding.

Tree diseases play important roles in the ecology of the boreal forest (Table 13), particularly in wet areas where the interval between fires is long. These roles include 1) the creation of canopy gaps and small-scale successions; 2) the creation of individual trees with specific habitat attributes; and 3) nutrient cycling.

In the North American boreal forest, small canopy gaps are characteristic features of spruce-dominated forests. Gaps can be formed by individual agents of tree mortality, such as root diseases, or by several factors operating together or in sequence. Small gaps are frequently caused by windthrow and root disease. The root rot fungus *Inonotus tomentosus* is widespread throughout the boreal forest and creates small clumps of dead trees and sometimes causes windthrow. Spruce and pine are the primary hosts of this fungus, although in pine it tends to cause cambial necrosis and less root decay than in spruce. In eastern Canada, *Armillaria ostoyae, Scytinostroma galactinum,* and *Coniophora puteana* are common root-rotting fungi of white and black spruce and balsam fir.

Small gaps are also formed by breakage of trees with heartrot. Recent observations in Alaska and central British Columbia suggest that heartrot and root and butt rot fungi are very important natural agents of disturbance that have a cumulative and significant impact on landscape patterning in these forests. Important decay species in spruce include the root rots listed above plus *Phellinus pini*. On true fir, *Poria subacida, Polyporus abietinus,* and *Echinodontium tinctorium* cause the most common butt and trunk rots.

Several common species of true heartrot fungi (e.g., *E. tinctorium* in true fir and *P. pini* in spruce) provide conditions suitable for cavity nesters. These fungi infect trees through small twigs or branch scars and decay the heartwood, leaving a sound ring of sapwood. Primary cavity nesters create an entrance hole into the softer, decayed heartwood where they then excavate a nest.

The importance of decay fungi in nutrient cycling is well known. Some pathogenic fungi in the boreal forest play key roles in either initiating the decay process by killing the tree or participating in some sequence of events that ultimately leads to total degradation of woody material. In the boreal forest, tree death is most commonly caused by abiotic events such as wind storms and fires. The most common primary biotic agents of spruce death are the spruce beetle (*Dendroctonus rufipennis*) and root disease. In the Alaskan boreal forest, death of spruce caused by beetles is resulting in significant changes in species composition and plant diversity. Other organisms may cause or contribute to mortality but occur much less frequently. In pine, biotic agents of mortality include the stem rusts (*Cronartium* spp.), Atropellis canker (*Atropellis piniphila*), and less commonly, root disease.

Once the tree is dead, sap rots, in conjunction with other organisms that assist the decay process, are the next agents of nutrient cycling and include many species of both white and brown rot fungi. Basidiomycetous fungi are the main decomposing organisms in boreal ecosystems. Populations of bacterial and soil fauna are smaller than those in temperate areas.

Much of the soil in boreal ecosystems is composed of brown rot residues. Brown rot fungi are found primarily in coniferous ecosystems and are considered circumboreal in distribution. Brown rot residues are important factors in water-holding capacity, ectomycorrhizal development, nitrogen fixation, and cation exchange capacity.

The roles of biotic and abiotic agents decline and increase, respectively, with increasing latitude. In the boreal forest, diseases caused by biotic factors are less economically important than in temperate forests, whereas damage from abiotic factors is relatively greater. Economically important biotic agents of disease in the boreal forest include root rots, especially those caused by *I. tomentosus* and *A. ostoyae*, which result in tree death, windthrow, butt cull, and growth reduction. The heartrot fungi cause significant volume losses in older stands, particularly in spruce and true fir. In mature pine, dwarf mistletoe (*Arceuthobium americanum*) is common throughout the range

TABLE 13. Important Conifer Pathogens in North American Boreal Forests

Pathogen	Hosts
Inonotus tomentosus	*Picea* spp., *Pinus contorta, P. banksiana*
Armillaria ostoyae	*Picea* spp., *Abies balsamea*
Scytinostroma galactinum	*Picea* spp., *Abies balsamea*
Coniophora puteana	*Picea* spp., *Abies balsamea*
Phellinus pini	*Pinus contorta, P. banksiana, Picea* spp.
Echinodontium tinctorium	*Abies* spp.
Endocronartium harknessii	*Pinus contorta, P. banksiana*
Arceuthobium	
americanum	*Pinus contorta*
pusillum	*Picea mariana*

of lodgepole and jack pine and in heavily infested areas can cause significant growth losses. *A. pusillum* causes significant losses in black spruce in eastern Canada. In young pine stands, either planted or naturally regenerated, stem rusts (*Cronartium* spp.) and western gall rust (*Endocronartium harknessii*) can cause high mortality rates in localized areas.

It is thought that the incidence of both root diseases and windthrow has increased because of forest management. Artificial regeneration immediately after clear-cutting and site treatment provides new susceptible host tissue for root rot inoculum in stumps. In the natural regeneration system, host species are frequently slow to colonize a site opened by root disease. Furthermore, intensive management practices, such as thinning, brushing, and weeding, have been shown to increase the incidence of *A. ostoyae*.

Health management in the boreal forests of North America has for years centered around prevention of massive bark beetle epidemics (spruce beetle and mountain pine beetle) and salvage or sanitation of the areas attacked. Several approaches have been used, most of them based on some method of detection followed by control and/or salvage operations. In contrast, diseases, both abiotic and biotic, are considered primarily during the development of silviculture prescriptions. Consideration is given to site factors that dictate the species selected for stand reestablishment. Generally, these reflect soil moisture and nutrient status. Prescriptions should also consider site susceptibility to frost and drought, windthrow, and root disease.

Depending on the incidence of root disease at a particular site, less susceptible species will be prescribed or some form of inoculum removal recommended. Push-over falling, in which an excavator is used, is becoming a relatively common technique for sites with *A. ostoyae*. Inoculum removal on sites with *I. tomentosus* is not a common operational prescription. Alternative species such as pine or a mixture of conifers and hardwoods are more common prescriptions.

Little is done to minimize damage by decays, except where a "pathological rotation age" (i.e., the age at which decay losses equal the annual increment) has been established. Harvest of these stands may occur at an age younger than normal to minimize losses to decay. Provincial forest regulations in British Columbia also prohibit damage to residual trees during logging operations to reduce the likelihood of infection by wound pathogens.

Forest management in Canada is moving toward an ecosystem-based approach. A good understanding of the normal range of variation in the different ecosystems and of the disturbance processes that drive them are necessary for this type of management. Mixed-species management (several conifer species or conifers and deciduous trees together) is becoming more accepted as a result of some of this work. These mixed-species forests often more closely resemble natural forests than the single-species, conifer-dominated forests that reflect current management objectives. Mixed-wood forests are expected to have less root disease because of the lower density and less uniform distribution of preferred host species. These stands are also more resistant to spruce weevil (*Pissodes strobi*) and bark beetle infestations.

(Prepared by K. Lewis)

Eastern Siberia and the Russian Far East

Russia has the largest forest resources of any country in the world. Forests cover more than 800 million hectares, i.e., one-third of the total land area. The regions of eastern Siberia and the Far East account for one-third of the forested land of the country and are for the most part sparsely populated and relatively undeveloped. More than 24 species of the family Pinaceae are present in the forests of these regions, characterized by four main genera: *Abies*, *Larix*, *Picea*, and *Pinus*. The boreal coniferous forests of eastern Siberia and the Far East are in the "taiga" zone, which is characterized by cold winters, relatively warm summers (July mean temperatures of 10–20°C), frequent penetration by cold arctic air masses, moderate amounts of precipitation, and a layer of permafrost near the surface in much of the area. Throughout the taiga zone, there are considerable variations in climatic conditions that determine the composition of forest vegetation.

Within central Siberia, three distinguishable regions surround major river systems: the Yenisei taiga, the Tunguska taiga, and the Angara taiga. The forests types of the Yenisei taiga include spruce-fir-birch in the south, spruce-Siberian pine (*Pinus siberica*)-birch in the central regions, and spruce-larch-Siberian pine-birch in the north. The Tunguska taiga is characterized by the predominance of Siberian larch (*Larix siberica*) with an admixture of spruce, pine, and fir. Birch-aspen stands develop after disturbances, followed by pine-larch forests and ultimately fir, spruce, and Siberian pine. The Angara taiga contains pine forests of fairly high productivity with admixtures of larch, Siberian pine, and spruce.

Much of eastern Siberia is covered by the Yakut taiga, which is dominated by Dahurian larch (*L. dahurica*) forests with admixtures of pine, birch, and aspen. This region is penetrated by cold air masses, and as a result, permafrost lies near the soil surface. The Altai-Sayan mountain taiga in south-central Siberia forms a distinct floristic region because of its continental climate with harsh, long winters and short, hot summers. Forest types are arranged along an elevation gradient with pine, larch, aspen, and birch forests in the lower elevations of the foothills; mountain forests of conifers, including pine, fir, Siberian pine and larch; and mountain-tundra with sparse stands of larch and Siberian pine in the higher elevations.

The Transbaikalian mountain taiga is an extensive region of mountain ranges near and east of Lake Baikal. The forests of this region are influenced by the mountainous relief, the harsh continental climate, and the permafrost layer. The northern part of this region consists of continuous mountain taiga of larch in the lower elevations with sparse subalpine forests in higher elevations with thickets of Japanese stone pine (*Pinus pumila*). In the southern and eastern parts of this region, the low elevations have predominantly larch-pine forests with an admixture of hardwoods, the middle elevations have mountain taiga of primarily Dahurian larch, and the high subalpine forests have Japanese stone pine.

The climate of the Far Eastern mountain taiga is more moderate than that of the other taiga zones. There is not much snow during the winter; 90% of the yearly precipitation falls during April through October. The northwestern part of this zone is primarily larch taiga, with spruce-fir-larch in the central part and larch and spruce-fir-cedar in the Sikhote Alin mountains. The mountain taiga of northeastern Siberia differs sharply from the other silvicultural zones. Here the climate is the most severe in Russia with a short growing season and permafrost at a very shallow depth. Forest vegetation is limited to low elevations on mountains and consists primarily of Dahurian larch with some Jeddo spruce (*Picea ajanensis*) and Khingan fir (*Abies nephrolepis*). Forest stands are extremely sparse and of low quality because of the short growing season, high winds, and shallow permafrost.

The conifer forests east of the Ural Mountains have not been harvested as much as the forests of the Uralo-European part of the country. Until recently, most of the harvesting had taken place along the Trans-Siberian Railway, completed in the early 1900s, and in the drainage areas of the Yenisei, Amur, and Ussuri Rivers. The recently completed Baykal-Amur Mainline has lead to expansion of logging adjacent to the railroad.

Another railroad, the Amur-Yakutsk Mainline is under construction and will open forest resources north of current logging operations. Forest products from Siberia and the Far East regions have been exported mainly to Japan and China and used regionally because of the excessive transport costs.

Although there have been several mycological surveys of various Siberian regions (Table 14), little is known of the pathology of these forests. The following are short characterizations of the biology, ecology, and regional peculiarities of pathogens most frequently encountered in Siberia and the Far East, those of particularly high pathogenicity, or those of quarantine importance.

Several canker fungi are known. *Lachnellula pini* is widespread in the Transbaikal region and causes a canker disease of fir, Siberian pine, and dwarf Siberian pine. In forests weakened by industrial pollution, the amount of damage caused by this species increases. *L. calyciformis* affects fir, spruce, Siberian pine, and dwarf Siberian pine. On irreversibly weakened trees, the canker disease will spread to the entire bole.

Gremmeniella abietina was recorded on pine in the Transbaikal region. Information on the distribution of this species in other regions of Siberia is currently lacking.

Dasyscyphus willkommii is recorded on larch in the Transbaikal and Postbaikal regions. *Delphinella balsameae* is widely distributed in Transbaikal. This is the first report for Russia. It damages needles and shoots of the current year. It is most damaging to seedlings and saplings. Perennial centers of injury can be found in the dark taiga and mountain taiga zones of Transbaikal.

Biatorella difformis causes a stem and branch canker in grass- and shrub-grass-pine forests. *Bothrodiscus berenice* is a widespread species in Transbaikal. It has only recently been reported as a pathogen in Russia. It causes cankers of branches and stems of fir. Centers of reproduction appear in forests of the dark taiga and mountain taiga zones. *Sclerophoma pythiophila* appears on many conifer species, causing cankers. Various factors contribute to the development of the disease, including deteriorating growth of stands, unfavorable climatic conditions, and damage from industrial emissions.

Among the needle fungi, *Hartigiella laricis*, which is widely distributed in Transbaikal, causes a needle disease of larch.

Leptostroma pinastri most commonly affects needles of Siberian pine and is rarely found on Scot's pine. *Lophodermium pinastri* affects needles of Siberian pine and dwarf Siberian pine. High percentages of needle infections are recorded in forests clouded with sulfur-containing compounds. *L. macrosporium* infects needles of spruce on the previous year's shoots, causing needle blight. The greatest amount of damage is found in plantations. *L. nervisequium* causes blight of fir and infects 2- and 3-year-old needles. The disease is found in Siberian pine-fir and whortleberry-grass forests. *Rhizosphaera pini* causes reddening of fir needles. The pathogenicity of this species is greater in regions with high anthropogenic loads than in healthy stands.

Herpotrichia nigra causes brown felt of spruce, fir, Siberian pine, dwarf Siberian pine, and juniper and damages needles and branches. It causes death of seedlings and saplings; in older trees, only the lower branches are affected. The amount of damage caused by this species depends on snow depth and the rate of snow melt in the spring. *Phacidium infestans* also causes snow blight of pine.

Phellinus pini is the most common wood-decay fungus of conifers in the Far East. The most significant impact is on spruce and Korean pine (*Pinus koraiensis*); larch, Scot's pine, and *Pinus funebris* are affected to a lesser extent. *P. hartigii* causes a stem decay of fir. Infection levels can be significant in fir species of the Far East including *Abies nephrolepis*, *A. sachalinensis*, and *A. holophylla*. *Fomitopsis officinalis* causes a stem decay of larch. The percentage of trees infected in the southern regions is considerably higher than that in the northern regions of Siberia.

Root pathogens are poorly understood. *Heterobasidion annosum* causes root decay in many regions of Siberia. It is reported to affect all the important conifer species of the Far East, especially spruce, larch, and fir. Root decay caused by this fungus leads to windthrow of spruce, Dahurian larch (*Larix dahurica*), and whitebark fir (*A. nephrolepis*).

Phaeolus schweinitzii is widely distributed and infects Siberian pine and larch. It causes stem and root decay and is a common cause of windthrow, especially of large Dahurian larch in the Amur region. Infected trees are often killed by insects. *Armillariella mellea* sensu latu is a species with a wide

TABLE 14. Fungal Pathogens of Conifers in Siberia and the Far East

Foliage Fungi
Delphinella balsameae (A. M. Waterman) E. Müller in E. Müller & Arx
Meria laricis Vuill.
Herpotrichia nigra R. Hartig
Hypodermella laricis Tub.
Hypodermella sulcigena (Rostr.) Tub.
Lophodermella conigenum
Lophodermium pinastri (Schrad.:Fr.) Chev.
Lophodermium macrosporium (Hart.) Rehm.
Lophodermium nervisequium Rehm.
Micropera pinastri Sacc.
Phacidium infestans P. Karst.
Rhizosphaera pini (Corda) Maubl.

Decay Fungi
Armillariella mellea (Vahl:Fr.) P. Karst
Fomes fomentarius (L.:Fr.) J. Kickx fil.
Fomitopsis cajanderi (P. Karst.) Kotlaba & Pouzar
Fomitopsis officinalis (Villars:Fr.) Bondartsev & Singer
Fomitopsis pinicola (Sw.:Fr.) P. Karst.
Ganoderma applanatum (Pers.) Pat.
Gloeophyllum abietinum (Fr.)
Heterobasidion annosum (Fr.:Fr.) Bref.
Inonotus weirii (Murrill) Kotlaba & Pouzar
Ischnoderma resinosum (Schrad.:Fr.) P. Karst.
Laetiporus sulphureus (Bull.:Fr.) Murrill
Phaeolus schweinitzii (Fr.:Fr.) Pat.
Phellinus hartigii (Allesch. & Schnabl) Pat.
Phellinus pini (Thore:Fr.) A. Ames

Rust Fungi
Calyptospora goeppertiana Kühn
Chrysomyxa ledi de Bary
Chrysomyxa woroninii Tranzschel
Cronartium ribicola J. C. Fisch.
Cronartium flaccidum (Alb. & Schwein.) Wint.
Gymnosporangium juniperi Link
Melampsora pinitorqua (de Bary) Rostr.
Melampsoridium betulinum Kleb.
Melampsorella caryophyllacearum J. Schröt.
Melampsorella symphyti Bul.
Pucciniastrum pustulatum Dietel in Engl. & Prantl

Canker Fungi
Ascocalyx abietis (Pers.) Rehm.
Bothrodiscus berenice (Berk. & M. A. Curt.) Groves
Cytospora abietis Sacc.
Cytospora kunzei Sacc.
Dasyscyphus willkommii (R. Hartig) Rehm
Dothiorella pithya Sacc.
Gremmeniella abietina (Lagerberg) Morelet
Lachnellula calyciformis (Batsch) Dharne
Lachnellula pini (Brunch.) Dennis
Nectria cucurbitula (Tode:Fr.) Fr.
Phoma abietella-siberica Schwarzman
Phoma acicola (Lév.) Sacc
Sclerophoma pythiophila (Corda) Höhn.
Zythiostroma pinastri Karst

trophic spectrum and is subject to quarantine objectives. *Inonotus weirii* is included in the list according to information provided by A. S. Bondartseva, who noted this species on larch in the Russian region of Irkutsk Oblast. It too is subject to quarantine surveillance.

Melampsorella caryophyllacearum, widely distributed in Siberia, causes a canker disease of branches and stems of fir and forms witches'-brooms. Damage is inflicted in areas with considerable sulfurous pollution. *Melampsora pinitorqua* causes rust of pine shoots. *Cronartium ribicola* is the cause of blister rust of branches and stems of pine, Siberian pine, and dwarf Siberian pine.

Melampsoridium betulinum causes rust of needles and shoots of larch. Uredinial and telial stages occur on leaves of birch. *Pucciniastrum pustulatum* is the cause of rust on the current year's needles of fir. The uredial stage is found on willow herb. The percentage of infection on fir increases significantly in relation to the distance from epicenters of sulfurous compound pollution. *Chrysomyxa woroninii* causes rust of needles and shoots of the current year on spruce. Infected shoots perish, resulting in retarded growth of damaged juveniles. Teliospores are found on leaves of wild rosemary. The highest percentage of infection is found in flood plain spruce forests. *Chrysomyxa pirolata* causes rust of spruce cones in flood plain spruce forests. The telial stage is found on the previous year's leaves of pyrola, and the urediospores are found on the current year's leaves.

(Prepared by T. I. Morozova and B. Tkacz)

Coastal Western North America

The region along the Pacific coast of North America has characteristics that favor maximal development of coniferous forests. It has a maritime climate; both mild winters and dry summers appear to provide evergreens with definite advantages over deciduous tree species. A preponderance of conifers, high stand densities over extensive areas, phenomenal growth rates, and great size and longevity of dominant tree species are typical of western coastal forests and make them unique among the forests of the world.

Western hemlock, an extremely shade-tolerant tree, is the climax species in most coastal forests. Pure climax stands are relatively rare, however, because seral species are so long lived that they remain important stand components between even widely separated disturbance events. Douglas-fir is the dominant seral species and the preferred species for management over much of the region, although Sitka spruce is gaining importance in the north.

Pacific coast conifer forests have been premiere, high-quality wood producers from the time of first settlement to the present. Clear-cutting was preferred because of its economic advantages and because the favored tree species for management, Douglas-fir, required relatively open conditions for establishment and rapid juvenile growth. A typical management regime involved clear-cut harvest, site preparation by broadcast burning, planting of Douglas-fir at relatively close spacing to ensure full stocking, precommercial thinning at 12–15 years, one or more commercial thinnings timed appropriately, and final harvest at a rotation age varying from 40 to 150 years. Such regimes are still in use where timber production is the principal management objective.

During recent years, other management objectives, including preservation of wildlife habitat, development of recreational areas, maintenance of visual quality, and protection of the watershed, have increased markedly in importance in coastal forests, especially on publicly owned lands, resulting in modifications to management prescriptions and the development of a new philosophy that is more holistic, systems oriented, and ecologically sensitive than that of the past. This ecosystem-management approach is the policy for public land managers in coastal forests in the United States.

Laminated root rot is probably the most important tree disease and certainly the most significant root disease of coastal forests (Table 15). When wildlife habitat, visual quality, or watershed protection are major management objectives, laminated root rot is often considered beneficial or innocuous. It is very destructive when it occurs in stands where timber production is a major management objective, however. It reduces harvestable volume by as much as 50–55% at rotation age in affected areas. In such situations, it is managed, often aggressively, by mechanical removal of inoculum (stump removal) or, more commonly, by tree species manipulation.

Armillaria root disease can be found in virtually any stand, but usually mortality levels are much lower in coastal forests than in the interior. Mortality caused by Armillaria root disease is most common in Douglas-fir plantations or heavily stocked patches of natural Douglas-fir or true fir regeneration. *A. ostoyae*, like *Phellinus weirii*, is capable of surviving for many years in roots of stumps and dead trees and generally causes scattered deaths or formation of small infection centers around large, old stumps. Mortality caused by Armillaria root disease often decreases substantially in plantations when they reach 20–30 years of age. Ecologically, *A. ostoyae* acts as a natural thinning agent and a winnower of weak or suppressed trees.

Black-stain root disease is distributed throughout the range of Douglas-fir in the Pacific coastal region; levels are highest in northwestern California and southwestern Oregon. Occurrence and associated mortality have increased substantially during the 1980s and 1990s. Surveys in southern Oregon indicate that the disease occurs in up to 25% of the Douglas-fir plantations there. New infection centers tend to appear in areas with compacted soils or wounded Douglas-firs, probably reflecting vector preference for stressed trees. Vectors of the black stain pathogen are also attracted to fresh precommercial thinning slash and stumps.

Current management strategies for black-stain root disease are primarily preventative, aimed at reducing or eliminating establishment of new disease centers by avoiding creation of conditions favorable for the insect vectors. This involves minimizing site disturbance, avoiding tractor logging and associated soil compaction, not wounding trees during harvest operations or during road building or maintenance, and timing precommercial thinnings so that slash and stumps are not fresh at the time that vectors are flying.

S-type *Heterobasidion annosum* is common in forest stands in all but the northern part of the coastal region. Climatic conditions are particularly favorable for the pathogen, stump and wound infection courts are plentiful, and long-distance spread by windborne spores occurs virtually throughout the year. Although most conifers can be infected on occasion, Pacific silver fir and especially western hemlock are the principal hosts. Infection levels in these species tend to be very high,

TABLE 15. Important Pathogens of Conifers in Coastal Forests of Western North America

Pathogen	Hosts
Phellinus weirii	*Pseudotsuga menziesii*, *Abies* spp.
Armillaria ostoyae	*Pseudotsuga menziesii*
Leptographium wageneri	*Pseudotsuga menziesii*
Heterobasidion annosum	*Tsuga heterophylla*, *Abies amabilis*
Phytophthora lateralis	*Chamaecyparis lawsoniana*
Cronartium ribicola	*Pinus monticola*, *P. lambertiana*
Phellinus pini	Most conifers
Phaeolus schweinitzii	*Pseudotsuga* spp. and other conifers
Echinodontium tinctorium	*Tsuga heterophylla*, *Abies* spp.
Arceuthobium tsugense	*Tsuga heterophylla*

often 20% of the trees or more. *H. annosum* infection can result in extensive decay and associated mortality caused by stem breakage. However, such effects are usually not manifested until trees are 180 years old or older. Although infection levels are high, effects in younger stands tend to be quite minor.

The effects of *H. annosum* are minimized where timber production is the main objective by managing damage-prone species on short rotations (40–120 years), by favoring more resistant tree species on appropriate sites, and by limiting tree wounding. Stump or wound treatments with borax or other chemicals or biological agents do not appear to be nearly as useful in coastal forests as they are in the interior. With the emphasis on longer rotations and management of more late successional forests under ecosystem-management approaches, the impact of annosum root disease can be expected to increase.

White pine blister rust occurs throughout the Pacific coast region wherever hosts occur. *Cronartium ribicola* was introduced into coastal forests in 1910 at Vancouver, British Columbia, and spread especially rapidly in coastal forests because of the frequently ideal climatic conditions and the abundance of *Ribes* spp. Five-needle pines have been all but eliminated in some coastal stands and seriously reduced in others. It is estimated that 80–95% of the western white pine and sugar pine have been affected.

Currently, planting of five-needle pines with various levels of resistance to *C. ribicola* is accelerating on appropriate sites in coastal forests. Tree-improvement programs to screen apparently resistant pines and to breed for higher levels of resistance have existed in both Canada and the United States for some time.

Fungi that cause decay of wood in living conifers are widely distributed in coastal forests. Many species of fungi are involved, but those that have the greatest effects are *Phellinus pini, Echinodontium tinctorium, Phaeolus schweinitzii,* and *Fomitopsis pinicola*. Most stem-decay fungi are associated with tree wounds, and most take a number of years to develop substantial decay columns. Thus, although the effects of decay fungi are encountered in almost all stands, they are greatest in old stands where many trees are wounded.

In stands managed entirely for timber-production purposes, decays have been viewed as very damaging. In such situations, wood losses are minimized by short rotations (120 years or less), by favoring the more decay-resistant species where possible, and by preventing wounding in thinning entries or partial-cut harvests. Wounding has been greatly reduced by use of modified harvest preparation procedures and sale administration techniques.

In stands managed for a variety of objectives under an ecosystem-management approach, the virtues of stem decays in providing important wildlife habitat and enhancing long-term site productivity are recognized. Active programs to systematically create snags, dead tops, or wounds for decay entry are commonplace. Actual inoculations of individual trees with decay fungi are occurring on a limited basis.

Hemlock dwarf mistletoe (*Arceuthobium tsugense*) is the only dwarf mistletoe encountered in most coastal forests, and its impact on its host is generally not as severe as that of many other dwarf mistletoe species. Nevertheless, effects can be very significant on heavily infected, old trees. Young trees may also be severely impacted if they have numerous infections in their very tops. Where the disease is severe, it contributes to a "pathological rotation" for western hemlock by causing substantial decline and mortality of old, infected trees. It limits development of western hemlock understories in infected stands and gives nonhosts distinct growth advantages over hemlocks. Dwarf mistletoe-induced witches'-brooms and large, infected branches also provide preferred habitat for several wildlife species, including marbled murrelets.

Where timber production is the principal objective, managers either eliminate dwarf mistletoe from infected stands via regeneration harvest, favor nonhosts in such areas, or live with the disease if it is not severe by managing on short rotations (40–120 years). Where an ecosystem-management approach is being used, some level of dwarf mistletoe infection is often desired or at least accepted. Dwarf mistletoes thus represent a significant planning challenge. How can stands be managed so that some mistletoe is maintained but not so much that long-term impacts are too severe?

(Prepared by D. J. Goheen)

Inland West of North America

The Inland West encompasses a vast area that includes the Rocky Mountains west to the Cascade and Sierra Nevada mountain ranges. The area contains diverse landforms, climates, and forests. The range of forest conditions includes the warm, moist, productive cedar-hemlock-white pine forests of the interior northwest; the hot, dry ponderosa pine forests at lower elevations; the cold, spruce-fir-whitebark pine forests at high elevations; and a diverse group of mixed-conifer forests that occur where the climate is moderate.

Conifer diseases impact forest resources on millions of acres throughout the Inland West each year (Table 16). Both commodities and amenities are affected. Root diseases and dwarf mistletoes, for example, each kill and reduce the growth of timber measured in the hundreds of millions of cubic feet. Root and stem diseases cause tree failures and lead to personal and property damage.

Pathogens are also important for their ecological functions. Forest disease and insect management is undergoing its biggest change since the 1960s, when controversy about the use of pesticides led to the adoption of IPM (integrated pest management) as a new philosophy. But even with the change to IPM, pathogens have been evaluated almost exclusively on the basis of the damage they do to resources, whether the resource is timber, recreation, wildlife, or other commodity or amenity values. The new emphasis on ecosystem management has caused us to reconsider the "nonpest" roles of pathogens in forest ecosystems, i.e., the positive as well as negative effects on essential ecosystem processes such as succession, nutrient recycling, evolution, and wildlife population dynamics.

A very large number of pathogens significantly affect conifer forests in the Inland West. All forest types and conifer species are affected to some degree. Dwarf mistletoes are the most abundant of all pathogen groups in the Inland West, infesting more than 22 million acres. Many different species are present, each with its preferred host or hosts. Growth loss is estimated at 393 million cubic feet per year.

Dwarf mistletoes may also be among the most important pathogens ecologically. Several studies have shown that dwarf mistletoe brooms are used for nesting and cover for various

TABLE 16. Important Conifer Pathogens of the Inland West

Pathogen	Hosts
Arceuthobium	
americanum	*Pinus contorta*
campylopodum	*Pinus ponderosa*
vaginatum	*Pinus ponderosa*
laricina	*Larix occidentalis*
douglasii	*Pseudotsuga menziesii*
Armillaria ostoyae	*Pseudotsuga menziesii, Abies grandis*
Heterobasidion annosum	*Abies* spp.
Cronartium ribicola	*Pinus monticola*
Phellinus pini	Many conifers
Phaeolus schweinitzii	Many conifers
Echinodontium tinctorium	*Abies* spp.

bird and animal species and that birds and mammals serve as vectors. Insect-dwarf mistletoe interactions result in pollination; insects feed on dwarf mistletoes; and dwarf mistletoes predispose trees to insect attack.

Dwarf mistletoes not only reduce tree growth, but also cause tree death. The rate of mortality tends to be slow, but given the large area infested by these agents, mortality levels can be quite significant over long time periods. The interaction of dwarf mistletoes with fire has been only partially explored but appears to be significant in terms of forest development and dwarf mistletoe distribution. Since these parasites require a living host, stand-replacement fires that kill all hosts eliminate them. Mixed severity fires and underburns perpetuate them by promoting susceptible understories. Dwarf mistletoes increase stand susceptibility to wildfires. Heavily broomed understories act as fuel ladders, carrying fire into the overstory and increasing the probability of stand replacement.

Dwarf mistletoe distribution and severity has been altered greatly by management in individual stands, but changes over the landscape have been less significant. The trend has been mixed. Dwarf mistletoes have been eliminated from stands, sometimes over large areas, when timber management has been the primary objective. But in other cases, the severity of the dwarf mistletoe problem has been increased by various forms of partial cutting, by control of wildfire, and by other forms of management. These obligate parasites have only moderate effects on the growth of young, single-story stands of vigorous trees. Severe growth loss and tree mortality occur in aging forests and dense, multistoried stands. Evidence suggests that the acreage of such forests has increased.

Root diseases are considered to be among the most significant agents in terms of resource impacts and are among the most difficult problems to diagnose and treat. They are of management concern on about 10 million acres in both interior and coastal forests of the west. They have been increasing in importance during the past decades because of management practices that have increased host composition of forests and increased spread of the pathogens.

Root diseases also serve many important ecological functions. They form openings or gaps in forests, which are valuable to wildlife. They kill trees, which provides standing and down wildlife snags, and they decay trees, which recycles carbon and nutrients. In some areas, they cause extensive, broad-scale tree mortality that alters patterns of vegetation on the landscape and changes successional trends. The "weeding" effect of selective tree killing in mixed-species stands changes the forest type to disease-tolerant species. Removal of overstories in stands of susceptible species accelerates succession to climax species or results in self-perpetuating root disease patches, depending on the site and the understory composition.

The two most significant root pathogens are probably *Armillaria ostoyae* and *Heterobasidion annosum*. Both are widespread in mixed-conifer forests of the Interior West, and both are capable of causing significant tree mortality. *Phellinus weirii* is a virulent pathogen, but its distribution is limited. Reductions of 50% or more in timber volumes are common for stands affected by Phellinus and Armillaria root diseases in the inland northwest.

The spatial distribution of root disease deserves comment, because current concepts lead to underestimation of disease significance. Distribution of the pathogens, and the resultant disease expression, can be localized or general. Localized distribution, the common concept of disease distribution, implies that the pathogen is present in patches or centers of variable but limited size and is not present outside these patches. However, surveys and observations in the Inland Empire and elsewhere indicate that certain pathogens, including *A. ostoyae*, *P. weirii*, and *H. annosum*, can be generally distributed throughout stands and landscapes, causing levels of disease that vary with the stage of stand development, species composition, and stand history. Disease is expressed as the killing of a single tree or of small centers in young stands of susceptible species or in mixed-species stands and thus appears to be localized. But observation reveals infected trees throughout the stand, and within one to a few decades, the entire canopy can be lost, an outcome much different from that forecast for localized disease, which tends to increase very slowly by comparison.

The evidence suggests that root disease has been increased and its roles altered by management practices. For example, wildfire control, economic selection and other forms of partial cutting, reduction of white pines by blister rust, and other human-caused changes have greatly increased the abundance of susceptible hosts in intermountain and other areas, and Phellinus and Armillaria root diseases have responded. They now convert susceptible forests to climax forests and to persistent root-diseased areas consisting of brush and small trees on large areas where their prior function was the creation of gaps and the weeding of mixed-species stands. Annosum root disease has intensified as a result of harvesting, which has increased spread by providing stumps for infection. Ecologically, annosum disease in ponderosa pine might be considered very significant, even though the area involved is limited. One or a few harvests can increase the number of disease centers many fold, and pines can be eliminated over the decades as these centers enlarge. In areas where pine is the only native conifer, the forest cover can be lost for an indefinite time period.

Stem rusts are common on pine species in the Inland West, but except for the introduced white pine blister rust, they tend to be of localized importance. White pine blister rust has had a severe and lasting impact on resource production and ecosystem function throughout the range of western white pine and sugar pine. Ecological impacts are severe in whitebark pine forests over much of the inland northwest and northern Rocky Mountains. Southwest white pine is threatened.

Western white pine and sugar pine are two of the most valuable timber species. Forests composed primarily of white pine were once found in northern Idaho and the vicinity, and both species were an important component of mixed-conifer forests of the inland northwest. Few white pine forests now exist, and the white pine component of many mixed forests has been reduced by 90%. Rust-resistance research and development programs have provided the means to successfully reforest with these species. But reestablishment of these valuable species for the long term will require a major reforestation effort and may depend on continued research and development to maintain levels of resistance.

Ecological impacts of blister rust are also very significant. Extensive, old-growth pine forests existed a century ago, and the replacement forests of Douglas-fir, grand fir, subalpine fir, and other species now are very different in character. The replacement forests are also very susceptible to root diseases, which are now causing a transition to climax forests of western hemlock and western red cedar on a scale that likely never occurred before. In whitebark pine forests of the glacier ecosystem, loss of old-growth whitebark pine to blister rust and other agents has greatly reduced the availability of whitebark cones, considered an important food source for grizzly bears.

Stem decays were considered the most important group of diseases early in this century when pathologists first visited the west. The early investigations were made to determine the amount of decay in merchantable forests and to develop indicators for use in harvesting. Decay was extensive in old-growth forests, a condition that persisted into the middle of the century. Quinine conks, for example, were reported shipped from Missoula, Montana, by the train boxcar. Stem decays are still common in the remaining old-growth forests, but they have been much reduced. Some decays are still common, for example, those caused by *Phellinus pini* and *Phaeolis schweinitzii*. Others can be locally abundant. Quinine conks are now rarely found in areas where they were once plentiful.

With the drastic reduction in stem decays, we can assume some loss of function in the systems. The decays provide sites for cavity-nesting birds and animals, they recycle biomass, and they weaken trees, making them susceptible to breakage by wind and snow. Observations indicate that over the long term, stem decay-related breakage may have had a significant effect on stand structure.

(Prepared by J. W. Byler)

Southern United States

Land ownership in the south is dominated by the private sector. There are large numbers of nonindustrial, privately owned areas of relatively limited size and a more limited number of large, often contiguous forested tracts managed by industry. Consequently, private lands furnish the vast majority of the economic products obtained from southern forests. The importance of these lands as a source of renewable forest products is likely to expand in coming years because changing priorities for public land management, especially in the western states, are placing less emphasis on commodity production.

The expanding role of southern forests as major suppliers of economic products is supported by the rapid growth of southern pines, their responsiveness to intensive culture, and the suitability of their wood for diverse products ranging from pulp and paper to composites, dimension lumber, and plywood. In terms of distribution, economic value, and ecological importance, the southern pinery is dominated by only a few species. Loblolly and slash pine are the primary economic species; longleaf pine is of growing ecological importance; shortleaf and sand pines are less widely distributed but of considerable importance at the local level. The intolerant, seral nature of southern pines makes them well suited for even-aged silvicultural systems. Regeneration harvesting, in which clear-cutting is followed by the planting of bare-root 1-0 stock on prepared sites, is the system of choice on most private lands. Site preparation can entail any combination of mechanical or chemical procedures, with or without the use of prescribed fire. Most seedlings are machine planted, although hand planting is needed or prescribed on very rough sites or by landowners seeking high rates of seedling survival. Intermediate cultural techniques often include nutrient enhancement, use of prescribed fire, and precommercial and commercial thinning.

Rotation ages vary according to product and management objectives. Pulpwood rotations are usually less than 25–30 years; early saw-timber rotations require an additional 10 years. Harvesting is fully mechanized on most sites. Recent economic incentive programs have resulted in the reforestation of substantial acreage of marginal or highly erodible croplands throughout the south.

Relatively few pathogens have major impacts on the physical, social, and environmental products of southern pine forests (Table 17). Many diseases have strong anthropocentric linkages (e.g., soil erosion and littleleaf, fire suppression leading to increased oak abundance and fusiform rust, and thinning and annosum root rot) or are strongly influenced by general environmental conditions (e.g., moisture stress and pitch canker and soil texture and annosus root rot). Recognition of these types of relationships is important in the formulation of forest management plans. Many of the diseases of southern pines also have important direct or indirect linkages to insects (e.g., vectors, wounding agents, and predisposing influences). Consequently, disease management must often address multiple agents in an integrated approach.

Because of its effects on product quality, tree mortality, and stocking reduction, fusiform rust is clearly the major concern of forest managers throughout much of the southern pine ecosystem. This is especially true on lands managed for wood products. The major impact area ranges from South Carolina through Georgia, Florida, Alabama, and Mississippi. Within and beyond this area, incidence and severity traverse the spectrum from low levels that provide an often-beneficial thinning effect to localized epiphytotics that leave few if any healthy survivors. Affected trees suffer early death and stem deformation and are predisposed to pitch canker, stem breakage, and attack by the pitch moth.

In many areas, fusiform rust management is highly integrated into land management activities. Economic viability requires rapid growth rates, which in turn require cultural activities that often lead to increases in disease incidence. Managers frequently delay fertilization until after 5 years of age to reduce infection during the most vulnerable years. Genetically resistant planting stock is the major disease management tool used to minimize incidence and severity of fusiform rust. This stock also retains the ability for rapid early growth through the application of other silivicultural practices.

Since the 1960s, the cooperative work of forest pathologists has brought the use of genetic resistance from a desired goal to a routine practice in reforestation of slash and loblolly pine forests in the south. Culturally induced increases in growth favor increased levels of rust infection across the spectrum of resistance levels. However, the more resistant sources maintain lower and usually acceptable infection rates. At the genetic level, the absence of a linkage between rust resistance or susceptibility and growth rate allows the simultaneous utilization of genetically faster-growing trees that also possess enhanced resistance.

Littleleaf disease reduces growth and increases mortality and predisposition to bark beetle attack. Facing limited opportunities to eradicate the pathogen (*Phytophthora cinnamomi*) and given its widespread occurrence in highly eroded soils throughout the Piedmont region where shortleaf pine is common, forest managers frequently avoid future problems with littleleaf by converting affected stands to the more disease-tolerant loblolly pine. In situations where harvest is not possible or desirable and where fire is excluded, littleleaf disease, alone or in combination with the southern pine beetle, will hasten succession to the tolerant hardwood climax.

Delineation of site hazard according to edaphic conditions has allowed managers to adjust harvest scheduling to avoid anticipated littleleaf-related losses and to identify sites where short rotations of shortleaf pine are feasible. Additionally, site risk ratings allow managers to target shortleaf regeneration efforts on sites with higher probabilities for success.

The strong link between annosum root rot losses and edaphic (low water table and coarse-textured surface soils) and atmospheric (temperature) conditions and the cultural practice of thinning provides the opportunity for development of effective management strategies. Annosum root rot is of little concern on lands managed on short rotations in which thinning is not a component of the silvicultural prescription. Similarly, managers do not commonly confront this disease on sites with a high water table in the lower coastal plain or on most sites lacking coarse-textured surface soils.

TABLE 17. Important Pathogens of Southern Pines

Pathogen	Hosts
Phytophthora cinnamomi	*Pinus echinata*
Heterobasidion annosum	*Pinus* spp.
Phellinus pini	*Pinus* spp.
Cronartium quercuum f. sp. *fusiforme*	*Pinus elliottii, P. taeda*, other *Pinus* spp.
Fusarium subglutinans	*Pinus* spp.
Mycosphaerella dearnessii	*Pinus palustris*
Bursaphelenchus xylophilus	*Pinus* spp.

Chemical stump treatment is routinely used in high-value situations (e.g., roguing of seed orchards) but is used infrequently in normal operational thinnings. Until such time as thinning becomes a more frequently utilized silivicultural practice and annosum root rot becomes a more commonly encountered disease, investments in stump treatment will probably be limited to those organizations and individuals with prior experience with this disease and those seeking to avoid the acquaintance.

Losses caused by heartrot are of little concern in most contemporary southern pine forests managed under the relatively short rotations used for production of wood products. In forests managed for other objectives, heartrot (principally caused by *Phellinus pini*) in living trees can represent a valuable forest resource as the preferred nesting habitat for the red-cockaded woodpecker, an endangered species. The scarcity of old, heartrot-affected southern pines, especially longleaf, limits the reproductive success of this cavity-nesting bird. Changes in land-management objectives, especially on public lands, will increase forest acreage managed under the extended rotation lengths designed in part to favor suitable habitat for species favoring old-growth forests. Those attempting to artificially accelerate heartrot by deliberate inoculation could benefit by an increased understanding of the decay process in living southern pines.

The pinewood nematode is ubiquitous in southern pine ecosystems and is effectively vectored by several cerambycid beetles that utilize severely stressed, dying trees and recently harvested trees for oviposition and brood rearing. Despite this widespread occurrence, only in very rare instances has the pinewood nematode demonstrated pathogenic capability on native pines growing within their native ranges and on sites typical of their occurrence. Because wood chips produced from colonized trees regularly contain living nematodes, however, export markets for southern pine wood chips have been restricted. The resultant economic impacts to the forest products sector have been significant, despite the minimal effects of this organism in the forest itself.

The pitch canker pathogen occurs throughout the south, can infect all species of southern pines, and can impact all components of the silvicultural system from seed to mature trees. Through its effects on growth reduction, stem deformity, and mortality, pitch canker can impact productivity and limit product and management options. Integration of pitch canker management into silvicultural prescriptions follows two major paths: utilization of genetic resistance and stress management. Overall, loblolly is less affected by pitch canker than slash pine. On sites suitable for both species and in areas where pitch canker is a concern, this species-level differential in susceptibility often favors loblolly as the species of choice.

The effects of pitch canker in mature slash pine stands have focused additional attention on regulation of stocking levels to minimize competitive stress and to help ensure the adequacy of site resources. Accordingly, planting densities are lowered and thinning is more frequent in order to manage stress. A reduced number of stressed trees may also limit the availability of brood trees for the eastern pine weevil, a known vector and wounding agent associated with pitch canker in slash pine. Precommercial thinning of pitch canker-affected loblolly and early commercial thinning of pitch canker-affected slash pine have reduced postthinning disease and favored crop tree recovery and growth. Operational use of these tactics has been limited, primarily because of reservations about making additional investments in stands that have demonstrated genetic susceptibility to the disease.

Brown spot needle blight can be a serious problem in longleaf pine forests in the western part of the species range. Given the increasing interest in this species, especially on public lands, brown spot needle blight is likely to assume a more dominant position in management considerations.

Cultural practices such as herbaceous weed control and fertilization are effective in promoting rapid early growth and reducing the duration of the infection-prone grass stage. Forest managers are increasingly using genetic sources of longleaf pine that tend to leave the grass stage at an early age or that show resistance to the brown spot fungus. Prescribed fire is useful as a site-preparation tool for the reduction of potential competing vegetation and eradication of any on-site inoculum. Cool winter burns are also used in infected grass-stage plantings to destroy inoculum. Additionally, preplant treatment with a fungicidal root dip has proved effective in reducing postplanting infections and is used in situations where this extra effort and cost is warranted by the hazard of the planting locale.

(Prepared by G. M. Blakeslee)

India

The coniferous forests of the Indian Himalayas are composed of species of *Pinus, Cedrus, Picea,* and *Abies*. Chir pine, blue pine, and deodar are the most important. Of these, chir pine (*Pinus roxburghii*) is most widely grown. Chir is a tall tree with a spreading crown. It occurs in the outer hills and valleys of the Himalayas and does not usually extend beyond the range of monsoon rains. Chir pine forms pure forests over extensive areas, although it also occurs in mixtures with deodar and blue pine at the upper elevation limits of its range and with various hardwoods at the lower elevation limits. The chir pine is one of the least exacting of Indian conifers, but the tree demands full light in all but the hottest situations. Chir pine is now worked mostly as even-aged forests with rotations of 90–160 years and regeneration periods of 20–35 years.

Blue pine (*Pinus wallichiana*), or kail, is a tall, five-needle pine with spreading or drooping branches. It occurs throughout the temperate Himalayas, extending to higher elevations than other Indian conifers. In the western Himalayas, it grows in pure stands at between 1,800 and 2,500 m. Blue pine is light demanding, although it may persist in shade for some years while young.

Deodar (*Cedrus deodara*) is a large tree that often reaches 60 m in height and 10 m in girth. Deodar forests are common in Kashmir, especially between elevations of 1,650 and 2,400 m, and form the bulk of the vegetation. Deodar is grown as a rather dense stand to produce a clean bole. It may be grown in mixtures with spruce to produce the highest quality timbers. Deodar, because of its economic importance, was the first Indian conifer to be raised artificially. Plantations have been established on a large scale, even beyond the natural range of the tree.

Diseases are major determinants of forest productivity (Table 18). In natural forests, the diseases are generally endemic and do not cause much damage to the trees because of the genetic variation in forest composition. However, diseases

TABLE 18. Important Pathogens of Conifers in India

Pathogen	Hosts
Heterobasidion annosum	*Cedrus deodara, Pinus wallichiana,* other conifers
Armillaria mellea	*Cedrus deodara, Picea* and *Abies* spp.
Inonotus tomentosus	*Cedrus deodara, Pinus wallichiana, Picea* spp.
Phellinus pini	*Pinus wallichiana,* other conifers
Cronartium himalayense	*Pinus roxburghii*
Arceuthobium minutissimum	*Pinus wallichiana*
Diplodia pinea	*Pinus kesiya,* other *Pinus* spp.
Peridermium cedri	*Cedrus deodara*

are likely to pose serious problems in man-made forests, particularly when forest tree species are raised as pure crops or attempts are made to provide green cover to the degraded forest areas beset with one or more stress problems. Because of the heavy denudation of the forests in the Himalayas since the 1950s, a challenging situation has arisen that is posing a serious threat in the form of flash floods, silting of rivers and dams, adverse climatic changes, and shortages of fuel wood and timber. To check and reverse this trend, the government of India has imposed a moratorium on green felling on the hills above 1,000 m. This ban on tree felling is likely to succeed if it is coupled with a massive planting program in the Himalayas to provide green cover and thus restore the otherwise fragile ecosystem. Protection of forests from diseases is therefore very necessary to mitigate damage to tree crops and thereby augment forest productivity in the Himalayas.

Heterobasidion annosum attacks deodar, blue pine, fir (*Abies pindrow*), and spruce (*Picea smithiana*), causing root and butt rot in many areas of the Himalayas. Although endemic in natural forests, the disease may cause serious damage to plantations on unsuitable sites, particularly those with poor drainage. Tree wounds in the butt region and on exposed roots caused during felling operations appear to constitute the infection court for the pathogen. However, in India, root rot caused by *H. annosum* is not yet as important as it is in Europe and the United States.

The disease can be managed through an integrated approach involving measures such as silvicultural and management practices, chemotherapy, and biological control. The suitability of the site is very important in growing disease-free crops; therefore, consideration should be given to elevation, aspect, and slopes of the hills when a plantation sight is selected. During thinning and improvement fellings, diseased trees should be removed to reduce the fungal inoculum. Since the disease normally becomes a problem after thinning, the freshly cut stumps should be treated with urea (20%) or sodium nitrite (10%).

Armillaria mellea causes root rot in deodar plantations in the Himalayas where the original conifer forests have been clear-cut. The pathogen is also known to attack spruce, fir, and oak. Ring barking of trees before felling minimizes the disease in the subsequent plantation by allowing innocuous soil fungi to colonize the roots, exhausting the food reserves and making them no longer suitable for infection by the pathogen. Treatment of stumps with 40% ammonium sulfamate also favors rapid colonization by fast-growing saprophytic microorganisms, which limit invasion by *A. mellea*. Removal of colonized stumps and trenching around areas of infection also help minimize the disease in plantations.

Phellinus pini attacks the heartwood of blue pine and causes white pocket rot. The incidence of heartrot is high in the outer Himalayas where annual rainfall is 1,000 mm or more. Disease levels gradually decline with decreasing rainfall below 1,000 mm in the inner ranges of the Himalayas. This fungus is also observed occasionally on spruce, fir, and deodar in coniferous forests at higher elevations and on chilgoza pine in dry zones. It is seen rarely on chir pine after resin tapping. In some areas of the Himalayas, villagers are allowed to lop the limbs of blue pine, and the fungus enters the tree through branch stubs, open knots, or wounds.

Peniophora luna causes a serious brown cubical butt and trunk rot in living deodar and has been recorded in pure plantations clear-cut in Himachal Pradesh. About 56% of the trees and 24% of the basal area were affected. It is likely that the butt rot fungus enters through root injuries.

Cronartium himalayense alternates between chir pine and *Swertia* spp., which come up as weeds during the rains. The disease was recorded in a severe form that caused large-scale mortality during the 1930s and 1940s. However, it later declined and was almost absent during the 1960s. Recently, the presence of the rust in pure chir pine plantations has caused great concern, because the afforestation program in the area is likely to be disrupted. The disease causes death in young plantations up to 10–15 years of age. Susceptibility to rust attack is greatly reduced in the pole crop. Mature trees are fairly resistant, although branches may occasionally exhibit symptoms of infection. This indigenous rust is also reported to attack and cause death of *P. canariensis*, an exotic pine, and it is no longer planted in India. Eradication of *Swertia* spp. from chir pine forests for 3–5 years, up to a distance of 300 m from young plantations, may help minimize the disease. Systemic fungicides have been reported to be effective against the disease in young chir pine plantations in Uttar Pradesh.

Cronartium ribicola attacks blue pine but is not as damaging as it is to white pine in Europe and to the majority of five-needle white pines in the United States, and therefore no control measures are called for. *C. quercuum* causes gall rust on khasi pine (*P. kesiya*), and telia develop on *Quercus griffithii*. Khasi pine should not be raised in close proximity to oak to avoid the gall rust infection.

Dwarf mistletoe (*Arceuthobium minutissimum*) is common and widespread on blue pine in areas with low rainfall. Its incidence is low in mixed stands but high in pure blue pine stands. Heavily infected trees are stunted and eventually killed, resulting in stand openings that reduce the aesthetic value of the forests, particularly in health resorts such as Pahalgam and Gulmarg in Kashmir.

Control measures based on the degree of infection of the stand by the parasite have been recommended. In locations where infection is not widespread and is mainly confined to branches, pruning of branches bearing mistletoe is an effective method of control. However, in heavily infected areas, the stand should be clear-cut and regenerated. It is advisable to burn the area after felling, because fire normally results in profuse regeneration.

The exotic cypresses, *Cupressus arizonica*, *C. lusitanica*, and *C. sempervirens*, and *Juniperus procera* are attacked by *Monochaetia unicornis*. Indigenous *C. torulosa* planted in the same locality is apparently free from disease, and *C. cashmeriana*, possibly indigenous to Tibet and now naturalized in the eastern Himalayas, is also not attacked by the pathogen. Systematic removal of diseased parts and plants over the years may lead to a significant drop in disease incidence.

Diplodia dieback, caused by *Diplodia pinea*, has been recorded on khasi pine from Meghalaya and also on several exotic pines from Uttar Pradesh. Avoiding unfavorable locations for planting both exotic pines and *P. kesiya* and restricting movement of planting materials from infected areas within the country are helpful in minimizing the disease. For individual trees, pruning and burning of infected twigs during the autumn followed by three sprays of Bordeaux mixture (the first when new shoots just begin to emerge, the second when needles begin to develop, and the third when needles are partly grown) control the disease.

Peridermium needle and broom rust of deodar is caused by *Peridermium cedri*, an autoecious fungus that attacks young needles of the current year's shoots. The infected needles are shorter than normal and curve backward. Witches'-brooms are produced on the affected branches, and in extreme cases, the whole tree may be converted into witches'-brooms. The pathogen is believed to perennate as mycelium in the needles during the winter to infect the needles that appear the following spring.

The attack on saplings and young pole crops may result in arrested height growth or in extreme cases, even death of branches and the tree. Large-scale mortality in planted deodar has been reported from Himachal Pradesh and in natural pole crops in Uttar Pradesh, where the infected trees had to be removed during thinnings in favor of healthy trees. The disease occurs in a serious form in damp valleys, so planting of deodar in such areas should be avoided.

Dothistroma needle blight of pines was first recorded on *P. radiata* and *P. elliottii* trial plantations in Tamil Nadu. Because of the highly destructive nature of the disease, planting of *P. radiata* has been abandoned in the country. However, the disease is of no consequence on indigenous pines and warrants no control.

(Prepared by M. D. Mehrotra)

Australia and New Zealand

The native coniferous flora of Australia and New Zealand includes at least 44 species in the families Cupressaceae, Araucariaceae, Taxodiaceae, and Podocarpaceae. These species range from small, alpine shrubs to large forest trees. Natural stands of a number of species have been exploited for timber in the past, and some continue to be harvested in commercial quantities (e.g., species of *Callitris,* or cypress pine, in Australia). Most species, with the exception of *Araucaria cunninghamii* (hoop pine) are too slow growing or lack appropriate silvicultural characteristics to be suitable for economic production in plantations. *A. cunninghamii* has been used to develop a plantation resource in Queensland and to a small extent (47,000 ha) in New South Wales. The exotic coniferous flora introduced since European settlement is numerically extensive in terms of species, but commercial forestry is dominated by species of Pinaceae. Predominant among these has been *Pinus radiata* in the temperate and Mediterranean zones (1,899,000 ha) and on a more moderate scale, *Pinus pinaster* in Western Australia (32,000 ha) and *Pseudotsuga menziesii* in New Zealand in cool, wet environments (67,000 ha). In the subtropics, *Pinus caribaea* var. *hondurensis* (56,000 ha) and *Pinus elliottii* var. *elliottii* (84,000 ha) have been widely planted, and more recently, a hybrid (slash × Honduras pine) has been selected for planting on poorly drained sites. In temperate New Zealand, *Cupressus* spp. and *Chamaecyparis lawsoniana* are being increasingly planted. Plantations are normally grown on rotations of 25–60 years, and mean annual increments are typically in the range of 15–30 m^3/ha/year. Plantations of exotic species in New Zealand provide a source of general-purpose softwood for solid timber, reconstituted wood products, and pulp and paper, which meets domestic needs with a significant surplus for export.

Diseases of introduced species are caused by pathogens likely to have been introduced with their hosts or at subsequent times, indigenous organisms, and other non-host-specific pathogens such as *Armillaria* spp. and *Phytophthora cinnamomi* (Table 19). Several of the diseases are of sufficient economic importance to warrant control measures.

TABLE 19. Conifer Diseases in Australia and New Zealand

Plant Part Pathogen	Major host
Needles	
Dothistroma septospora	*Pinus radiata*
Lophodermium spp.	*Pinus radiata*
Cyclaneusma minus	*Pinus radiata*
Phaeocryptopus gaeumannii	*Pseudotsuga menziesii*
Shoot or stem	
Sphaeropsis sapinea	*Pinus radiata*
Amylostereum areolatum	*Pinus radiata*
Seiridium unicorne	*Cupressus* spp.
Seiridium cardinale	*Cupressus* spp.
Roots	
Phytophthora cinnamomi	*Pinus radiata*
Phellinus noxius	*Araucaria cunninghamii*
Junghuhnia (Poria) vincta	*Araucaria cunninghamii*
Armillaria novae-zelandiae	*Pinus radiata*
Armillaria limonea	*Pinus radiata*

Dothistroma needle blight, caused by *Dothistroma septospora* (a combination not accepted in New Zealand, where the binomial *D. pini* is used), is the most important foliage disease in terms of its potential impact on yield and on control costs. Moderate temperatures (annual mean, <20°C) and a wet and humid environment, particularly with high summer rainfall levels, favor disease development. Severe infection is less widespread in Australia than in New Zealand, where a larger proportion of the plantation estate is located in areas with high rainfall levels (>1,000 mm per year). The disease was first noticed in New Zealand in 1962 and had spread to most plantations in both the North and South Islands by 1987. The disease was first recorded in Australia in 1975 and has extended its range to include most of the areas of eastern Australia where radiata pine is grown. Reduction in height and diameter growth escalates as crown infection increases. An approximate linear relationship exists between the percentage of the crown infected and growth: for each 10% increase in crown infection there is a decrease in increment growth of about 10%. Repeated and severe infections may result in death.

Cyclaneusma minus is one of the more pathogenic of a number of needle-cast fungi recorded in Australia and New Zealand. The fungus is ubiquitous and causes considerable growth loss. A reduction in volume increment of 60% at a disease severity level of 80% has been recorded. Disease severity is variable and most severe in cool, wet environments.

In Tasmania, spring needle cast is common in some of the most highly productive plantations in wet areas and may result in growth losses comparable to those caused by Dothistroma needle blight and Cyclaneusma needle cast. The etiology of spring needle cast is uncertain but may involve several needle-cast fungi, including *C. minus,* that act in association with physiological or nutritional stress. *Sirex*-induced mortality, which results from the combined action of a toxic mucus and the fungus *Amylostereum areolatum* (introduced into the tree trunk at the time of oviposition by *Sirex noctilio*), is a problem principally of overcrowded or drought-stressed pine stands. It is not normally a problem in tended stands.

Since its discovery in New Zealand in 1959, *Phaeocryptopus gaeumannii* has spread to most areas of New Zealand and Australia where *Pseudotsuga menziesii* is grown. The fungus causes foliar chlorosis and reduced needle retention, but the effects of infection on tree growth have not been fully quantified. H. Beekhuis, who compared the predicted growth of uninfected trees with the actual growth of infected trees, calculated that there was a loss of 2.7 m^3/ha in the mean annual increment at age 46.

Sphaeropsis sapinea is a common pathogen throughout the pine-growing regions of Australia and New Zealand and causes locally severe shoot and leader dieback, stem malformations, and blue staining in *P. radiata*. Although the fungus can infect undamaged tissue, infection is typically most severe after shoot or stem injury caused by hail, insects, pruning, stand thinning, or drought stress. Early stand infections may reduce stocking, and stem malformation may reduce merchantable volume. A reduction of 6–12% of recoverable volume has been reported.

Root diseases vary in importance by region. *P. cinnamomi* has caused death and growth loss in *P. radiata* on poorly drained, sandy or lateritic soils of low fertility in Western Australia and in some shelterbelts on heavy soils in New Zealand. *Phellinus noxius* is the most widespread and destructive root disease of *Araucaria cunninghamii* plantations in Queensland, where it causes locally severe root rot and death of trees of all ages. *Junghuhnia (Poria) vincta* also causes root rot in *A. cunninghamii* plantations and is responsible for low levels of mortality in *P. radiata* on sites cleared of indigenous forest in the Bay of Plenty region of New Zealand. *Armillaria novae-zelandiae* and *A. limonea* are important root rot pathogens in New Zealand's pine forests. On first-rotation sites cleared from indigenous forest, mortality may be severe, and although the incidence of tree death decreases with increasing stand age, infection persists throughout the rotation in a nonlethal, chronic form. Low to moderate levels of mortality may also occur in second-rotation sites, and chronic infection may be widespread in some forests. Growth losses caused by *Armillaria* infection were estimated to be 6–13% of potential volume for a 28-year sawlog rotation.

At the forefront of disease-management strategies in Australia and New Zealand is the prevention of further introductions of pathogens from the natural range of the hosts or elsewhere. Diseases of particular concern include western gall rust, pitch canker, and those caused by root-disease fungi such as *Armillaria* spp., *Heterobasidion annosum,* and *Phellinus weirii*. Intracountry quarantine has also been practiced in Australia to limit the spread of pathogens such as *D. septospora*. The substantial investment in plantations of exotic species has ensured an active interest in the assessment of potential threats and potential modes of introduction, the effectiveness of pest-detection activities, and quarantine policy.

Aerial applications of copper fungicides have been regularly used to control *D. septospora* in New Zealand on a large scale and in Australia on an intermittent and more localized basis since the mid-1970s. Stands in the age classes susceptible to needle blight (<16 years for *P. radiata*) are surveyed, and those with more than 15% crown infection in high-hazard areas and more than 25% in other areas are sprayed once or twice per summer, depending on the initial level of crown infection. In Kinleith Forest (area, 145,000 ha; annual rainfall, 1,500 mm) in the North Island of New Zealand, total costs of control from 1967 to 1988 were estimated to be 18.4 million New Zealand dollars (1988 dollars) and total costs in New Zealand approximately $35 million during the same period.

The intensive management of conifer plantations in Australia and New Zealand, with its combination of wide spacing (usually <1,500 stems per hectare at planting, with a final stocking of <400 stems per hectare), heavy thinning, and pruning, tends to reduce the incidence and severity of certain diseases. Dothistroma needle blight is less severe in pruned and thinned stands because pruning reduces the amount of available inoculum and thinning improves stand ventilation. A similar effect of thinning on spring needle cast has been reported. It was shown that by delaying thinning to the age (about 6 years) at which individual trees susceptible to Cyclaneusma needle cast can be identified and removed, the impact of the disease can be minimized. Thinning also reduces the adverse effect of *S. pinea* infection by improving stand vigor. In New Zealand, incidence of Armillaria root disease has been reduced on first-rotation sites cleared from indigenous forests by physical removal of the stumps and large roots (the inoculum). Despite the effectiveness of the method, it is not widely practiced because of the expense and the slope limitations on where destumping can be conducted.

In Western Australia, material selected for resistance to *P. cinnamomi* has been deployed onto high-hazard sites via cuttings and a controlled crossing program developed to produce resistant genotypes for future plantings. In both Australia and New Zealand, selection and breeding programs for resistance to *D. septospora* are being pursued as a cost-effective option for planting on high-hazard sites and to reduce the dependence on chemical control.

(Prepared by G. A. Kile, K. M. Old, and P. D. Gadgil)

Africa

Native conifers are not particularly dominant in Africa, and in terms of timber production, they are relatively unimportant. By far the most important conifers propagated for timber are exotic species of *Pinus,* which are important components of the forest industries in numerous sub-Saharan African countries. Because of incomplete records and the relatively poor communication between pathologists in Africa, this review cannot attempt to provide a complete and thorough treatment of the topic. It is therefore intended merely to highlight some of the more important diseases known in the area.

In sub-Saharan Africa, *Pinus* spp. are grown in intensively managed plantations and have thus been separated from their natural enemies. Intensive propagation and the possibility of introducing new pathogens into the area have heightened the potential threat of diseases to these trees. Some diseases have been intensively studied and are known to have had a significant impact on forestry operations (Table 20). Perhaps the most damaging and best known of these is Dothistroma needle blight, caused by *Dothistroma septospora*.

Dothistroma needle blight was first recorded in Tanzania and then spread to exotic pine plantations throughout central and southern Africa. It is known on a number of *Pinus* spp., but its impact on *P. radiata* has been most serious because this species is extensively planted. Dothistroma needle blight has caused the virtual cessation of *P. radiata* propagation in most of Africa. In South Africa, this tree species is still grown relatively extensively in the southern and western cape provinces that have a Mediterranean climate. *D. septospora* occurs in this area but does not result in substantial defoliation of trees, which is enigmatic.

One of the most important diseases of pines in southern Africa is dieback caused by *Sphaeropsis sapinea*. This pathogen is known to be associated with a wide variety of symptoms on numerous *Pinus* spp., but the most serious losses are associated with infection after hail damage.

The two *Pinus* spp. most seriously affected by hail-associated infection by *S. sapinea* are *P. radiata* and *P. patula*. The primary strategy to avoid this disease has been to limit plantings of susceptible species to areas where hail damage does not occur. Susceptible trees growing in valleys are more seriously affected by this disease than are those growing on more exposed sites. Infestation by bark beetles and weevils tends to hasten disease development. Even though they might show serious signs of disease, trees often recover after hail damage. Clear-cutting should thus begin with trees in high-risk sites (valleys) and with trees that have become infested with insects.

Fusarium subglutinans f. sp. *pini* causes a serious disease of pines known as pitch canker, and indications are that this disease is beginning to acquire a worldwide significance. The

TABLE 20. Major Pathogens of Pine in Africa

Needle Pathogens	Stem and Shoot Pathogens	Root Pathogens
Cyclanuesma minus	Amylostereum areolatum (associated with Sirex noctilio)	Armillaria spp. (A. heimii)
Cercoseptoria pini-densiflorae	Fusarium subglutinans f. sp. pini	Leptographium spp.
Dothistroma septospora	Sphaeropsis sapinea	Pseudophaeolus baudoni
Lophodermium australe		Phytophthora cinnamomi
		Phaeolus schweinitzii
		Phellinus noxius
		Rhizina undulata

pathogen has recently been discovered in South Africa, although it occurs mainly in seedling nurseries. However, inoculations of mature trees have produced symptoms typical of pitch canker. Indications are that *F. subglutinans* f. sp. *pini* was introduced into the area on seed, and it is beginning to appear naturally in plantations. Of particular importance is that *P. patula* is highly susceptible to this pathogen and that this tree species, which is extensively planted in Africa, has been considered to be relatively disease free.

Armillaria root rot is a well-known disease of *Pinus* spp. in Africa. The disease is similar to those associated with species of *Armillaria* in other parts of the world: trees die in patches, and a thick, white, mycelial mat develops under the bark of dying trees. Comparatively little attention has been given to this disease in Africa, and until relatively recently, the species of *Armillaria* involved was not clear. More recently, however, a number of isolates from various parts of Africa have been identified as *A. heimii*, but more than one species is believed to be present.

Sporocarps of *Armillaria* on pines in Africa are seldom seen, and rhizomorphs are apparently absent. It has also been found that infection centers in pine plantations tend to disappear after a number of rotations of pine, evidently because trees do not attain particularly large diameters before felling and the relatively small, softwood stumps do not afford the pathogen an adequate food base for infections in progressive rotations.

Rhizina root disease, caused by *Rhizina undulata*, is one of the most important causes of tree death in southern Africa. One of the most serious problems associated with this pathogen is death of trees newly planted after fire on sites previously planted to pine. This problem became particularly acute when slash burning was used as a routine approach to site preparation. Burning of pine slash is now avoided, but accidental fires and subsequent *R. undulata* infections are still common.

The most practical means of reducing losses from infection by *R. undulata* is to avoid replanting after fire until risks of infection are minimal. Other than fire, factors associated with infection have not been clearly defined, and the periods of highest risk after fire tend to differ depending on time of the year and available moisture. A common recommendation, therefore, is that foresters plant small, randomly distributed test plots on previously burned sites at monthly intervals until seedlings in these plots cease to die. Large-scale planting is then considered to be safe.

Cypress canker is a serious stem disease of various species of *Cupressus* that have been relatively commonly grown in various parts of Africa. The disease has been associated with three species of *Seiridium*, *S. cardinale*, *S. unicorne*, and *S. cupressi*, although recent evidence suggests that these might all represent a single taxon. Cypress canker causes serious stem malformation on infected trees and eventually leads to tree death. This disease has therefore resulted in the failure of plantings of *Cupressus* spp. and the lack of attractiveness of this tree for forestry purposes. *Cupressus* spp. remain important shade and ornamental trees, and certain species appear to be less susceptible to infection than others.

The forestry industries of Africa, particularly those that depend on plantations of exotic *Pinus* spp., have been relatively fortunate in terms of losses caused by disease. A number of pathogens, such as *Dothistroma*, *Sphaeropsis*, and *Rhizina* spp., have resulted in, and in some cases continue to cause, serious diseases. However, the number of pathogens that have as yet not appeared in the area and that potentially threaten these trees is vast. While there is a great need to study diseases that are currently causing losses to these trees, intensive quarantine strategies to prevent the introduction of additional pine pathogens are also required. The recent appearance in southern Africa of the pitch canker pathogen and of the *Sirex* wood wasp-*Amylostereum areolatum* complex suggests that the strict quarantine that has been applied is not entirely effective. It is therefore predicted that losses to disease will increase in severity in the future.

(Prepared by M. J. Wingfield and T. Coutinho)

Part III. Diseases in Special Settings

Forest Tree Nurseries

Forest tree nurseries are an integral and sensitive link in most forest-management enterprises, but they have much more in common with intensive, high-input agriculture than with forestry. Forest tree nurseries are highly modified and simplified ecosystems and can be maintained only with high energy inputs in the form of cultivation, fertilization, and pest management. The same environmental conditions that make a successful forest nursery (high seedling density, favorable temperature and moisture, and intensive agricultural crop-management techniques) are also ideal for many of the pests that attack tree seedlings. Most of the pathogens of seedling conifers in nurseries are also serious pathogens of agricultural crops. The following are key features in successful nurseries.

Careful culture and handling of seedlings. Constant attention to the vigor of seedlings at all stages from sowing to lifting and storage before outplanting assures trees with the best chance of resisting pathogen attack and reaching the field in optimal physiological condition (Plate 163).

Quality seed. Emphasis on the collection of clean seed and its careful handling during storage and stratification have improved germination vigor and greatly reduced losses from seed decay and damping-off.

Careful soil management. Many nursery problems result from poor soil management. Bare-root nurseries must be sited on good agricultural ground with very well-drained soils. Nursery management must concentrate on preventing the compaction, depletion, or contamination of the soil. Containerized nurseries face a similar challenge to ensure a reliable supply of pathogen-free planting mix and clean containers and to prevent subsequent contamination of soil or containers during the growing cycle.

Modern pesticides and application procedures. Success in forest nursery operations, as in most intensive agricultural operations, depends on the timely application of disease (and insect and weed) control agents (Plate 164).

Damping-off (Plate 165) and the various seedling root rots are the most common and potentially destructive of the nursery diseases. The root pathogens of seedlings in nurseries are for the most part common agricultural pathogens, not forest dwellers. Thus, the species of *Rhizoctonia, Fusarium, Pythium,* and *Phytophthora* (Plates 166 and 167) that cause damping-off and root rot cause similar diseases in many crops. Indeed, these four genera of soil fungi invariably rank at the tops of lists of pathogens in forest tree nurseries, whatever the host and wherever in the world the nursery is located. In warm climates, *Macrophomina* joins the list, and in colder regions, *Caloscypha fulgens*, the seed fungus, and *Phacidium* and other snow molds are constant threats.

The most widespread pathogens of aboveground seedling parts are *Fusarium* and *Phoma* spp. and *Botrytis cinerea*. The first two are spread from the soil by rain splash; *B. cinerea* is airborne. All three normally require wounds or natural openings in the stem to initiate infection.

Selected References

Cordell, C. E., Anderson, R. L., Hoffard, W. H., Landis, T. D., Smith, R. S., Jr., and Toko, H. V. 1989. Forest nursery pests. U.S. Dep. Agric. For. Serv. Agric. Handb. 680.

Sutherland, J. R., and Glover, S. G., eds. 1991. Diseases and insects in forest nurseries. Proc. IUFRO Working Party S2.07-09. For. Can. Pac. Yukon Reg. Inf. Rep. BC-X-331.

(Prepared by E. Hansen)

Christmas Tree Plantations

The modern Christmas tree originated in Germany during the seventeenth century. The custom spread to England during the early nineteenth century and was popularized by Queen Victoria's German husband, Prince Albert. The tradition of using a Christmas tree as part of the Christmas celebration was brought to North America by German settlers and had become very popular by the nineteenth century. Today, Christmas trees and other cut conifer foliage are utilized in a number of Christian and non-Christian celebrations during late December and early January throughout the world.

Historically, Christmas trees were obtained from naturally forested areas. Even as late as 1960, approximately 75% of the Christmas trees produced in the United States reportedly came from wild or natural stands. However, by 1990, only 10% of the trees came from such stands, while 90% came from farms or plantations. A number of different species of conifers are grown as Christmas trees. The top-selling Christmas trees in North America include balsam fir (*Abies balsamea*), Douglas-fir (*Pseudotsuga menziesii*), Fraser's fir (*A. fraseri*), noble fir (*A. procera*), and Scotch or Scot's pine (*Pinus sylvestris*). In European and Scandinavian countries, the most common trees are Norway spruce (*Picea abies*), Scot's pine, and Nordmann fir (*A. nordmanniana*).

TABLE 21. Root Diseases of Conifers Commonly Grown as Christmas Trees

Disease Causal Organism	Hosts
Annosum root rot *Heterobasidion annosum*	Many conifers, but most common in second and third rotations of *Abies* spp.
Armillaria root rot *Armillaria* spp.	Many conifers; particularly troublesome in areas where stump culture is practiced
Phytophthora root rot *Phytophthora* spp.	Mainly *Abies* spp., especially *A. fraseri, A. procera, A. concolor,* and *A. magnifica* var. *shastensis*
Pinewood nematode *Bursaphelenchus xylophilus*	*Pinus* spp., especially *P. sylvestris*
Procerum root disease *Leptographium procerum*	*Pinus strobus*

Diseases can be a limiting factor in the production of high-quality Christmas trees (Tables 21, 22, and 23). Although some diseases, such as Phytophthora and Armillaria root rot, can kill trees prior to harvest, most diseases adversely affect the aesthetic quality and grade of a tree by causing shoot and branch dieback or premature casting of needles.

The types of diseases that are important and the disease-management options in Christmas tree plantations can differ significantly from those of conifers growing in forest settings. In general, needle cast diseases are much more important on Christmas trees than on conifers grown for a variety of forest product uses. Off-site seed sources or species may be much more susceptible to certain diseases that are favored by environmental conditions in local areas. For example, the coastal or green forms of Douglas-fir that are typically grown along the Pacific coastal areas of North America are fairly resistant to Rhabdocline needle cast. However, planting Rocky Mountain seed sources of Douglas-fir in these areas generally results in severe *Rhabdocline* problems.

Planting conifer species that are endemic to the area but that do not naturally grow on the sites that are typically utilized for Christmas tree production can also increase disease problems. Phytophthora root and stem canker disease and the physiological needle disorder known as current-season needle necrosis on noble fir grown as Christmas trees in the northwestern portions of North America are rarely seen in areas within these states where noble fir naturally occurs.

Cultural practices used in the production of Christmas trees can also increase some disease problems. Increased branch density associated with shearing to shape trees and plant spacing of 1.5–1.7 m can create conditions that prolong periods of moisture on needles and thus favor disease development (Plate 168). For example, Lecanosticta needle blight is a relatively minor problem on open-grown Scot's pine in landscape situations but can be a limiting problem when Scot's pine are grown as Christmas trees (Plate 169).

While there are some practices associated with Christmas tree production that may increase disease problems, the relatively high value of the crop, the types of sites on which many Christmas trees are grown, and the short rotations between planting and harvesting provide Christmas tree growers with many more disease-management options than are available to foresters. For example, growers typically spray fungicides to control locally important needle cast diseases. They might also install drain systems to alleviate soil drainage problems and thus reduce the potential for Phytophthora root rot.

TABLE 22. Stem and Branch Diseases of Conifers Commonly Grown as Christmas Trees

Disease Causal Organism	Common Hosts
Atropellis canker *Atropellis* spp.	*Pinus sylvestris*
Cytospora canker *Cytospora kunzei*	*Picea* spp., especially *P. pungens* and *P. abies*; *Abies*, *Pinus*, and *Pseudotsuga* spp.
Diplodia shoot blight and canker *Sphaeropsis sapinea*	*Pinus resinosa*, *P. sylvestris*, *P. nigra*, and *P. radiata*
Grovesiella canker *Grovesiella abieticola*	*Abies* spp., especially *A. magnifica* var. *shastensis* and *A. concolor*
Phomopsis canker *Phomopsis* spp.	*Pseudotsuga menziesii*
Phytophthora stem canker and shoot blight *Phytophthora* spp.	Mainly *Abies* spp., especially *A. fraseri*, *A. concolor*, *A. magnifica* var. *shastensis*, *A. magnifica*, *A. balsamea*, *A. lasiocarpa*, and *A. numidica*
Pine-pine gall rust *Endocronartium harknessii*	*Pinus sylvestris* and *P. radiata*
Pine shoot twisting rust *Melampsora pinitorqua*	*Pinus sylvestris*
Pitch canker *Fusarium subglutinans*	*Pinus radiata*
Scleroderris canker *Gremmeniella abietina*	*Pinus* spp., *Picea* spp., and *Pseudotsuga menziesii*
White pine blister rust *Cronartium ribicola*	*Pinus strobus*

TABLE 23. Needle Diseases of Conifers Commonly Grown as Christmas Trees

Disease Causal Organism	Common Host
Brown spot needle blight *Mycosphaerella dearnessii*	*Pinus sylvestris*
Current-season needle necrosis (physiological)	Mainly *Abies procera* and *A. grandis*
Cyclaneusma needle cast *Cyclaneusma minus*	Mainly *Pinus sylvestris*
Dothistroma needle blight (red band) *Mycosphaerella pini*	*Pinus nigra*
Fir needle rusts *Pucciniastrum* and *Uredinopsis* spp.	*Abies* spp.
Lophodermium needle cast *Lophodermium seditiosum* and *L. abietis*	Various *Pinus* spp., especially *P. sylvestris* and *P. nigra*; *Abies* spp., *A. alba*
Pine needle rust *Coleosporium* spp.	*Pinus sylvestris* and *P. resinosa*
Rhabdocline needle cast *Rhabdocline* spp.	*Pseudotsuga menziesii*
Rhizosphaera needle blight *Rhizosphaera kalkhoffii*	*Picea pungens*, occasionally *P. glauca*
Spruce needle rusts *Chrysomyxa* spp.	*Picea mariana*, *P. pungens*, *P. glauca*, and especially *P. abies*
Swiss needle cast *Phaeocryptopus gaeumannii*	*Pseudotsuga menziesii*

Selected References

Benyus, J. M. 1983. Christmas Tree Manual. U.S. Department of Agriculture, Forest Service, North Central Forest Experiment Station, St. Paul, MN.

Merrill, W., and Cameron, E. A. 1986. Christmas tree pests and pest management in the northeast. Pa. State Univ. Coll. Agric. Agric. Exp. Stn. Prog. Rep. 388.

Nicholls, T. H., and Wray, R. D. 1996. Pocket Guide to Christmas Tree Diseases. U.S. Dep. Agric. For. Serv. North Cent. For. Exp. Stn. http://www.ncfes.umn.edu/pubs/html/pgctree.html

(Prepared by G. A. Chastagner)

Horticultural Landscapes

The diseases that are most damaging to a species in horticultural landscapes are seldom the most damaging in its natural habitat. For example, pines and other conifers planted as ornamentals sustain shoot blight, dieback, and cankers caused by the cosmopolitan fungus *Sphaeropsis sapinea*, but this pathogen is inconsequential in natural forests. Root rot caused by *Heterobasidion annosum* is a major disease in coniferous forests but is uncommon in horticultural landscapes. Most diseases that are important in both forests and landscapes represent new host-pathogen combinations that occur after intercontinental dispersal. Seiridium canker of cypress in the Mediterranean countries is an apparent example. The host, Italian cypress, is native to the region, but the pathogen, *Seiridium cardinale*, was unknown there until the mid-twentieth century.

Published research on diseases of conifers reflects differences between horticultural and other settings in the real or perceived importance of particular diseases. Foliar diseases, shoot blights, cankers, and decline syndromes have received the most attention in horticultural settings, while trunk decay, root diseases, stem rusts, and dwarf mistletoes have been the main subjects of forest disease research. These distinctions are related in part to differing bases of tree value: contribution to shade or landscape beauty versus contribution to timber volume. They also reflect the differing epidemiology of diseases in populations of scattered versus aggregated trees. Some root pathogens of conifers in forests spread from tree to tree via root contact, but this mode of spread is inoperable or less important in scattered landscape specimens. If the vitality of a landscape tree is reduced by pathogens of feeder roots (e.g., nematodes, oomycetes, and fungi), this effect may not be separable from the influence of local environmental factors. For these reasons and because the subjects are scattered on diverse sites, root pathology of landscape conifers tends to be obscure.

Effective disease management in woody ornamentals is a matter mainly of prevention, with emphasis on site selection and preparation, species or cultivar selection, and routine care that promotes or preserves plant vitality. Benefits of plant care such as covering root zones with chip mulch, fertilizing (lightly in occasional years), and watering (thoroughly) during dry weather have rarely been measured in terms of disease incidence and severity but are recommended on the basis of conventional wisdom. Measures for suppression of specific diseases are usually not prescribed until after diagnosis indicates the likelihood that valuable specimens will be disfigured or killed.

An annotated list of diseases and noninfectious disorders that commonly affect coniferous plants in landscape settings appears below. In the list, the term "hard pines" refers to species having two or three needles per fascicle, except piñon.

Foliar Disorders

Cercospora blights. Cypresses and junipers are damaged and sometimes killed by the conidial fungi *Asperisporium sequoiae* (syn. *Cercospora sequoiae*) and *Pseudocercospora juniperi* (syn. *Cercospora sequoiae* var. *juniperi*).

A. sequoiae is favored by mild, moist climate. Diseased leaves and often entire shoots affected by either pathogen turn brown and eventually grayish. Brown stromata break through the epidermis and produce conidia soon after leaves die. Conidia, dispersed by water, may cause new infections throughout the growing season. Symptoms appear first on low branches and spread upward if favorable weather prevails. The pathogens overwinter in diseased leaves on trees. For new plantings in areas of high disease hazard, the most highly susceptible species should be avoided; e.g., consider eastern red cedar instead of Rocky Mountain juniper in the central Great Plains. Fungicidal sprays may be useful to protect susceptible plants in localities where disease has been severe.

Ploioderma needle cast. Ploioderma needle cast affects hard pines in the eastern United States. *Ploioderma lethale* is the most damaging species in horticultural settings. Scattered, highly susceptible trees are attacked. Severe disease results in loss of 1-year-old needles. Individual trees of Austrian, loblolly, and slash pines vary markedly in needle cast severity.

Rhizosphaera needle cast. Rhizosphaera needle cast, caused by *Rhizosphaera kalkhoffii* Bubák, affects spruce, especially Colorado blue and Engelmann's spruce, in the northern hemisphere. Infection usually occurs during the spring, and symptoms arise late in the summer as yellow mottling. Diseased needles turn brown (purplish brown in Colorado blue spruce) during late winter and early spring and are cast during summer and autumn. Conidia are dispersed by water and may infect healthy, young needles or stressed, older needles. The disease is common on trees growing beyond their natural range or weakened by other factors. Protection is possible with fungicides. Norway spruce is somewhat resistant.

Sooty molds. Sooty molds are caused by diverse, dark, saprophytic fungi and can be found on many plants, especially pines among the conifers. Sooty molds are not parasitic; they grow primarily on plant and insect secretions, e.g., honeydew from soft scales and aphids, and form a dark brown to black coating of hyphae and spores on plants and other surfaces. Dense deposits of sooty mold can suppress photosynthesis and may require control for aesthetic reasons. Suppression is possible through control of insects that secrete honeydew.

Dothistroma needle blight. Dothistroma needle blight, caused by *Mycosphaerella pini*, is discussed in the section on needle diseases.

Tip Blights, Diebacks, and Cankers

Botryosphaeria dieback. *Botryosphaeria stevensii* Shoemaker causes a dieback of junipers. It is especially common on Rocky Mountain juniper and is found less often on eastern red cedar in the central Great Plains of the United States. It causes cankers and branch dieback and may kill junipers weakened by heat and drought. Eastern red cedar sustains less damage than Rocky Mountain juniper and may be the species of choice on sites where this disease has been severe.

Cenangium dieback. Cenangium dieback of pines, caused by *Cenangium ferruginosum* Fr.:Fr. or *Crumenulopsis atropurpurea* (syn. *Cenangium atropurpureum* Cash & R. W. Davidson) is usually found on species that are exotic in a given region, e.g., Austrian and Japanese red pines in the eastern United States. *C. ferruginosum* occurs around the northern hemisphere.

Girdling cankers that result in dieback develop on twigs and branches in occasional years when host defenses are impaired by winter injury, pests or pathogens, or natural senescence. Foliage on affected parts turns brown during late winter and spring. Tiny, cup-shaped apothecia become conspicuous on killed bark early in the summer. *C. ferruginosum* also occurs as an endophyte in green needles of Scot's pine. Thus, this fungus may be a cryptic resident of apparently healthy trees, and needles possibly provide a route by which it can invade predisposed bark. Trees that show symptoms every year are exceptional. This disease can be avoided through selection of adapted species. Pruning diseased parts improves trees' appearance but does not prevent new infections.

Leucostoma and Valsa cankers. Cankers caused by *Leucostoma kunzei* and *Valsa* spp. can be found on all conifers, especially fir, hemlock, and spruce. *Leucostoma* and *Valsa* spp. attack diverse gymnosperms. These fungi are early colonists of dying bark, regardless of the cause. Pathogenic strains cause cankers and dieback primarily on trees stressed by factors such as drought, salt, and freezing and those planted beyond their natural range. Resin often exudes from the edges of lesions and hardens into a white crust. Spores are dispersed primarily by dripping and splashing rain but are also airborne. *L. kunzei* is the most important pathogen of the group because of the branch dieback it causes on ornamental spruces, especially Colorado blue spruce planted east of its natural range. Low branches are usually attacked first.

Damage can be delayed or suppressed by practices that minimize stress, e.g., proper site selection and regular care such as chip mulch and water during drought, but Leucostoma canker of blue spruce may develop regardless of tree care. Pruning improves the appearance of diseased tees but probably has no effect on disease progress, because the pathogen is established in branches before they show symptoms.

Phomopsis, Kabatina, and Sclerophoma blights. Blights caused by *Phomopsis juniperovora*, *Kabatina juniperi*, and *Sclerophoma pithyophila* (Corda) Höhn. may be found on junipers and occasionally on other gymnosperms. *P. juniperovora* is cosmopolitan; the other two fungi are widespread in North America and Eurasia. *P. juniperovora* causes shoot blight, twig cankers, and dieback. Landscape and nursery plantings of susceptible junipers may be severely disfigured, but plants in natural stands sustain no significant damage. *K. juniperi* and *S. pithyophila* kill 1-year-old juniper twigs in the spring, and *S. pithyophila* sometimes attacks older plant parts. These two fungi enter through insect wounds or winter injuries and do not require preventive control. *S. pithyophila* also colonizes other conifers. *Phomopsis*-resistant juniper cultivars are available. Overhead irrigation of beds of cultivars highly susceptible to Phomopsis blight should be avoided. The appearance of valuable individual junipers or small beds can be improved by pruning diseased parts when dry. Fungicidal protection against *P. juniperovora* is appropriate for valuable plantings of susceptible cultivars.

Seiridium cankers and dieback. Cankers and dieback of cypresses, caused by *Seiridium cardinale, S. cupressi,* and *S. unicorne,* are lethal on several cypresses, notably Italian and Monterey cypresses. Some authors consider all three taxa to represent one variable species for which the name *Seiridium cardinale* is appropriate. *S. unicorne* causes cankers and dieback of various members of Cupressaceae in the southern Great Plains of the United States, in East Africa, and in other areas subject to summer drought.

S. cardinale commonly causes lethal damage to Italian, Leyland, and Monterey cypresses. The majority of Monterey cypresses on inland sites in central and southern California have been destroyed by *S. cardinale*, while trees of the same species in native coastal groves with moderate temperature and frequent fog have remained unaffected. *S. cardinale* appeared in Europe during the 1940s and now is destroying populations of Italian cypress in Mediterranean countries. Girdling cankers commonly start in wounds on twig or branch bases. Death of diseased trees is hastened by hot weather and sometimes by insects, especially bark beetles. Water stress per se may not promote high susceptibility. However, Seiridium cankers in well-watered Italian cypress were found to elongate more rapidly than those in water-deprived plants. Planting highly susceptible species in areas with hot, dry summers should be avoided. Watering during dry weather may be useful. Resistance occurs in some provenances of Italian cypress.

Sphaeropsis blight. Blight caused by *Sphaeropsis sapinea* and *S. sapinea* f. sp. *cupressi* is discussed more fully in the section on cankers and twig blights. Minimizing stress by regular tree care is essential to avoiding disease. Fungicidal protection may be needed for cone-bearing pines on which shoot blight has been noted. Trees with branch dieback may be pruned to improve their appearance.

Pitch canker. Pitch canker of pines, caused by *Fusarium subglutinans* f. sp. *pini*, is discussed in the section on cankers and twig blights.

Rusts

Gymnosporangium rusts. Rusts caused by *Gymnosporangium* spp. are cosmopolitan on *Juniperus* spp. and pomaceous plants. These rusts cause galls, stem swellings, witches'-brooms, and dieback of twigs and branches on their evergreen hosts. Death of diseased twigs and small branches over years may slowly disfigure a tree. Eastern red cedar and Rocky mountain juniper growing as ornamentals often display such damage where alternate hosts such as hawthorn or serviceberry are common.

Control of Gymnosporangium rusts on ornamentals is impractical in most situations, because rust alternates between junipers and pomaceous ornamental hosts. Therefore, rust-induced twig dieback of junipers is unavoidable. Witches'-brooms can be pruned off.

Pine-pine gall rust (western gall rust). Rust caused by *Peridermium harknessii* is discussed in the section on stem rusts. Pruning dead or dying branches may improve the appearance of trees but will not suppress incidence of new infections.

Fusiform rust. Rust caused by *Cronartium quercuum* f. sp. *fusiforme* is discussed in the section on stem rusts. Trees with branch galls can be preserved by pruning.

White pine blister rust. Rust caused by *Cronartium ribicola* is discussed more fully in the section on stem rusts. Susceptible currants or gooseberries should not be planted near susceptible pines in areas of known rust hazard. Infection is halted by pruning diseased branches before the rust reaches the trunk. Rust growth in the trunk can be halted by excising living bark to approximately 8 cm above and below and 6 cm beyond the sides of the obviously diseased tissues. Stem wounding for disease control results in messy resin exudation, however, and may attract insect pests such as bark beetles or pitch nodule makers. Resistant species include *Pinus armandii, P. griffithii, P. cembra,* and *P. koraiensis*.

Trunk Diseases

Trunk rots, caused by many basidiomycetes, may damage any species of conifer. Wood-decaying fungi colonize sapwood and/or heartwood exposed by limb or top breakage, cankers, and wounds made by construction equipment, lightning, errant vehicles, or falling trees. Many diseased trees lack obvious symptoms. The hazard of breakage or tree-fall can be reduced by monitoring the extent of decay and removing hazardous trees. A tree with a basidiocarp or other structure of a wood-decay fungus on the trunk should be removed if it is located where its uncontrolled fall could cause property damage or harm to people. Trees with other decay indicators, especially large, old wounds, should be inspected periodically for the development of trunk decay sufficient to make the tree a hazard. Adequate inspection may require boring to ascertain the location of decay and the thickness of sound wood.

Root Diseases

Annosum root rot. Rot caused by *Heterobasidion annosum* is discussed in the section on root rots.

Armillaria root rots. Rots caused by *Armillaria* spp. are discussed more fully in the section on root rots. Factors that predispose trees to infection should be avoided or mitigated. Site selection and cultural measures that promote tree vigor and reduction or avoidance of *Armillaria*-colonized substrates in soil minimize the hazard of lethal attack. If a tree or shrub is killed by *Armillaria*, removal of the stump with as many roots as possible may reduce further losses.

Other root decays. Decays caused by holobasidiomycetes may be encountered on a variety of conifers. The most notable, *Phaeolus schweinitzii*, is cosmopolitan in coniferous forest regions. Brown root and butt rot caused by *P. schweinitzii* may pose hazard problems on campuses and in cemeteries, parks, and campgrounds. Decay develops after root and butt injuries or, in absence of wounds, begins in deep roots killed by hypoxia on intermittently wet sites. Long-affected trees may fall during wind storms. In the northeastern United States, *P. schweinitzii* commonly infects roots of eastern white pine growing in shallow soils subject to periodic waterlogging and summer drying. Preventive measures other than avoidance of basal wounding are unavailable. The crown condition of potentially hazardous trees should be monitored and declining trees removed.

Phytophthora root rots. Rots caused by *Phytophthora* spp., notably *P. cinnamomi*, affect hundreds of plants, both angiosperms and gymnosperms, killing roots (but not decaying the wood) and in some hosts killing the inner bark and cambium of the butt. Damage is common on poorly drained sites. The causal fungi have been dispersed with nursery stock. English and Japanese yews and their hybrids are attacked by *P. cinnamomi* in North America and Europe, but this fungus does not persist where soils regularly freeze deeply during the winter. *P. citricola* and *P. citrophthora* also attack yews.

Root rot of Port Orford cedar, caused by *P. lateralis,* is severe in forests and landscapes from northern California to southern British Columbia, especially along waterways and roads. This disease has greatly diminished the indigenous population of

Port Orford cedar. It can girdle and kill large trees within 2–4 years. Other Phytophthora root rots are known mainly in nurseries and young plantations.

The hazard of damage can be reduced by selecting well-drained planting sites, using pathogen-free planting stock, and planting species tolerant or resistant to the *Phytophthora* species known to occur in the locality. For example, most conifers, except for Alaska cedar and Pacific yew, are resistant to *P. lateralis* and may be planted where Port Orford cedar has died.

Procerum root disease. Root disease caused by *Leptographium procerum* (Kendrick) M. J. Wingfield is found on pines in North America, Europe, and New Zealand. *L. procerum* infects the bark and sapwood of roots and the stem bases of various pines and occasionally other conifers, causing root lesions, basal cankers, and decline or wilting and death of the most susceptible species, notably eastern white pine. This fungus is vectored by weevils and seems to be an opportunist that causes the most damage to plants stressed by other agents. The disease is often associated with heavy, shallow soils that are water-saturated during the spring and become dry for prolonged periods during the summer. White pines should not be planted on conducive sites or where this disease has been diagnosed previously.

Systemic Infections

Pine wilt, caused by the pinewood nematode, *Bursaphelenchus xylophilus*, is discussed more fully in a separate section. Regular care and watering during drought may reduce predisposing stress. Pines adapted to the locality should be planted.

Damage by Abiotic Stimuli or Deficiencies

Air pollutants. Pollution-induced chlorosis or needle browning, slow growth, and decline are confined to areas in or near polluted air basins. Damage to individual plants is difficult to verify, except near point sources of pollutants.

Needle blight. Blight of semimature needles of eastern white pine is considered by some to be caused by ozone and by others to result from intense solar radiation. It occurs on east ern white pine in eastern North America. Severely affected plants should be replaced.

Deicing salt. Most conifers are sensitive to salt. Damage occurs along northern highways within approximately 50 m of the pavement. Salty water causes foliar browning, bud death, twig dieback, and decline of plants when sprayed repeatedly on aerial parts or applied in sufficient quantity to roots. Dry deposition of salt dust augments spray deposition. Ion accumulation in tissues is facilitated by minute injuries to leaves as twigs abrade one another. Symptoms appear late in the winter and intensify during early spring. Conifer needles turn brown, beginning at the tips and progressing toward the bases. Branches injured during successive years slowly become barren and die. Conifers vary in salt spray tolerance; Austrian pine often remains green where other conifers turn brown. Eastern white pine and eastern hemlock are highly sensitive.

Decline. Decline, the progressive loss of vitality caused by various factors, affects all conifers. It may be caused by particular pathogens or other pests but in their absence is commonly associated with adverse soil or site factors such as clay soils with restricted drainage, high pH, nutrient imbalances, compacted soil, or restricted rooting space. It may also be caused by drought and planting outside the natural range of the species. Symptoms often appear several years after planting when restricted or damaged roots can no longer sustain top growth. Decline can be delayed by choosing species adapted to the locality and site, adhering to planting guidelines as presented in arboricultural manuals, and providing regular care to young trees.

Selected References

Phillips, D. H., and Burdekin, D. A. 1982. Diseases of Forest and Ornamental Trees. Macmillan, London.

Riffle, J. W., and Peterson, G. W., tech. coords. 1986. Diseases of trees in the Great Plains. U.S. Dep. Agric. For. Serv. Gen. Tech. Rep. RM-129.

Sinclair, W. A., Lyon, H. H., and Johnson, W. T. 1987. Diseases of Trees and Shrubs. Cornell University Press, Ithaca, NY.

(Prepared by W. A. Sinclair)

Host Index

Host	Disease / Causal Agent	Pages
Abies	Cankers	
alba	*Cytospora abietis*	47
amabilis	*Phellinus canchriformans*	24, 49
balsamea	Mistletoes	
concolor	*Arceuthobium abietinum*	36
grandis	*A. tsugense*	79, 80
magnifica	*Phoradendron* spp.	38–39
procera	*Psittacanthus* spp.	39–40
	Viscum abietis	39
	Needle blights and casts	
	Herpotrichia parasitica	63
	Needle and broom rusts	
	Calyptospora goeppertiana	52
	Melampsorella caryophyllacearum	53, 73, 74
	Pucciniastrum epilobii	52
	Uredinopsis pteridis	53
	Root rots	
	Armillaria ostoyae	13–14
	Coniophora puteana	76
	Heterobasidion annosum	12–13
	Inonotus dryadeus	18
	Oligoporus balsameus	18
	Perenniporia subacida	18
	Phaeolus schweinitzii	16
	Phellinus weirii	14–15
	Scytinostroma glactinum	76
	Seed and cone infection	
	Caloscypha fulgens	51, 88
	Stem decays	
	Echinodontium tinctorium	25–26
	Fomitopsis pinicola	23
	Hydnum abietis	24
	Resinicium bicolor	23
	Stereum sanguinolentum	21
Araucaria	Root rots	
cunninghamii	*Heterobasidion annosum*	12–13
	Phellinus noxius	17–18
	Rigidoporus vinctus	18
Chamaecyparis	Root disease	
lawsoniana	*Phytophthora lateralis*	6–7
Cedrus	Broom rust	
	Melampsora cedri	53
	Root rots	
	Armillaria ostoyae	13–14
	Heterobasidion annosum	12–13
	Inonotus tomentosus	16–17
	Phaeolus schweinitzii	16
	Stem decay	
	Phellinus pini	24–25
Cupressus	Cankers	
	Seiridium spp.	47–48
	Mistletoes	
	Psittacanthus spp.	39–40
	Needle blight and cast	
	Asperisporium sequoiae	62–63
	Root rot	
	Phellinus noxius	17–18
	Stem decay	
	Phellinus pini	24–25
Juniperus	Cankers	
	Kabatina juniperi	48, 90
	Phomopsis juniperovora	48, 90
	Mistletoes	
	Arceuthobium oxycedri	36, 37
	Phoradendron spp.	38–39
	Needle blights and casts	
	Asperisporium sequoiae	62–63
	Herpotrichia juniperi	63
	Pseudocercospora juniperi	63, 90
	Root rot	
	Heterobasidion annosum	12–13
	Stem rusts	
	Gymnosporangium spp.	34–35
Larix	Cankers	
decidua	*Lachnellula willkommii*	41–42
laricina	Mistletoes	
occidentalis	*Arceuthobium laricis*	36
	Needle blights and casts	
	Hypodermella laricis	62, 78
	Meria laricis	62
	Mycosphaerella laricina	62
	Root rots	
	Armillaria ostoyae	13–14
	Heterobasidion annosum	12–13
	Phaeolus schweinitzii	16
	Stem decays	
	Amylostereum areolatum	21–22
	A. chailletii	21–22
	Phellinus pini	24–25
	Resinicium bicolor	23
	Stereum sanguinolentum	21
Picea	Blue stain	
abies	*Ophiostoma polonicum*	18–19
engelmannii	Bud blight	
glauca	*Gemmamyces piceae*	62
mariana	Cankers	
pungens	*Cytospora kunzei*	47, 78, 89
sitchensis	*Gremmeniella abietina*	43–45
	Sirococcus conigenus	48, 51
	Mistletoes	
	Arceuthobium microcarpum	36
	A. pusillum	36, 76, 77
	Needle and broom rusts	
	Chrysomyxa abietis	52
	C. arctostaphyli	53
	C. deformans	52, 53
	C. ledicola	52
	C. weirii	52
	C. woroninii	53, 78, 79
	Needle blights and casts	
	Lirula macrospora	61–62
	Rhizosphaera kalkoffii	62, 89, 90
	Root disease	
	Rhizina undulata	7–8
	Root rots	
	Armillaria ostoyae	13–14
	Coniophora puteana	76
	Heterobasidion annosum	12–13
	Inonotus tomentosus	16–17
	Oligoporus balsameus	18
	Phaeolus schweinitzii	16
	Scytinostroma galactinum	76
	Seed and cone infection	
	Caloscypha fulgens	51, 88
	Chrysomyxa pirolata	49–50
	Thekopsora aureolatum	52
	Stem decays	
	Amylostereum areolatum	21–22
	A. chailletii	21–22
	Cylindrobasidium evolvens	23
	Echinodontium tinctorium	25–26
	Fomitopsis pinicola	23
	Phellinus pini	24–25
	Resinicium bicolor	23
	Stereum sanguinolentum	21
Pinus	Blue stain	
banksiana	*Ophiostoma clavigerum*	18–19
contorta	*O. minus*	18–19
densiflora	Cankers	
echinata	*Atropellis piniphila*	46–47
edulis	*Cenangium ferruginosum*	48, 90
elliotii	*Ramicloridium pini*	48
halepensis	*Sirococcus conigenus*	48, 51
jeffreyi	*Fusarium subglutinans*	45–46
kesiya	*Phacideum coniferarum*	47, 75

(continued on next page)

Host Index (*continued*)

Host	Disease / Causal Agent	Pages
Pinus (continued)		
lambertiana	*Gremmeniella abietina*	43–45
monophyla	*Sphaeropsis sapinea*	42–43
monticola	Mistletoes	
mugo	*Arceuthobium* spp.	36–38
nigra	*A. americanum*	32, 76, 80
palustris	*A. campylopodum*	80
patula	*A. minutissimum*	83, 84
pinaster	*A. vaginatum*	80
pinea	*Psittacanthus* spp.	39–40
ponderosa	*Viscum laxum*	39
pumila	Needle blights and casts	
radiata	*Cyclaneusma minus*	59
resinosa	*Davisomycella ampla*	61
roxberghii	*Elytroderma deformans*	60
strobus	*Herpotrichia coulteri*	63
sylvestris	*Lophodermella arcuata*	60
taeda	*L. cerina*	60
thunbergii	*L. concolor*	60
wallichiana	*L. montivaga*	60
	L. morbida	60
	L. sulcigena	60, 73, 75
	Lophodermium seditiosum	61
	Mycosphaerella dearnessii	57
	M. pini	57–59
	Phacidium infestans	63, 75, 78
	Ploioderma hedgcockii	59, 60
	P. lethale	59, 60, 90
	Needle rusts	
	Coleosporium spp.	52–53, 89
	Pine wilt	
	Bursaphelenchus xylophilus	19–20
	Root diseases	
	Leptographium procerum	10, 88, 92
	L. serpens	11
	L. terebrantis	10, 11
	L. wageneri	8–9, 10
	Phytophthora cinnamomi	4–6
	Rhizina undulata	7–8
	Root rots	
	Armillaria limonea	85
	A. novae-zelandiae	85
	A. ostoyae	13–14
	Heterobasidion annosum	12–13
	Inonotus tomentosus	16–17
	Phaeolus schweinitzii	16
	Phellinus noxius	17–18
	Rigidoporus lineatus	18
	Seed and cone infection	
	Cronartium conigenum	32, 50
	C. strobilinum	50
	Rigidoporus vinctus	18
	Stem decays	
	Amylostereum areolatum	21–22
	Phellinus pini	24–25
	Resinicium bicolor	23
	Stereum sanguinolentum	21
	Stem rusts	
	Cronartium arizonicum	32, 33
	C. coleosporioides	31, 32, 33
	C. comandrae	31
	C. comptoniae	32
	C. conigenum	32, 50
	C. flaccidum (and *Peridermium pini*)	30–31
	C. himalayense	83, 84
	C. occidentale	27
	C. quercuum f. sp. *banksianae*	29
	C. quercuum f. sp. *fusiforme*	27–29
	C. ribicola	26–27

Host	Disease / Causal Agent	Pages
Pinus (continued)	Stem rusts (continued)	
	Endocronartium harknessii	29–30
	E. sahoanum	27
	Melampsora pinitorqua	33–34, 53
	Peridermium yamabense	27
Pseudotsuga menziesii	Cankers	
	Cytospora abietis	47, 78
	Dermea pseudotsugae	48
	Phomopsis lokoyae	48
	Sirococcus conigenus	48, 51
	Mistletoe	
	Arceuthobium douglasii	80
	Needle blights and casts	
	Phaeocryptopus gaeumannii	55–56
	Rhabdocline spp.	54–55
	Needle rust	
	Melampsora occidentalis	53
	Root diseases	
	Leptographium wageneri	8–9, 10
	Rhizina undulata	7–8
	Root rots	
	Armillaria ostoyae	13–14
	Heterobasidion annosum	12–13
	Phaeolus schweinitzii	16
	Phellinus weirii	14–15
	Sparassis crispa	16, 18
	Stem decays	
	Fomitopsis cajanderi	22–23
	Phellinus pini	24–25
	Resinicium bicolor	23
	Stereum sanguinolentum	21
Thuja plicata	Needle blights and casts	
	Asperisporium sequoiae	62–63
	Didymascella thujina	62
	Root rots	
	Ceriporiopsis rivulosa	18
	Oligoporus sericeomollis	18
	Phellinus weirii	14–15
	Resinicium bicolor	23
	Stem decay	
	Phellinus pini	24–25
Tsuga canadensis heterophylla mertensiana	Cankers	
	Botryosphaeria tsugae	48
	Sirococcus conigenus	48, 51
	Mistletoe	
	Arceuthobium tsugense	79, 80
	Needle blight and cast	
	Fabrella tsugae	63
	Root diseases	
	Leptographium wageneri	8–9, 10
	Rhizina undulata	7–8
	Root rots	
	Armillaria ostoyae	13–14
	Heterobasidion annosum	12–13
	Inonotus tomentosus	16–17
	Perenniporia subacida	18
	Phellinus weirii	14–15
	Rust	
	Melampsora farlowii	53
	Stem decays	
	Echinodontium tinctorium	25–26
	Fomitopsis pinicola	23
	Hydnum abietis	24
	Phellinus pini	24–25
	Resinicium bicolor	23
Sequoia sempervirens	Root rot	
	Ceriporiopsis rivulosa	18

Index

Abies, 40, 79, 88
    cone and seed diseases of, 51
    leafy mistletoes on, 39–40
    needle and broom rust of, 52, 53
    Phacidium canker of, 47
    red-brown butt rot of, 16
    red rot of, 21
    rust-red stringy white rot of, 25–26;
        Pls. 51, 54
    Scleroderris canker of, 43–45
    Sphaeropsis shoot blight and canker of, 42–43
    white stringy rot of, 23
Abies alba, 73
    annosum root rot of, 12; Pl. 16
    broom rust of, 53; Pl. 126
    decline of, in Europe, 67
    felt blight of, 63
    Lophodermium needle cast of, 89
    Viscum abietis on, 39; Pl. 86
    Viscum laxum on, 39
Abies amabilis, 61, 79
Abies balsamea, 18, 76, 88, 89
    Lachnellula canker of, 47
    Scleroderris canker of, 43
Abies cephalonica, 73
Abies concolor, 47, 88, 89
    and *Hydnum abietis,* 24
    leafy mistletoes on, 38–39
    Lirulla needle blight of, Pl. 150
    Nectria fuckeliana infection of, 49
    Phellinus cancriformans infection of, 24, 49;
        Pl. 50
    Virgella needle blight of, 61
Abies durangensis
    dwarf mistletoes on, 36–38
    leafy mistletoes on, 38–39
Abies fraseri, 88, 89
    Lachnellula canker of, 47
    Lirulla needle blight of, 61
Abies grandis, 47, 80
    black mildew on, 63; Pl. 157
    current-season needle necrosis of, 89
    laminated root rot of, 14–15
    Lirulla needle blight of, 61
Abies holophylla, 78
Abies lasiocarpa, 76, 89
    broom rust of, Pl. 127
    cork-bark disease of, 49; Pl. 114
    Isthmiella needle blight of, 61
    Lirulla needle blight of, 61
Abies magnifica
    dwarf mistletoes on, 36–38
    Virgella needle blight of, 61
    var. *shastensis,* 88, 89
Abies nephrolepis, 77, 78
Abies nordmanniana, 88
Abies numidica, 89
Abies procera, 88
    current-season needle necrosis of, 89
    dwarf mistletoes on, 36–38

    Lirulla needle blight of, 61
    Virgella needle blight of, 61
Abies religiosa, 36–38
Abies sachalinensis, 43, 78
Abies vejarii, 36–38
Acidic deposition, 67, 79
Adelges larici, 41
Adelopus gaeumannii, 55
Agalinis, 40
Agathis
    annosum root rot of, 12
    brown root rot of, 17–18
Air pollution, 65–66
    and Waldsterben, 67–69
Alaska-cedar decline, 70
Allantophomopsis, 47
Alternaria, 51
Amelanchier, 35
Amylostereum
    areolatum, 21–22, 85, 86
    chailletii, 21–22
Amylostereum rot, 21–22
Annosum root rot, 12–13, 82, 83;
    Pls. 12–16
Apostrasseria, 47
Appalachian blister rust, 32
Araucaria
    annosum root rot of, 12
    brown root rot of, 17–18; Pls. 29, 31
    needle rust of, 52, 53
    Sphaeropsis shoot blight and canker of, 42–43
Araucaria cunninghamii, 18, 85; Pl. 33
Arceuthobium, 36–38; Pls. 78–84
    americanum, 32, 76, 80
    campylopodum, 80
    douglasii, 80
    laricina, 80
    minutissimum, 83, 84
    oxycedri, 36, 37
    pusillum, 69, 76, 77
    tsugense, 79, 80
    vaginatum, 80
Arctostaphylos, 53
Areolaria, 40
Argyrodendron trifoliatum, Pl. 30
Armillaria, 72
    heimii, 86, 87
    limonea, 85
    mellea, 14, 83, 84
        var. *obscura,* 13
    novae-zelandiae, 85
    obscura, 13
    ostoyae, 13–14; Pls. 17–19
        in Europe, 73, 74
        in Fennoscandia, 75
        in North America, 76, 77, 79, 80, 81
        and red-brown butt rot, 16
Armillaria root disease, 13–14, 87;
    Pls. 17–19

Armillariella
    mellea, 78
    obscura, 16
    ostoyae, 13
Ascocalyx
    abietina, 43
    abietis, 78
Ash yellows, 66
Aspergillus, 51
Asperisporium
    juniperinum, 63
    sequoiae, 62, 63, 90
Asteridiella pitya, 63
Atichia glomerulosa, 63
Atropellis
    apiculata, 46
    arizonica, 46–47
    piniphila, 46–47, 76; Pls. 102, 103
    tingens, 46
    treleasei, 48
Atropellis cankers, 46–47, 89; Pls. 102, 103
Austrocedrus, 52, 53
Austrocedrus chilensis, decline of, 72; Pls. 161, 162

Bacteria, 1
Bark beetles, 17, 33, 65, 86
    and blue-stain fungi, 2, 18–19; Pl. 35
    and forest management, 77
    and Lophodermella needle cast, 60
    and Port Orford cedar root disease, 6
    as vectors of *Leptographium* diseases, 8–11;
        Pls. 10, 11
    and yellow-cedar decline, 71
Betula, 53
    papyrifera, 76
    pendula, 75
Biatorella difformis, 78
Bifusella, 61
    linearis, 60
    pini, 60
    saccata, 60
Bifusella needle blight, 60–61
Black mildews, 63; Pl. 157
Black-stain root disease, 8–9, 79; Pls. 7–10
Bleeding fungus, 21
Blister rust, 74
Blue stain, 2, 9, 18–19; Pl. 35
Bothrodiscus berenice, 78
Botryosphaeria
    dothidea, 18
    piceae, 48; Pl. 113
    stevensii, 90
    tsugae, 48
Botrytis, 51
    cinerea, 34, 88
Brown crumbly rot, 23; Pls. 42–44
Brown cubical rot, 22, 84
Brown felt, 63; Pl. 158
Brown root rot, 17–18; Pls. 29–31

Brown rots, 11, 18, 20, 76
  brown crumbly rot, 23; Pls. 42–44
  red-brown butt rot, 16; Pls. 23–25
  yellow brown top rot, 22–23; Pls. 40, 41
Brown snow mold, 63
Brown spot needle blight, 57, 58, 83, 89;
  Pls. 134, 135
Brunchorstia, 48
  pinea, 43
Buchnera, 40
Buckleya distichophylla, 32
Bursaphelenchus
  mucronatus, 19
  xylophilus, 19–20, 82, 83, 88; Pls. 36, 37

Caeoma
  mexicanum, 32
  pinitorquum, 33
Caliciopsis arceuthobii, Pl. 84
Callitris, 85
Calocedrus, 35
Calocera viscosa, 74
Caloscypha fulgens, 51, 88; Pls. 120, 121
Calyptospora goeppertiana, 52, 78
Camarosporium strobilinum, 62
Castilleja, 33
Cedar root rot, 6–7; Pls. 1–3
Cedrus, 48
  red-brown butt rot of, 16
  Sphaeropsis shoot blight and canker of, 42–43
Cedrus atlantica, 73
Cedrus deodara, 53, 83
Cenangium
  atropurpureum, 90
  ferruginosum, 48, 90; Pl. 111
  piniphilum, 46–47
Cerastium, 52, 53
Ceratocystis, 2
  fagacearum, 66
  polonica, 18–19
  wageneri, 8
Cercoseptoria, 62
  pini-densiflorae, 86
Cercospora sequoiae, 62, 63, 90
  var. *juniperi*, 63, 90
Cercospora needle blight, 62, 63, 90
Ceriporiopsis rivulosa, 18
Ceropsora, 52
Chaenomeles, 35
Chamaecyparis, 35, 63
  Seiridium cankers of, 47
  Sphaeropsis shoot blight and canker of, 42–43
Chamaecyparis lawsoniana, 79, 85, 91
  Port Orford cedar root disease of, 6–7;
    Pls. 1–3
Chamaecyparis nootkatensis
  Kabatina tip blight of, 48; Pl. 109
  yellow-cedar decline of, 70–71; Pls. 159, 160
Chamaecyparis thyoides, 62
Christmas trees, 88–89; Pls. 116, 168, 169
Chrysomyxa, 52, 89
  abietis, 52
  arctostaphyli, 53; Pl. 125
  deformans, 52, 53
  ledi, 78
  ledicola, 52; Pl. 122
  monesis, 50
  pirolata, 49–50, 79
  weirii, 52
  woroninii, 53, 78, 79
Cladocolea cupulata, 40
Climate, 64–65, 66
Coccomyces pseudotsugae, 48
Coleosporium, 52–53, 89; Pl. 124
  tussilaginis, 52

Collar rot, 42
Colpoma crispum, 49
Coltricia tomentosa, 16
Comandra, 31; Pl. 70
Comandra blister rust, 30, 31; Pls. 68–70
Comptonia, 32, 35
Coniophora puteana, 76
Coniothyrium faulii, 61
Conk rot, 24
Contarinia pseudotsugae, 55
Cordylanthus, 33
Cork bark, 49; Pl. 114
Coronado Peridermium, 33
Corticium
  evolvens, 23
  laeve, 23
Coryneum, 47
Crataegus, 35
Cronartium, 76, 77
  appalachianum, 32
  arizonicum, 32, 33
  asclepiadeum, 30
  cerebrum, 28
  coleosporioides, 31, 32, 33; Pl. 71
  comandrae, 31, 32; Pls. 68–70
  comptoniae, 32; Pl. 72
  conigenum, 32, 50
  flaccidum, 30–31, 33, 73, 74, 75, 78
  fusiforme, 28
  himalayense, 83, 84
  occidentale, 27
  pyriforme, 31
  quercuum, 28, 84
    f. sp. *banksianae*, 29, 30; Pl. 61
    f. sp. *densiflorae*, 29
    f. sp. *echinatae*, 29
    f. sp. *fusiforme*, 2, 27–29, 45, 82, 91;
      Pls. 59, 60
    f. sp. *virginianae*, 29
  ribicola, 26–27, 78, 89, 91; Pls. 55–58
    in Europe, 74
    in India, 84
    in North America, 79, 80
    and pole blight, 71
  stalactiforme, 31
  strobilinum, 50
Crumenulopsis
  atropurpurea, 90
  sororia, 48
Cryptomeria, 47
Cryptomeria japonica, 63
Cupressocyparis, 47
Cupressus, 34, 35, 40, 48, 85, 87
  brown root rot of, 17–18
  leafy mistletoes on, 39–40
  Seiridium canker of, 47–48; Pl. 106
  Sphaeropsis shoot blight and canker of, 42–43
Cupressus arizonica, 63, 84
Cupressus cashmeriana, 84
Cupressus lusitanica, 38–39, 84
Cupressus macrocarpa, 47
Cupressus sempervirens, 47, 73, 84
Cupressus torulosa, 84
Current-season needle necrosis, 89
Cyclaneusma minus, 59, 85, 86, 89
Cyclaneusma needle cast, 59, 89
Cydonia, 35
Cylindrobasidium evolvens, 23
Cylindrocarpon cylindrioides, 49
Cypress canker, 87
Cytospora
  abietis, 47, 78; Pl. 104
  kunzei, 47, 78, 89
Cytospora cankers, 47, 89; Pl. 104

Damping-off, 42, 51
  in nurseries, 4, 88; Pl. 165
Dasistoma, 40
Dasyscyphus, 47
  willkommii, 41, 78
Davisomycella
  ampla, 61
  medusa, 61
Davisomycella needle casts, 61
Deladenus
  siricidicola, 22
  wilsoni, 22
Delphinella balsameae, 78
Dendroctonus. See also Bark beetles
  frontalis, 18–19
  ponderosae, 18–19; Pl. 35
  rufipennis, 76, 77
  terebrans, 11
  valens, 11
Dendrophthoe falcata, 40
Dermea
  balsamea, 49
  pseudotsugae, 48; Pl. 110
  rhytidiformans, 49; Pl. 114
  tetrasperma, 49
Diaporthe lokoyae, 48
Dichomera gemmicola, 62
Didymascella
  chamaecyparidis, 62
  tetramicrospora, 62
  thujina, 62; Pl. 156
Diedickea piceae, 62
Dioryctria, 45
Diplodia
  gossypina, 51
  pinea, 42, 51, 83, 84
Diplodia blight, 42, 89
Discocainia treleasei, 48
Dothidea acicola, 57
Dothiorella pithya, 78
Dothistroma
  pini, 2, 58, 85
  septospora, 21, 58, 85, 86
Dothistroma needle blight, 57–58, 85, 86, 89;
  Pls. 136–138
Drought, 64
Durandiella pseudotsugae, 48
Dwarf mistletoes, 1, 23, 36–38; Pls. 78–84
  and Cytospora cankers, 47
  in India, 84
  in North America, 80, 81

Eastern gall rust, 29; Pl. 61
Echinodontium tinctorium, 25–26, 76, 79, 80;
  Pls. 51–54
Elytroderma
  deformans, 59, 60, 61; Pls. 141, 142
  torres-juanii, 60
Elytroderma needle disease, 60; Pls. 141, 142
Endocronartium
  harknessii, 29–30, 76, 77, 89; Pls. 62–64
  pini, 30, 73, 74, 75
  sahoanum, 27
Entoloma arbortivum, 14
Epilobium, 52
Epipolaeum abietis, 63
Eucalyptus, 16
Eupelte farrae, 63
Eurasian spruce bark beetle, 18–19. See also
  Bark beetles

Fabrella tsugae, 63
Fairy rings, 7
Feather rot, 18
Felt blights, 63–64; Pl. 158

Fendlera, 35
Figwort, 36, 40
Fir hydnum, 24; Pl. 46
Fire
    and dwarf mistletoe, 38, 81
    and Rhizina root disease, 7, 8
Fomes
    annosus, 12
    cajanderi, 22
    fomentarius, 78
    noxius, 17
    pini, 24
    pinicola, 23
    subroseus, 22
Fomitiporia weirii, 14
Fomitopsis
    annosa, 12
    cajanderi, 22–23, 78; Pls. 40, 41
    officinalis, 78
    pinicola, 22, 23, 78, 80; Pls. 42–44
    rosea, 22
Fraxinus, 16
Freeze-drying, 65
Frost damage, 64–65
Fusarium, 88
    lateritium f. sp. *pini*, 45
    moniliforme var. *subglutinans*, 45, 51
    subglutinans, 45–46, 82, 89; Pls. 98–101
        f. sp. *pini*, 45, 86, 87
Fusiform rust, 27–29, 82; Pls. 59, 60

Ganoderma
    applanatum, 78
    australe, 18
Gelatinosporium, 49
Gemmamyces piceae, 62; Pl. 152
Geocaulon, 31
Gillenia, 35
Global climate change, 66
Globose gall rust, 29
Gloeophyllum abietinum, 78
Gremmeniella
    abietina, 43–45; Pls. 96, 97
        and abiotic disease symptoms,
            compared, 64, 65
        in Europe, 73, 74
        in the Far East, 78
        in Fennoscandia, 75
        var. *abietina*, 44, 45
        var. *balsamea*, 44
        var. *cembrae*, 44, 45
    laricina, 48
Griphosphaeria corticola, 49
Group dying, 7–8; Pls. 4–6
Grovesiella, 49, 89; Pl. 116
Gymnosporangium, 34–35, 52, 91; Pls. 75–77
    bermudianum, 35
    clavariiforme, Pl. 76
    clavipes, 34, 35
    juniperi, 78
    kernianum, 34, 35; Pl. 77
    libocedri, 34
    nidus-avis, 34, 35
    sabinae, 34, 35
Gymnosporangium stem rusts, 34–35;
    Pls. 75–77

Haematostereum sanguinolentum, 21
Hartigiella laricis, 78
Heartrots, 2, 20, 83
    annosum root rot, 12–13; Pls. 12–16
    brown crumbly rot, 23; Pls. 42–44
    red ring rot, 24–25; Pls. 47–50
    rust-red stringy white rot, 25–26; Pls. 51–54
    yellow brown top rot, 22–23

Helicotylenchus
    dihystera, 5
    erythrinae, 5
Hemicycliophora vidua, 5
Hemiphacidium
    longisporum, 60
    planum, 63
Hemlock, 18; Pl. 107
Hendersonia
    acicola, 60
    pinicola, 60
Hericium abietis, 24, 50
Herpotrichia, Pl. 158
    coulteri, 63
    juniperi, 63
    nigra, 78
    parasitica, 63
Herpotrichia brown felt blight, 63; Pl. 158
Heterobasidion
    annosum, 12–13, 88, 89; Pls. 12–16
        in Australia and New Zealand, 86
        in Europe, 73, 74
        in Fennoscandia, 75
        in India, 83, 84
        in North America, 79, 80, 81, 82, 83
        and *Resinicium bicolor*, 24
        in Siberia and the Far East, 78
    araucariae, 12
Honey mushroom, 13
Horticultural landscapes, 89–92
Hyalopsora, 52, 53
Hydnum abietis, 24; Pl. 46
Hylastes, 11
    nigrinus, 9; Pl. 10
Hylobius
    abietis, 41
    pales, 11
Hylurgus, 11
Hypodermella
    laricis, 62, 78
    sulcigena, 78
Hypodermella needle blight, 62

Ibalia leucospoides, 22
Ileostylus micracanthus, 40
Indian paint fungus, 24
Inonotus
    dryadeus, 18
    sulphurascens, 14
    tomentosus, 16–17, 20, 76, 77, 83;
        Pls. 26–28
    weirii, 14, 78, 79
Insect vectors, 2, 6, 17, 48
    of Amylostereum rot, 21–22
    of *Leptographium* diseases, 8–11; Pls. 10, 11
    of pinewood nematodes, 19–20
Inyo Peridermium, 33
Ips typographus, 18–19. *See also* Bark beetles
Ischnoderma resinosum, 78
Isthmiella
    abietis, 61
    crepidiformis, 62
    faullii, 61
    quadrispora, 61

Junghuhnia vincta, 85
Juniperus, 48, 90, 91; Pl. 77
    annosum root rot of, 12
    leafy mistletoes on, 38–39
Juniperus brevifolia, 36–38
Juniperus communis, 63; Pl. 76
Juniperus occidentalis, Pl. 85
Juniperus procera, 84
Juniperus standlyei, Pl. 75
Juniperus virginiana, 47, 63

Kabatina
    juniperi, 48, 90
    thujae, 48; Pl. 109
Keithia blight, 62; Pl. 156
Keteleeria, 36–38
Korfia tsugae, 63
Korthalsella dacrydii, 40

Lachnellula
    abietis, 47
    agassizii, 47
    arida, 47
    calyciformis, 47, 78
    gallica, 47
    laricis, 47
    occidentalis, 47
    pini, 47, 78
    pseudotsugae, 47
    willkommii, 41–42, 47, 73, 75;
        Pls. 89–91
Lachnellula cankers, 47; Pl. 105
Laetiporus sulphureus, 78
Laminated root rot, 14–15, 79; Pls. 20–22
Lamourouxia, 33
Larch canker, 41–42; Pls. 89–91
Larch needle cast, 62; Pls. 154, 155
Larix
    Amylostereum rot of, 21–22
    annosum root rot of, 12
    dwarf mistletoes on, 36–38
    Gremmeniella canker on, 48
    Hypodermella needle blight of, 62
    Lachnellula canker of, 47
    larch needle cast of, 62; Pls. 154, 155
    needle rust of, 52, 53
    Phacidium canker of, 47
    red-brown butt rot of, 16
    red rot of, 21
    Sphaeropsis shoot blight and canker of,
        42–43
    white stringy rot of, 23
Larix dahurica, 77, 78
Larix decidua, 34, 73, 75
    larch canker of, 41, 42; Pls. 89, 90
Larix × *eurolepis*, 41, 42, 73
Larix gmelini, 41
Larix kaempferi, 41, 73
Larix laricina, 41
Larix leptolepis, 41
Larix occidentalis, 80
Larix siberica, 41, 77
Lasiodiplodia theobromae, 18, 51
Leafy mistletoes, 38–40; Pls. 85–88
Lecanosticta acicola, 57
Lecanosticta needle blight, 57, 89; Pl. 169
Ledum, 52, 53
Lepteutypa cupressi, 47
Leptographium, 72, 86
    procerum, 10, 88, 92; Pl. 11
    serpens, 11
    terebrantis, 10, 11
    wageneri, 2, 8–9, 10, 79; Pls. 7–10
        var. *ponderosum*, 8–9; Pl. 7
        var. *pseudotsugae*, 8–9; Pl. 7
        var. *wageneri*, 8–9; Pl. 7
Leptosphaeria faulii, 61
Leptostroma, 61
    pinastri, 78
Leptothyrella laricis, 62
Leucocytospora kunzei, 47
Leucostoma kunzei, 47, 49, 90
Libocedrus decurrens, 38–39
Limacinia alaskensis, 63
Limb rusts, 33; Pls. 73, 74
Lindleya, 35

Lirula, Pl. 150
   abietis-concoloris, 61
   macrospora, 61–62; Pl. 151
   punctata, 61
Lirula needle blight, 61–62; Pls. 150, 151
Littleleaf disease, 4–6, 82
Live oak decline, 66
Long-pitted rot, 24
Long-pocket rot, 24
Lophodermella, 59, 61
   arcuata, 60
   cerina, 60; Pl. 147
   concolor, 60; Pl. 143
   conigenum, 78
   montivaga, 60
   morbida, 2, 60; Pl. 146
   sulcigena, 60, 73, 75; Pl. 144
Lophodermella needle casts, 60;
   Pls. 143–147
Lophodermium, 59, 61, 85
   abietis, 89
   australe, 61, 86
   conigenum, 61
   macrosporium, 78
   molitoris, 61
   nervisequium, 78
   nitens, 61; Pl. 149
   pinastri, 78; Pl. 148
   seditiosum, 61, 73, 75, 89; Pl. 148
Lophodermium needle blight, 61–62
Lophodermium needle casts, 61, 89;
   Pls. 148, 149
Lophomerum autumnale, 61
Lophophacidium hyperboreum, 63, 64

Macranthera, 40
Macrophoma
   pinea, 42
   sapinea, 42
Macrophomina, 88
Magnesium deficiency, 68
Malus, 35
Maurodothina farrae, 63; Pl. 157
Melampsora, 52
   cedri, 53
   farlowii, 53
   occidentalis, 53
   pinitorqua, 33–34, 53, 89
     in Europe, 73, 74
     in Fennoscandia, 75
     in Siberia and the Far East, 78, 79
   populnea, 33, 74
   tremulae, 33
Melampsorella, 52
   caryophyllacearum, 53, 73, 74, 78, 79;
     Pl. 126
   symphyti, 78
Melampsoridium, 52, 53
   betulinum, 78, 79
Meloderma desmaziersii, 60
Meloderma needle blight, 60–61
Meria laricis, 55, 62, 78; Pls. 154, 155
Mespilus, 35
Metrosideros polymorpha, 64
Micropera pinastri, 78
Mikronegeria, 52
   alba, 53
Milesina, 52, 53
Mistletoe blister rust, 32
Moisture stress, 5, 64; Pl. 167
Moneses uniflora, 49
Monochaetia unicornis, 47, 84
Monochamus scutellatus oregonensis, 19–20;
   Pl. 36
Mottled bark disease, 21

Mountain pine beetle, 77, 18–19; Pl. 35.
   See also Bark beetles
Mushroom root rot, 13
Mychorrhizae, 1, 5
Mycosphaerella
   dearnessii, 57, 82, 89; Pls. 134, 135, 169
   laricina, 62
   pini, 21, 57–59, 89; Pls. 136–138
Myrica, 32, 35

Naemacyclus minor, 59
Naohidemyces, 52
Nectria
   cucurbitula, 78
   fuckeliana, 49
Needle and broom rusts, 51–53; Pls. 122–127
Nematodes, 1, 5
   as biological control agents, 22
   and pine wilt, 19–20; Pls. 36, 37
Nipterella tsugae, 49
Nitschkia molnarii, 49
Northeastern subalpine red spruce decline,
   69–70
Nothofagus, 52, 53, 72; Pl. 162
Nurseries, 88; Pls. 163–167
Nutrient deficiency, 64, 68

Odontia bicolor, 23
Oligoporus
   balsameus, 18
   sericeomollis, 18; Pl. 32
Oligostroma acicola, 57
Onnia tomentosa, 16
Ophiostoma, 2, 10
   clavigerum, 18–19
   minus, 18–19
   polonicum, 18–19
   ulmi, 8
   wageneri, 8
Orobanche minor, 40
Orthocarpus, 33
Oxycedrus, 34, 35

Paeonia, 30
Pedicularis, 33
   white pine blister rust of, 26–27
Penicillium, 51
Peniophora luna, 84
Perenniporia subacida, 18
Peridermium, 52
   appalachianum, 32
   bethelii, 32
   cedri, 53, 83, 84
   cerebrum, 28
   comptoniae, 32
   filamentosum, 33
   fusiforme, 28
   harknessii, 29–30, 91
   pini, 30–31, 33; Pls. 65–67
   stalactiforme, 31
   yamabense, 27
Pestalopezia tsugae, 49
Pestalotia, 47
Pestalotiopsis funerea, 49
Pezicula livida, 49
Phacidiopycnis, 47
   pseudotsugae, 47
Phacidium, 88
   abietis, 63
   balsamicola, 47
   coniferarum, 47, 75
   dearnessii, 63
   infestans, 63, 75, 78
   sherwoodiae, 63
   taxicola, 63

Phacidium cankers, 47
Phaeocryptopus
   gaeumannii, 55–56, 74, 85, 89;
     Pls. 132, 133
   nudus, 62
Phaeolus schweinitzii, 16, 91; Pls. 23–25
   in Africa, 86
   in Europe, 73, 74
   in North America, 80
   in Siberia and the Far East, 78, 79
Phellinus
   cancriformans, 24, 49; Pl. 50
   chrysoloma, 24
   hartigii, 78
   heinrichii, 14
   noxius, 17–18, 85, 86; Pls. 29–31
   pini, 2, 24–25, 32; Pls. 47–49
     in Europe, 73, 74
     in Fennoscandia, 75
     in India, 83, 84
     in North America, 76, 82, 79, 80
     in Siberia and the Far East, 78
     var. *pini,* 49
   sulphurascens, 14
   weirii, 2, 14–15, 16, 18; Pls. 20–22
     in Australia and New Zealand, 86
     in North America, 79, 81
Philadelphus, 35
Phlebia gigantea, 74
Phlebiopsis gigantea, 13, 75
Phloeosinus, 6, 71
Phoma, 88
   abietella-siberica, 78
   acicola, 78
Phomopsis, 47, 89
   juniperovora, 48, 90
   lokoyae, 48; Pl. 118
   occulta, 49; Pl. 117
   porteri, 48
Phoradendron, 38–39; Pl. 85
Photinia, 35
Phragmopora pithya, 49
Phyllostictina hysterella, 49
Phytophthora, 1, 89; Pl. 166
   in nurseries, 88; Pl. 167
   cinnamomi, 4–6, 62, 85, 86, 91
   citricola, 91
   citrophthora, 91
   gonapodyides, 4, 71
   lateralis, 6–7, 79, 91; Pls. 1–3
Phytophthora root rot, 4–7, 88, 91;
   Pls. 1–3, 166, 167
Picea, 18, 40, 48, 76
   Amylostereum rot of, 21–22
   annosum root rot of, 12
   broom rust of, 53
   bud blight of, 62; Pl. 152
   needle and broom rust of, 52
   red-brown butt rot of, 16
   Sirococcus tip blight of, 48
   Sphaeropsis shoot blight and canker of,
     42–43
   spruce cone rust of, 49–50
   tomentosus root rot of, 16–17;
     Pls. 26, 27
   white stringy rot of, 23
Picea abies, 73, 75, 88, 89
   annosum root rot of, 12; Pls. 12, 14
   blue stain of, 18–19
   cone and seed diseases of, 51
   and *Cylindrobasidium evolvens,* 23
   decline of, in Europe, 67
   and larch canker, 41
   red rot of, 21; Pls. 38, 39
   Scleroderris canker of, 43–45; Pl. 97

Picea ajanensis, 77
Picea engelmannii
  broom rust of, Pl. 125
  brown felt blight of, Pl. 158
  dwarf mistletoes on, 36–38
  rust-red stringy white rot of, 25–26
  Stigmina verrucosa infection of, 62
Picea glauca, 76, 89
  rust-red stringy white rot of, 25–26
  Scleroderris canker of, 43
  Stigmina verrucosa infection of, 62
Picea mariana, 76, 89
  Scleroderris canker of, 43
  Stigmina verrucosa infection of, 62
Picea pungens, 47, 89, 90
  dwarf mistletoes on, 36–38
  Leucostoma canker of, 90
  Rhizosphaera needle blight of, 62, 89
  Sirococcus tip blight of, 48
Picea rubens
  decline of, 69–70
  Rhizina root disease of, 8
Picea sitchensis, 73
  Armillaria ostoyae on, Pl. 19
  Botryosphaeria piceae on, 48; Pl. 113
  Chrysomyxa monesis on, 50
  Fomitopsis pinicola on, Pl. 42
  Lirula needle blight of, 61–62; Pl. 151
  red-brown butt rot of, 16; Pls. 23–25
  and *Resinicium bicolor*, Pl. 45
  Rhizina root disease of, 8
  Sirococcus tip blight of, 48
Picea smithiana, 84
Pine needle rusts, 52–53, 89; Pl. 124
Pine twist rust, 33–34
Pine wilt, 19–20; Pls. 36, 37
Pine-oak gall rust, 29
Pine-pine gall rust, 29, 89
Pinus
  annosum root rot of, 12
  brown root rot of, 17–18
  Lachnellula canker of, 47
  leafy mistletoes on, 39–40
  Lophodermium needle cast of, 61;
    Pls. 148, 149
  needle rust of, 52, 53
  red-brown butt rot of, 16
  Sphaeropsis shoot blight and canker of, 42–43
  tomentosus root rot of, 16–17
  white pine blister rust of, 26–27;
    Pls. 55, 56, 58
  white stringy rot of, 23
Pinus albicaulis
  Bifusella needle blight of, 60
  Lophodermella needle cast of, 60
Pinus arizonica, 33, 45
Pinus armandii, 91
Pinus attenuata
  dwarf mistletoes on, 36–38
  Elytroderma needle disease of, 60
  Lophodermella needle cast of, 60
Pinus avacahuite, 36–38
Pinus banksiana, 76
  comandra blister rust of, 31
  Davisomycella needle cast of, 61
  eastern gall rust of, 29; Pl. 61
  Elytroderma needle disease of, 60
  Lophodermella needle cast of, 60
  Meloderma needle blight of, 60–61
  Scleroderris canker of, 44
  stalactiform blister rust of, 31
  sweetfern blister rust of, 32; Pl. 72
  western gall rust of, 29–30; Pl. 63
Pinus brutia, 60
Pinus canariensis, 84

Pinus caribaea, 18
  var. *hondurensis*, 36–38, 85
Pinus cembra, 44, 91
Pinus cembroides, 60
  subsp. *orizabensis*, 36–38
Pinus clausa, 59
Pinus contorta, 73, 75, 76, 80
  Atropellis cankers of, 46–47; Pl. 102
  Bifusella needle blight of, 60
  black-stain root disease of, 8–9
  blue stain of, 18–19; Pl. 35
  Cenangium ferruginosum on, 48; Pl. 111
  comandra blister rust of, 31; Pls. 68, 69
  Cyclaneusma needle cast of, 59
  Davisomycella needle cast of, 61
  dwarf mistletoes on, 36–38
  Elytroderma needle disease of, 60
  Lophodermella needle cast of, 60; Pl. 143
  mistletoe blister rust of, 32
  pine twist rust of, 33–34
  Ramichloridium dieback of, 48; Pl. 112
  Scleroderris canker of, 43–45
  Sirococcus tip blight of, 48
  stalactiform blister rust of, 31; Pl. 71
  sweetfern blister rust of, 32
  var. *latifolia*, western gall rust of, 29–30;
    Pls. 62, 64
Pinus cooperi, 33
Pinus densiflora, 19–20
Pinus discolor, 36–38
Pinus durangensis, 33, 36–38
Pinus echinata, 28, 82
  littleleaf disease of, 4–6
  pitch canker of, 45
  Ploioderma needle cast of, 59
Pinus edulis
  Bifusella needle blight of, 60
  Elytroderma needle disease of, 60
  pinyon blister rust of, 27
Pinus elliottii, 18, 40, 82
  cone and seed diseases of, 51
  fusiform rust of, 27–29
  Lophodermella needle cast of, 60
  Ploioderma needle cast of, 59
  var. *densa*, 28
  var. *elliottii*, 45–46, 50, 85; Pl. 101
Pinus engelmannii, 33
Pinus estevezi, 45
Pinus flexilis
  Bifusella needle blight of, 60
  Lophodermella needle cast of, 60
  Meloderma needle blight of, 60–61
Pinus funebris, 78
Pinus glabra, 59
Pinus griffithii, 36–38, 91
Pinus halepensis
  Elytroderma needle disease of, 60
  pine twist rust of, 33–34
  resin top disease of, 30–31
Pinus jeffreyi, 28
  Cyclaneusma needle cast of, 59
  Elytroderma needle disease of, 60
  limb rust of, 33
Pinus kesiya, 18, 40, 83, 84
Pinus koraiensis, 78, 91
Pinus lambertiana, 60, 79
Pinus lawsonii, 33
Pinus leiophylla var. *chihuahuana*, 50
Pinus luchuensis, 45
Pinus massoniana, Pl. 37
Pinus michoacana, 33
Pinus monophylla
  Bifusella needle blight of, 60
  black-stain root disease of, 8–9
  pinyon blister rust of, 27

Pinus montana, 33–34
Pinus montezumae, 36–38; Pl. 81
Pinus monticola, 79, 80
  Bifusella needle blight of, 60
  Dothistroma needle blight of, Pl. 137
  dwarf mistletoes on, 36–38
  Lophodermella needle cast of, 60
  Meloderma needle blight of, 60–61
  pole blight of, 71–72
  Scleroderris canker of, 43
Pinus mugo
  Cyclaneusma needle cast of, 59
  Lophodermella needle cast of, 60
  resin top disease of, 30–31
  Scleroderris canker of, 44
  western gall rust of, 29–30
Pinus muricata, 40
  cone and seed diseases of, 51
  western gall rust of, 29–30
Pinus nigra, 73, 89
  cone and seed diseases of, 51
  Cyclaneusma needle cast of, 59
  Dothistroma needle blight of, 57–58, 89;
    Pl. 136
  Lophodermella needle cast of, 60; Pl. 144
  Lophodermium needle cast of, 89
  pine twist rust of, 33–34
  resin top disease of, 30–31; Pl. 67
  Scleroderris canker of, 43–45
  western gall rust of, 29–30
Pinus occidentalis, 36–38
Pinus oocarpa, 36–38; Pl. 88
Pinus palustris, 28, 82
  brown spot needle blight of, 57; Pl. 135
  pitch canker of, 45
  Ploioderma needle cast of, 59
  southern pine cone rust of, 50
Pinus patula, 40, 86, 87
  pitch canker of, 45
  red rot of, 21
Pinus pedatum, 59
Pinus peuce, 16
Pinus pinaster, 73, 85
  pine twist rust of, 33–34
  resin top disease of, 30–31
  western gall rust of, 29–30
Pinus pinea
  pine twist rust of, 33–34
  resin top disease of, 30–31
Pinus ponderosa, 28, 80
  Armillaria ostoyae on, Pl. 18
  Atropellis cankers of, 46–47
  black-stain root disease of, 8–9
  comandra blister rust of, 31
  Cyclaneusma needle cast of, 59
  damping-off of, Pl. 165
  Davisomycella needle cast of, 61
  Dothistroma needle blight of, 57–58
  dwarf mistletoes on, 36–38; Pl. 78
  Elytroderma needle disease of, 60;
    Pls. 141, 142
  limb rust of, 33; Pls. 73, 74
  Lophodermella needle cast of, 60;
    Pls. 145, 146
  mistletoe blister rust of, 32
  Rhizosphaera needle blight of, 62
  Scleroderris canker of, 43
  western gall rust of, 29–30
Pinus pumila, 27, 77
Pinus radiata, 28, 85, 89
  Amylostereum rot of, 21–22
  Cyclaneusma needle cast of, 59
  Dothistroma needle blight of, 57–58, 85, 86;
    Pl. 138
  Meloderma needle blight of, 60–61

Pinus radiata (continued)
    pitch canker of, 45–46, 89; Pl. 99
    Ploioderma needle cast of, 59
    red rot of, 21
    western gall rust of, 29–30
Pinus resinosa, 89
    Meloderma needle blight of, 60–61
    Scleroderris canker of, 43–45
    Sirococcus tip blight of, 48
    Sphaeropsis shoot blight and canker of, 42–43; Pl. 92
    sweetfern blister rust of, 32
Pinus rigida, 28
    Meloderma needle blight of, 60–61
    Ploioderma needle cast of, 59
Pinus roxburghii, 83, 84
Pinus sabiniana, 51
Pinus serotina, 28, 59
Pinus strobiformis, 36–38
Pinus strobus, 11, 88, 89, 92
    Bifusella needle blight of, 60
    Meloderma needle blight of, 60–61
    pine twist rust of, 33–34
    pitch canker of, 45–46; Pl. 100
    red ring rot of, 24
Pinus sylvestris, 53, 73, 75, 88, 89
    annosum root rot of, 12; Pl. 13
    brown spot needle blight of, 57, 89
    cone and seed diseases of, 51
    Cyclaneusma needle cast of, 59, 89
    Lecanosticta needle blight of, 89; Pl. 169
    Lophodermella needle cast of, 60
    Lophodermium needle cast of, 89
    Phacidium canker of, 47
    and *Phellinus pini*, Pl. 48
    pine needle rust of, 89
    pine twist rust of, 33–34
    resin top disease of, 30–31; Pls. 65, 66
    Rhizina root disease of, Pl. 5
    Scleroderris canker of, 43–45; Pl. 96
    Viscum laxum on, Pl. 87
    western gall rust of, 29–30
Pinus tabulaeformis, 36–38
Pinus taeda, 18, 82
    fusiform rust of, 27–29
    Lophodermella needle cast of, 60
    pitch canker of, 45–46
    Ploioderma needle cast of, 59
Pinus taiwanensis, 40
Pinus thunbergiana, 19–20
Pinus thunbergii, 62
Pinus virginiana
    Appalachian blister rust of, 32
    pitch canker of, 45
    Ploioderma needle cast of, 59
Pinus wallichiana, 83
    dwarf mistletoe on, Pl. 82
    Meloderma needle blight of, 60–61
Pinyon blister rust, 27
Pissodes
    fasciatus, 9; Pl. 10
    nemorensis, 11
    strobi, 77
Pitch canker, 45–46, 51, 83, 86, 89; Pls. 98–101
Ploioderma
    hedgcockii, 59, 60
    lethale, 59, 60, 61, 90; Pls. 139, 140
    lowei, 59
Ploioderma needle casts, 59, 90; Pls. 139, 140
Podocarpus imbricata, 40
Pole blight, 71–72
Polyporus
    abietinus, 76
    annosus, 12

    balsameus, 18
    schweinitzii, 16
    tomentosus, 16
Populus, 53
    alba, 34
    × *tremula*, 34
Populus balsamifera, 76
Populus canescens, 34
Populus tremula, 34
Populus tremuloides, 76
Poria
    albipellucida, 18
    asiatica, 18
    subacida, 76
    vincta, 18, 85
    weirii, 14
Porodaedalea pini, 24
Port Orford cedar root disease, 6–7; Pls. 1–3
Postia balsameus, 18
Potebniamyces coniferarum, 47
Powell Peridermium, 33
Procerum root disease, 11, 92
Prunus, 22
    needle rust of, 52, 53
    red-brown butt rot of, 16
Pseudixus japonica, 40
Pseudocercospora juniperi, 63, 90
Pseudophaeolus baudoni, 18, 86
Pseudotsuga, 47, 53
    Amylostereum rot of, 21–22
    annosum root rot of, 12
    dwarf mistletoes on, 36–38
    red-brown butt rot of, 16
    red rot of, 21
    Sphaeropsis shoot blight and canker of, 42–43
    white stringy rot of, 23
Pseudotsuga macrocarpa, 55
Pseudotsuga menziesii, 34, 73, 79, 80, 85
    Armillaria root disease of, Pl. 17
    black-stain root disease of, 8–9; Pls. 8, 9
    cone and seed diseases of, 50
    Cytospora canker of, Pl. 104
    Dermea canker of, 48; Pl. 110
    dwarf mistletoes on, 36–38; Pls. 79, 83, 84
    Lachnellula canker of, 47; Pl. 105
    laminated root rot of, 14–15; Pls. 20–22
    Phacidium canker of, 47
    Phytophthora root rot of, 88; Pl. 166
    red-brown butt rot of, 16
    red ring rot of, 24; Pl. 47
    Rhabdocline needle casts of, 54–55, 89; Pls. 128–131
    Rhizina root disease of, 8
    rust-red stringy white rot of, 25–26
    Sclerophoma pithyophila on, Pl. 115
    Sirococcus tip blight of, 48
    Sparassis radicata sporocarp on, Pl. 34
    Swiss needle cast of, 55–56, 89; Pls. 132, 133, 168
    yellow brown top rot of, 22–23; Pls. 40, 41
    var. *glauca*, 55
Psittacanthus, 39–40; Pl. 88
    angustifolius, 40
Pucciniastrum, 52, 89
    epilobii, 52; Pl. 123
    pustulatum, 78, 79
Pyrola, 49–50
Pyrus, 35
Pythium, 1, 4, 88

Quercus, 50
Quercus griffithii, 84
Quercus kelloggii, 28

Quercus nigra, 27–29
Quercus phellos, 27–29

Ramichloridium pini, 48; Pl. 112
Rasutoria
    abietis, 63
    pseudotsugae, 63
    tsugae, 63
Red belt, 65
Red belt fungus, 23
Red-brown butt rot, 16; Pls. 23–25
Red heart, 21
Red pine decline, 11
Red ring rot, 24–25; Pls. 47–50
Red rot, 21; Pls. 38, 39
Red spruce decline, 69–70
Red streak, 21
Resin top disease, 30–31; Pls. 65–67
Resinicium bicolor, 23, 24; Pl. 45
Rhabdocline, 89
    parkeri, 55
    pseudotsugae, 74
        subsp. *epiphylla*, 54–55
    weirii, 54–55; Pls. 128–131
Rhabdocline needle casts, 54–55, 89; Pls. 128–131
Rhabdogloeum pseudotsugae, 54–55; Pl. 131
Rhizina, 10
    inflata, 7
    undulata, 7–8, 86, 87; Pls. 4–6
Rhizina root disease, 7–8; Pls. 4–6
Rhizoctonia, 88
Rhizopus stolonifer, 51
Rhizosphaera
    kalkhoffii, 62, 89, 90; Pl. 153
    pini, 78
Rhyssa persuasoria, 22
Ribes nigrum, 26–27; Pl. 57
Rigidoporus
    lineatus, 18
    vinctus, 18; Pl. 33
Roestelia, 34
    brucensis, 35
Rust-red stringy white rot, 25–26; Pls. 51–54
Rusts, 1
    Appalachian blister rust, 32
    blister rust, 74
    comandra blister rust, 31; Pls. 68–70
    eastern gall rust, 29; Pl. 61
    fusiform rust, 27–29, 91; Pls. 59, 60
    globose gall rust, 29
    Gymnosporangium stem rusts, 34–35, 91; Pls. 75–77
    limb rusts, 33; Pls. 73, 74
    mistletoe blister rust, 32
    needle and broom rusts, 51–53; Pls. 122–127
    pine twist rust, 33–34
    pine-oak gall rust, 29
    pine-pine gall rust, 29, 91
    pinyon blister rust, 27
    Southern pine cone rust, 50
    southwestern pine cone rust, 50
    spruce cone rust, 49–50
    stalactiform blister rust, 31, 33; Pl. 71
    sweetfern blister rust, 32; Pl. 72
    twisting rust, 74
    western gall rust, 29–30, 77; Pls. 62–64
    white pine blister rust, 26–27, 80, 81, 91; Pls. 55–58

Sabina, 34, 35
Sageria tsugae, 49
Salix, 53
Salt damage, 64, 66, 92
Sap stain, 42

Sarcotrochila
 balsameae, 63
 macrospora, 61
 piniperda, 62, 63
Schizophyllum commune, 50
Schwalbia, 40
Scirrhia
 acicola, 57
 pini, 58
Scleroderris canker, 43–46, 89; Pls. 96, 97
Scleroderris lagerbergii, 43
Sclerophoma, 49; Pl. 118
 pithyophila, 49, 65, 78, 90; Pl. 115
 semenospora, 49
Scurrula parasitica, 40
Scytinostroma galactinum, 76
Seiridium, Pl. 106
 cardinale, 47, 73, 85, 87, 89, 91
 cupressi, 87, 91
 unicorne, 47, 85, 87, 91
Seiridium cankers, 47–48, 91; Pl. 106
Sequoia, 63
Sequoiadendron, 48, 63
Seuratia millardetii, 63
Seymeria elliottii, 40
Shoestring root disease, 13
Sirex
 areolatus, 21
 californicus, 21
 cyaneus, 21
 imperialis, 21
 nitobie, 21
 noctilio, 21, 85, 86
Sirococcus
 conigenus, 48, 51; Pls. 107, 108, 119
 strobilinus, 48, 51
Sirococcus blight, 51; Pl. 119
Sirococcus tip blight, 48; Pls. 107, 108
Sirodothis, 49
Snow blights, 63–64
Snow damage, 65
Soil moisture, 64, 71; Pl. 167
Soil pH, 64
Soil temperature, 65
Sooty molds, 63, 90
Sorbus, 35
Southern pine beetle, 5, 18–19. *See also* Bark beetles
Southern pine cone rust, 50
Southwestern pine cone rust, 50
Sparassis
 crispa, 16, 18, 74
 radicata, 18; Pl. 34
Sphaeropsis
 ellisii, 42
 sapinea, 18, 42–43, 85, 86, 89, 91; Pls. 92–95
 f. sp. *cupressi*, 91
Sphaeropsis shoot blight and canker, 42–43, 91; Pls. 92–95
Spiniger meineckellus, 12
Spruce beetle, 76, 77
Spruce cone rust, 49–50
Spruce rusts, 52; Pl. 122
Spruce weevil, 77

Stalactiform blister rust, 31, 33; Pl. 71
Stegopezizella balsameae, 63
Stellaria, 52, 53
Steremnius carinatus, 9; Pl. 10
Stereum
 balsameum, 21
 chailletii, 21
 sanguinolentum, 21, 22, 23; Pls. 38, 39
Stigmina verrucosa, 62
Stomiopeltis pinastri, 63
Struthanthus
 deppeanus, 40
 palmeri, 40
 quercicola, 40
Sunscald, 65
Sweetfern blister rust, 32; Pl. 72
Swertia, 84
Swiss needle cast, 55–56, 89; Pls. 132, 133, 168
Systremma acicola, 57

Tar spot needle cast, 60
Taxillus
 kaempferi, 40
 matsudai, 40
Taxodium, 63
 mucronatum, 40
Taxus, 48
 brevifolia, 6–7
Tea break fungus, 7–8; Pls. 4–6
Thekopsora, 52
 aureolatum, 52
Therrya
 pseudotsugae, 49
 tsugae, 48, 49
Thuja, 63
 Seiridium cankers of, 47
 Sphaeropsis shoot blight and canker of, 42–43, 63
 white stringy rot of, 23
Thuja occidentalis, 62
Thuja plicata, 14, 18, 47; Pls. 32, 118
 Keithia blight of, 62; Pl. 156
Tip blight, 42
Tomentosus root rot, 16–17; Pls. 26–28
Trametes
 carnea, 22
 pini, 24
 radiciperda, 12
 subrosea, 22
Trichoderma, 51
Trichoscyphella willkommii, 41, 47
Trichothecium, 51
 roseum, 51
Tripodanthus acutifolius, 40
Tropaeolum, 30
Tsuga, 40, 47, 48
 annosum root rot of, 12
 and *Hydnum abietis*, 24
 needle rust of, 52, 53
 rust-red stringy white rot of, 25–26; Pls. 52, 53
 white stringy rot of, 23
Tsuga canadensis, 32, 63

Tsuga heterophylla, 48, 63, 71, 79
 dwarf mistletoes on, 36–38
 Rhizina root disease of, 8
 Sirococcus tip blight of, 48; Pls. 107, 108
Tsuga mertensiana, 63, 71
 black-stain root disease of, 8–9
 dwarf mistletoes on, 36–38
 Hericium abietis on, 50
 laminated root rot of, 14–15
Tussilago farfara, 52
Twisting rust, 74
Tympanis hyphopodia, 48

Ulmus, 16
Uredinopsis, 52, 89
 pteridis, 53
Uredo, 34
 apacheca, 35
 cupressicola, 35
Uroceras, 21–22
Ursus arctos, 71

Vaccinium, 52, 53
Valsa, 90
 abietis, 47
Vauquelinia, 35
Velvet top fungus, 16
Verticicladiella wageneri, 8
Vincetoxicum hirundinaria, 30
Virgella needle blight, 61
Virgella robusta, 61
Viruses, 1
Viscum
 abietis, 39; Pl. 86
 album, 39, 73, 74
 laxum, 39; Pl. 87

Waldsterben, 66, 67–69
Washoe Peridermium, 33
Weevils, 17, 24, 41, 42, 77, 86
 as vectors of *Leptographium* diseases, 8–11, 92; Pls. 10, 11
Western gall rust, 29–30, 76, 77; Pls. 62–64
 and southwestern pine cone rust, compared, 50
White pine blister rust, 26–27, 80, 81, 89; Pls. 55–58
White pine root decline, 11
White pine weevil, 24
White pocket rot, 24, 84
White rot, 11, 12–13, 14, 18, 20
White snow mold, 63
White stringy rot, 23
Winter drying, 65
Wood wasp, 21–22

Xenomeris abietis, 49; Pl. 118

Yellow brown top rot, 22–23; Pls. 40, 41
Yellow-cedar decline, 70–71; Pls. 159, 160
Yellow laminated root rot, 14

Zelleria haimbachi, 59
Zythiostroma pinastri, 78